The UN Watercourses Force

At the UN General Assembly in 1997, an overwhelming majority of states voted for the adoption of the UN Convention on the Law of the Non-Navigational Uses of International Watercourses (UN Watercourses Convention) – a global overarching framework governing the rights and duties of states sharing freshwater systems. Globally, there are 276 internationally shared watersheds, which drain the territories of 145 countries and represent more than forty percent of the Earth's land surface. Hence, interstate cooperation towards the sustainable management of transboundary waters in accordance with international law is a topic of crucial importance, especially in the context of the current global water crisis.

This volume provides an assessment of the role and relevance of the UN Watercourses Convention as a key component of transboundary water governance. To date, the Convention still requires further contracting states before it can enter into force. The authors describe the drafting and negotiation of the Convention and its relationship to other multilateral environmental agreements. A series of case studies assesses the role of the Convention at various levels: regional (West Africa, Central America), river basin (e.g. the Mekong, Amazon, Nile, Aral Sea and Congo) and national (e.g. Ethiopia and El Salvador). The book concludes by proposing how the Convention's future implementation might further strengthen international cooperation in the management, use and protection of shared water resources and their ecosystems.

Flavia Rocha Loures is a Senior Programme Officer, International Law and Policy, in the Freshwater Programme of WWF-US, based in Washington, DC.

Alistair Rieu-Clarke is a Reader in the Centre for Water Law, Policy & Science (under the auspices of UNESCO) at the University of Dundee, UK.

The UN Watercourses Convention in Force

Strengthening international law for transboundary water management

Edited by Flavia Rocha Loures and Alistair Rieu-Clarke

LONDON AND NEW YORK

First published 2013
by Routledge

2 Park Square, Milton Park, Abingdon, Oxon, OX14 4RN
711 Third Avenue, New York, NY 10017

First issued in paperback 2017

Routledge is an imprint of the Taylor and Francis Group, an informa business

British Library Cataloguing in Publication Data
A catalogue record for this book is available from the British Library

Library of Congress Cataloging-in-Publication Data
The UN Watercourses Convention in force : strengthening international law
for transboundary water management / edited by Flavia Rocha Loures, Alistair
Rieu-Clarke.
pages cm
"Earthscan from Routledge"
Includes bibliographical references and index.
1. United Nations Convention on the Law of International Watercourses
(1997) 2. Water rights (International law) 3. Water resources development—Law
and legislation. I. Loures, Flavia Rocha. II. Rieu-Clarke, Alistair. III. Title:
United Nations Watercourses Convention in force.
K3496.A41997U5 2013
341.4'4—dc23
2013006481

ISBN: 978-1-84971-446-4 (hbk)
ISBN: 978-1-138-57392-5 (pbk)

Typeset in Bembo by
FiSH Books, London

Contents

Figures and tables

Figures

Tables

Abbreviations

ACT	Amazon Cooperation Treaty
ACTO	Amazon Cooperation Treaty Organization
ASBP	Aral Sea Basin Programme
ASEAN	Association of Southeast Nations
CARU	[Administrative Commission on the River Uruguay]
CBD	Convention on Biological Diversity
CESCR	Committee on Economic, Social and Cultural Rights
CFA	Nile Basin Cooperative Framework Agreement
CILA	Comisión Internacional de Límites y Aguas [International Border and Waters Commission]
CIS	Commonwealth of Independent States
CoP	Conference of the Parties to the Convention on Biological Diversity
EBRD	European Bank for Reconstruction and Development
ECAGIRH	[Central American Strategy for the Integrated Management of Water Resources]
ECOSOC	UN Economic and Social Council
ECOWAS	Economic Community of West African States
EIA	environmental impact assessment
ENSAP	Eastern Nile Subsidiary Action Programme
EU	European Union
FAO	Food and Agriculture Organization of the UN
GEF	Global Environment Facility
GIZ	German International Cooperation
GMS	Greater Mekong Subregion
GOLD	General Organization for Land Development
GWh	gigawatt hours
IBWC	International Boundary and Water Commission
ICESCR	International Covenant on Economic, Social and Cultural Rights
ICJ	International Court of Justice
ICSD	Interstate Commission for Sustainable Development
ICWC	Interstate Commission for Water Coordination

IFAS	International Fund for Saving the Aral Sea
IHP-HELP	International Hydrological Programme: Hydrology for the Environment, Life and Policy
IIL	Institute of International Law
ILA	International Law Association
ILC	International Law Commission
IPCC	Intergovernmental Panel on Climate Change
IUCN	International Union for Conservation of Nature
IWRM	integrated water resources management
km	kilometers
LCBC	Lake Chad Basin Commission
M-POWER	Mekong Programme on Water, Environment and Resilience
MRC	Mekong River Commission
MW	megawatts
NBA	Niger Basin Authority
NBI	Nile Basin Initiative
Nile-COM	Nile Council of Ministers
Nile-TAC	Nile Technical Advisory Committee
NELSAP	Nile Equatorial Subsidiary Action Programme
NGO	non-governmental organization
OAS	Organization of American States
OECD	Organisation for Economic Co-operation and Development
OMVS	Organisation pour la mise en valeur du fleuve Sénégal [Senegal River Development Organization]
PACADIRH	Plan de Acción para el Manejo Integrado del Agua en el Istmo Centroamericano [Action Plan for the Joint Management of Water in the Central American Isthmus]
PACAGIRH	[Central American Action Plan for the Integrated Management of Water Resources]
PCIJ	Permanent Court of International Justice
PKK	Parti Karkerani Kurdistan
SADC	Southern African Development Community
SIWI	Stockholm International Water Institute
TECCONILE	Technical Cooperation Committee for the Promotion of the Development and Environmental Protection of the Nile Basin
UN	United Nations
UN DESA	UN Economic and Social Commission for the Asia and Pacific
UNCCD	UN Convention to Combat Desertification in Countries Experiencing Serious Drought and/or Desertification, Particularly in Africa
UNCED	UN Conference on Environment and Development (Rio)
UNCLOS	UN Convention on the Law of the Sea
UNCOD	UN Conference on Desertification
UNECE	UN Economic Commission for Europe
UNEP	UN Environment Programme

UNESCAP	UN Economic and Social Commission for the Asia and Pacific
UNESCO	UN Educational, Scientific and Cultural Organization
UNFCCC	UN Framework Convention on Climate Change
UNGA	UN General Assembly
UNSGAB	UN Secretary-General's Advisory Board on Water and Sanitation
UNWC	United Nations Convention on the Law of the Non-Navigational Uses of International Watercourses
VBA	Volta Basin Authority
WRI	World Resources Institute
WSAF	Water Security Analytical Framework
WTO	World Trade Organization
WWF	World Wide Fund for Nature

Notes on contributors

Teslim Abdul-Kareem is an energy and environment analyst. He holds a Bachelors degree in Materials Science and Engineering and recently concluded his Master's in Energy Studies with Specialization in Energy and the Environment. His research interests lie in the area of energy and environmental sustainability, health and safety, materials selection and design, climate change and water security.

Musa Mohammed Abseno is a PhD Scholar at the Centre for Water Law, Policy and Science. He holds LLB and LLM (Dean's Medal) degrees. He has worked as Head of Transboundary Rivers Department, Ministry of Water Resources, Ethiopia. He was a member of Nile-TAC, Panel of Experts and senior negotiator for the Nile River Basin Cooperative Framework Agreement.

Bennett L. Bearden is Special Counsel on Water Law and Policy in the Office of the State Geologist at the Geological Survey of Alabama (USA). His research focuses on international water law in the Mekong River Basin and domestic water law and policy in Alabama and the Southeastern United States.

Christian Behrmann, Attorney-at-Law, Policy Officer, Global and Multilateral Issues, European External Action Service. Before joining the diplomatic service, Dr Behrmann practiced law in private practice, European Union institutions and the UN. He holds a PhD in public international law and lectures at the University of Leuven.

Joseph W. Dellapenna is Professor of Law at Villanova University. He practices, teaches and writes about water. He serves as Director of the Model Water Code Project of the American Society of Civil Engineers and served as Rapporteur of the Water Resources Committee of the International Law Association.

Anton Earle is the Director of Capacity Building Programmes at the Stockholm International Water Institute (SIWI) and is completing a PhD in Peace and Development at the School of Global Studies at the University of Gothenburg. He specializes in transboundary integrated water resource management, facilitating the interaction between governments, basin organizations and other stakeholders in international river and lake basins.

Uschi Eid is Vice Chair of the UN Secretary General's Advisory Board on Water and Sanitation. As former deputy minister for development cooperation in Germany, she has initiated several policy initiatives in the water and sanitation sector. She is a vocal supporter to the campaign to help ratify the UN Watercourses Convention.

Amidou Garane is a Professor of Environmental and Natural Resources Law at l'Université Ouaga 2 in Burkina Faso. He holds a PhD in public international law and has written extensively on shared water resources and has served on expert panels on the topic. He has also conducted studies on transboundary basins on behalf of numerous African basin organizations and non-governmental organizations.

Nicole Kranz is policy advisor in International Water Policy and Infrastructure at German International Cooperation (GIZ). She holds a PhD in Political Science from Freie Universität Berlin and Master's degrees in environmental sciences from German and American universities. In her work, she mainly supports international processes targeted at ensuring access to water and sanitation.

Johan G. Lammers is a former Legal Adviser of the Netherlands Ministry of Foreign Affairs and a former Professor of International Law and International Environmental Law at the University of Amsterdam. He acted as Chairman of the Drafting Group during the negotiations in New York which led to the adoption of the UN Watercourses Convention.

Alexander López is currently director of the School for International Relations of the Universidad Naçional de Costa Rica. He holds a PhD from the University of Oslo, Norway. He has worked extensively in India, South Africa, Vietnam, the Brazilian Amazon, Mexico and Central America on issues related to water management and water conflict resolution.

Flavia Rocha Loures is currently a Senior Programme Officer with WWF, the global conservation organization. She holds a JD from the University of Parana, Brazil, and an LLM (summa cum laude) in environmental law, with a focus on international groundwater law, from Vermont Law School, USA. Her focus is on the codification and development of international water law, on the promotion of improved global and regional freshwater governance and on the implementation of water-related international conventions.

Daniel Malzbender is an environmental lawyer with law degrees from Germany and South Africa. He specializes in legal analysis and policy development, institutional and river basin strategy development, investment planning, the design of stakeholder participation processes as well as capacity building in transboundary natural resources management in Africa.

Stephen C. McCaffrey is Distinguished Professor and Scholar at the University of the Pacific, McGeorge School of Law. He was a member of the UN International Law Commission (ILC) from 1982 to 1991, chaired the ILC's

1987 Session and was its special rapporteur on international watercourses from 1985 to 1991, when the Commission adopted a complete first draft of articles on the topic. The final draft formed the basis for the negotiation of the 1997 UN Convention on the same subject.

Owen McIntyre is a Senior Lecturer in the Faculty of Law, University College Cork, National University of Ireland, where he teaches and researches on all aspects of environmental law. He regularly consults internationally on environmental law and international water law and serves as a panel member of the European Bank for Reconstruction and Development (EBRD) Project Complaint Mechanism (PCM) and as a member of the Scientific Committee of the European Environment Agency.

Naho Mirumachi is a Lecturer at the Department of Geography, King's College London. Trained in political science, international studies and human geography, she has research interests in the politics of water resources management, particularly in developing-country contexts.

Ruby Moynihan is a Researcher at the Centre for Environmental Research, Department for Environmental and Planning Law in Helmholtz, Germany, where her research and project consultancy work focuses on international, European and comparative national water law, climate adaptation and the interfaces between law, science and policy. Ruby is also doing a PhD with the University of Edinburgh School of Law, Scotland, and is a 2013 Scottish Government Hydro-Nation Scholar.

Joshua Newton is a PhD Candidate at the Fletcher School of Law and Diplomacy at Tufts University, whose work focuses on global water governance and, previously, on transboundary water issues. In the past, he has coordinated the political processes of the fifth and sixth World Water Forums on behalf of UNESCO and the World Water Council and has consulted for CONAGUA, the Organization of American States and the Universities Partnership for Transboundary Waters.

Meg Patterson is a Programme Officer with WWF, where she works on international water law and policy. She received her JD from Lewis and Clark Law School, with a certificate in Natural Resources and Environmental Law and a BA in international relations from The College of William and Mary.

Sokhem Pech is a full-time staff of the environmental company, Hatfield Consultants Partnership, based in North Vancouver, Canada. He is chair of the Mekong Programme on Water, Environment and Resilience (M-POWER), a regional research network, and also works for the Challenge Programme on Water and Food. He is a former deputy CEO, Director of the Technical Support Division and legal advisor of the Mekong River Commission (MRC) – an intergovernmental river basin management in the Lower Mekong Basin.

Guy Pegram is the managing director of Pegasys, with over 20 years professional experience in the water sector. He is a professionally registered civil engineer with a PhD in water resources planning and an MBA from University of Cape Town. He has worked extensively on strategic, institutional, financial and organizational aspects related to the water sector within the South African Development Community, Africa, and globally.

Jamie Pittock has been Programme Leader, Australia and US Climate, Energy and Water for the US Studies Centre, and also Director of International Programmes for the UNESCO Chair in Water Economics and Transboundary Water Governance since 2010. He is Senior Lecturer at the Fenner School of Environment and Society at The Australian National University. His research focuses on conflicts and synergies between biodiversity, climate, energy and water policies.

Alistair Rieu-Clarke is a Reader in International Law. He holds an LLB in Scots Law, an LLM in Natural Resources Law and Policy (with distinction), and a PhD in Sustainable Development, Water and International Law. His research and teaching activities focus on deepening knowledge and understanding of international law within the context of transboundary water cooperation.

Salman M. A. Salman is an academic researcher and consultant on water law and policy. Until December 2009, he worked as Lead Counsel and Water Law Adviser with the World Bank. He is the author/editor of ten books and more than 50 articles and book chapters on water law. Dr. Salman obtained his LLB from University of Khartoum and holds LLM and JSD degrees from Yale Law School.

Ricardo Sancho is a Professor at the School of International Relations at the Universidad Naçional, where he teaches comparative political systems, international law and international political economy. He holds Master's degrees in public administration (Harvard) and Political Science (Carleton University, Ottawa, Canada) and a graduated diploma in International Development and Cooperation from the University of Ottawa, Canada.

Ashok Swain is a Professor of Peace and Conflict Research at Uppsala University Sweden. He also serves as the Director of the Uppsala Center for Sustainable Development.

Attila Tanzi, PhD, is Professor of International Law at the University of Bologna. He has served as a Legal Consultant for the Ministry of the Environment (Italy) since 1999 and is the Chairman of the Legal Board of the 1992 UNECE Convention on the Protection and Use of Transboundary Watercourses and International Lakes. He has held numerous academic positions and has written extensively on state responsibility, international organizations, jurisdictional immunities, foreign investment law and environmental law.

Mara Tignino is a Senior Researcher at the Platform for International Water Law at the University of Geneva School of Law and Visting Professor at the Catholic University of Lille and the Libera Università Internazionale degli Studi Sociali in Rome. Dr Tignino also serves as a consultant to governments, international organizations, NGOs and the private sector. Dr Tignino received a PhD in international law from the Graduate Institute of International and Development Studies in Geneva. She has written articles on public participation, peaceful settlement of water disputes and related topics.

Jeroen Warner has a background in international relations and is currently Assistant Professor in Disaster Studies at the University of Wageningen, The Netherlands, where he also received his PhD degree. His publishing, teaching, lecturing and training centers on domestic and international hydropolitics, water conflict, participation and governance issues.

Boleslawa 'Lesha' M. Witmer is an independent senior advisor on sustainable development, with a focus on water management; human resources, including participatory approaches and gender issues; and vocational training. She holds a degree in human resource management and general (business) management from Nijenrode and studied law at the University of Amsterdam.

Patricia Wouters is the founding Director of the Centre for Water Law, Policy and Science. She has been named the IUCN Academy of International Environmental Law Distinguished Scholar (2011) and appointed as an expert at Xiamen Law School in China, where she will have a three-year joint appointment as Research Scholar. She holds a PhD in international law from the University of Geneva, Graduate Institute of Higher Education, and advanced degrees from schools in the USA, Canada and Switzerland. Her current research focuses on the rule of law (in international law) and how it influences regional peace and security through its normative framework governing the world's transboundary waters.

Mark Zeitoun is a Reader in the School of International Development, at the University of East Anglia, and Director of the UEA Water Security Research Centre. His primary research interests relate to the ways that power asymmetry influences transboundary water interaction, policy and negotiations.

Dinara R. Ziganshina is a Deputy Director of the Scientific Information Centre of Interstate Commission for Water Coordination in Central Asia. She has been working as a legal expert in water management at national and transboundary levels for more than 10 years. She earned her law degree (J.D. equivalent) from Tashkent State Institute of Law (2001), LL.M. in Environmental and Natural Resources Law from the University of Oregon School of Law (2008), and PhD in International Water Law from the University of Dundee (2012).

Foreword

Since the beginning of time, water has been shaping the face of the Earth, not only as a geological agent, but also as a major factor in the rise and fall of great civilizations and as a source of conflict and tension between nations. The first great civilizations arose on the banks of great rivers like the Nile in Egypt, the Tigris-Euphrates of Mesopotamia, the Indus in Pakistan and the Hwang Ho in China. While all these civilizations built large irrigation systems and made land productive, by the same token, civilizations collapsed when water supplies failed or were improperly managed.

Today, the way we think about water goes to the very heart of the increasing worldwide concerns about human health, security, food, ecosystem collapse, changing climate, and globalization of the economy. The water crisis was glossed over at Rio, gained some traction at Johannesburg and then fell off the screen again at Rio in 2012. With most continental countries in the world sharing transboundary surface basins or aquifers, most of the unfrozen freshwater on the planet is associated with these cross-border systems, and attempts to deal with a changing climate will have to begin with cooperation and collaboration among nations regarding droughts, floods, and infrastructure investments.

Enormous transaction costs are associated with managing these great rivers and aquifers. This volume provides a great service in assembling up-to-date papers and the latest thinking on these shared basins and the relevant international architecture of the UN Watercourses Convention (UNWC). These international rivers will have the best chance to have water in drought but many will be associated with destructive floods that can rob a country of two to three percent of its gross domestic product. The Convention represents a milestone in the international community, as many of the papers argue, yet so many basic challenges seem to remain, as a number of authors in this volume highlight.

But, in reality, nations do choose to work together on their shared water systems in light of all the water security issues that face them. In the interim since the adoption of the UNWC, the Global Environment Facility (GEF) grew from a pilot environmental fund to a 20-year-old permanent grant facility for addressing global environmental issues, including these international rivers. The GEF International Waters focal area has provided US$1.3 billion in grants to developing countries that share surface and subsurface waters, along with US$6 billion in co-financing for joint projects. The GEF Strategy was approved in 1995 by the GEF Council, which represents most countries in the world. It was specifically designed to provide grant funding consistent with the International Law Commission's Draft Articles on the Law of the Non-Navigational Uses of International Watercourses, which preceded the UNWC. GEF grants fund processes among countries sharing a basin and its watercourses to compile information, to undertake joint fact finding, to analyze their situation with models (including future scenarios), to consult with each other and civil society, and to work side by side in a participative manner towards a joint vision or action programme for the future. These GEF grants have helped to develop capacity and to undertake relaxed dialogue, so that trust and confidence can be built among nations to collaborate jointly on their shared watercourses by negotiating agreements, establishing joint bodies and using environmental issues as a common ground before the harder water allocation disputes are tackled.

While the Nile and Mekong Basins may be the most familiar, GEF International Waters projects undertaken through the World Bank and UN agencies have also been requested by the nations themselves and implemented for other great rivers such as the Danube, La Plata, Amazon, Orange, Okavango, Niger, Senegal and Volta, as well as Lakes Chad, Victoria, Tanganyika, Prespa/Ohrid, and Baikal. The papers in this volume note the opportunities that exist for nations riparian to international watercourses. More than 85 nations have asked for and received GEF assistance that should be considered as 'enabling activities' completed towards hopeful collaboration under the Convention. When it comes into force, the Convention's implementation will have had a 'head start', with assistance from the GEF, as more nations recognize that adaptive and joint management of these watercourses provides the only hope toward water security for the future. Enjoy this collection of papers!

Alfred M. Duda, Ph.D., Senior Advisor, International Waters (retired),
Global Environment Facility, Washington, DC.

Part 1

Background and evolution

1   Introduction

Alistair Rieu-Clarke and Flavia Rocha Loures

International law is clearly much more than a simple set of rules. It is a culture in the broadest sense in that it constitutes a method of communicating claims, counter-claims, expectations and anticipations as well as providing a framework for assessing and prioritizing such demands.[1]

The importance of strengthening international law related to transboundary water management boils down to certain fundamental aspects of water. Firstly, water has no substitute; water is 'a finite natural resource necessary for the sustenance of life and ecological systems and a key resource for social and economic development'.[2]

Secondly, water cannot be compartmentalized. As defined in the UN Watercourses Convention (UNWC), a 'watercourse' is 'a system of surface waters and groundwaters constituting by virtue of their physical relationship a unitary whole'.[3]

In addition, freshwater resources are often scarce, overexploited and polluted, and unevenly distributed across the globe. Freshwater ecosystems generate a diverse array of services for humanity but the enormous pressures they face are plain to see.[4] Large dams fragment 172 of the world's 292 longest rivers, affecting the ability of ecosystems to deliver valuable services such as fish and sediment transport – and more rivers will soon be dammed.[5] The world's 50,000 large dams have displaced around 40–80 million people and impacted the livelihoods of an

1   Shaw, M.N. *International Law* (6th ed., Cambridge University Press, 2008) at 67.
2   UN ACC Sub-Committee on Water Resources *Report: Water – A Key Resource for Sustainable Development*. Doc E/CN.17/2001/PC/17 (UN, 2001), at 2.
3   UNWC, at Article 2(a) [emphasis added].
4   Vörösmarty, C. J., Lévêque, C. and Revenga, C. 'Fresh Water' in Hassan, R., Scholes, R. and Ash, N. (eds) *Ecosystems and Human Well-being: Current State and Trends, Volume 1.* (Island Press, 2005), pp. 165–207, at 167.
5   Nilsson, C., Reidy, C. A., Dynesius, M. and Revenga, C. 'Fragmentation and Flow Regulation of the World's Large River Systems' *Science*, 2005; 308: 405–8. Available online at www.gwsp.org/fileadmin/downloads/Nilsson_Science2005.pdf (accessed March 14, 2013). See also BioFresh, 'DR Congo and South Africa Sign Pact to Implement 40,000-MW Grand Inga Dam', *Biofresh Blog*, November 23, 2011. Available online at http://biofreshblog.com/2011/11/23/dr-congo-and-south-africa-sign-pact-to-implement-40000-mw-grand-inga-dam(accessed March 14, 2013).

estimated 472 million people.[6] At the same time, demand for freshwater is increasing rapidly owing to population growth and shifting dietary habits as societies develop, raising consumer demand for thirstier products.[7] This is evident in the growing number of 'closed' river basins around the planet, where 1.4 billon people live and where nearly all available water is diverted for human consumption.[8] Even where the volume of water is not overexploited, poor catchment management and pollution degrade ecosystems and reduce water availability.[9] For example, aquifers and surface waters are often hydrologically linked, yet they are often managed as independent units, to the detriment of the sustainability of these resources.

Finally, climate change is already placing additional pressures on transboundary water resources. Through its impacts on water, climate change will progressively affect important triggers of conflict, including food, energy and land security, mass migration, poverty and social unrest and ecological disasters. In this sense, it is estimated that more than two billion people will be living in water-scarce countries by 2025.[10] Hence, even though water has commonly formed a basis for cooperation, climate change is progressively shaping a world in which peace and political stability could be less likely to prevail. Indeed, as the World Commission on Human Security stated back in 2003:

> water scarcity may... escalate tensions between nations. While the last outright war over water occurred 4500 years ago, historical precedent may not be an absolute guide in the case of water scarcity... [W]hile most international interactions over shared basins have been cooperative, tensions exist in many areas.[11]

It has become imperative, therefore, for the international community to acknowledge 'the global threat posed by water scarcity'.[12]

As a result of the above-mentioned pressures, global freshwater ecosystems, while being among the most diverse in biological terms, are being degraded at a faster rate than both marine and terrestrial biomes. Freshwater, along with wild fisheries, are two ecosystem services already exploited beyond their capacity to

6   Richter, B. D., Postel, S., Revenga, C., Scudder, T., Lehner, B., Churchill, A. and Chow, M. 'Lost in Development's Shadow: The Downstream Human Consequences of Dams', *Water Alternatives*, 2012; 3 (2): 14–42, at 38.

7   World Water Assessment Programme, *The UN World Water Development Report 3: Water in a Changing World*. (UNESCO and Earthscan, 2009). Available online at http://webworld.unesco.org/water/wwap/wwdr/wwdr3/tableofcontents.shtml (accessed March 14, 2013) pp. 29–40.

8   Falkenmark, M. and Molden, D. 'Wake Up to Realities of River Basin Closure', *International Journal of Water Resources Development*, 2008; 24 (2): 201–15.

9   WWDR (2009) (note 7) Part 3, at 160–236.

10   WWDR (2009) (note 7), at 36.

11   Commission on Human Security, *Human Security Now* (Commission on Human Security, 2003). Available online at www.policyinnovations.org/ideas/policy_library/data/01077/_res/id=sa_File1 (accessed March 14, 2013), at 15.

12   Ibid.

meet current, let alone future, needs.[13] The freshwater biome has the highest proportion of threatened species and is the least well conserved in nature reserves.[14]

Compounding these challenges, freshwater systems defy political boundaries. There are 145 countries sharing the world's 276 international river basins, which cover nearly half the Earth's land surface. These basins are home to 40 percent of the world's population and represent 60 percent of global freshwater flow.[15] In these basins, states must cooperate to reconcile existing and future competing uses, while ensuring the long-term viability of the resource. When states fail to do so, water disputes are likely to arise – and have happened frequently throughout history. And if the rules of the game are unclear, there is more scope for power imbalances among riparian states and thus for harmful unilateral measures and associated disagreements.

The essential link between the use of freshwater resources, livelihoods and economic activity means that the management of shared water resources is usually contested, with each jurisdiction seeking to maximize benefits for their own people.[16] As noted by Benvenisti, 'each state is interested in getting more out of the resource with minimal costs, and these interests conflict with those of other users. This conflict can lead the parties to a race to the bottom'.[17] In this context, international law plays an important role in leveling the playing field among co-riparian states, enabling ongoing dialogue and information exchange, as well as establishing clear standards for governments to be held accountable by the relevant stakeholders.

Recent times have seen a greater awareness and understanding of the pressures faced by freshwater ecosystems and, in particular, a heightened recognition of the need for stronger and better integrated governance frameworks at all levels.[18] In this sense, Agenda 21 identifies the need to strengthen and develop 'the appropriate institutional, legal and financial mechanisms' for water management.[19] This recommendation was taken further in numerous global policy statements over the last decade or so. For instance, the Ministerial Declaration of the 2nd World Water

13 Vörösmarty, *et al.* (note 4), at 167.
14 Ibid.
15 Giordano, M. A. and Wolf, A. T. 'The World's International Freshwater Agreements: Historical Developments and Future Opportunities' in UN Environment Programme, *Atlas of International Freshwater Agreements* (UNEP, 2002), at 2.
16 Wouters, P. and Ziganshina, D. 'Tackling the Global Water Crisis: Unlocking International Law as Fundamental to the Peaceful Management of the World's Shared Transboundary Waters – Introducing the H2O Paradigm' in Grafton, R. Q. and Hussey, K. (eds) *Water Resources Planning and Management* (Cambridge University Press, 2011), at 175.
17 Benvenisti, E. *Sharing Transboundary Resources: International Law and Optimal Resource Use* (Cambridge University Press, 2002), at 31–42. See also Hardin, G. 'The Tragedy of the Commons' *Science*, 1968; 162: 1243.
18 Conca, K. 'Transnational Dimensions of Freshwater Ecosystem Governance', in Turton, A. R., Hattingh, J. H., Maree, G. A., Roux, D. J., Claassen, M. and Strydom, W. F. (eds) *Governance as a Trialogue: Governance-Society-Science in Transition*, (Springer, 2007), at 101.
19 UN General Assembly, *Report of the UN Conference on Environment and Development, Rio de Janeiro, Brazil, 3–14 June 1992, Agenda 21: A Programme for Action for Sustainable Development. Volume 1: Resolutions Adopted by the Conference.* UN Doc A/CONF.151/26/Rev.1 (UN, 1992), at 277.

Forum identifies a number of key challenges in addressing water issues for the 21st century, including governing water wisely and promoting the peaceful cooperation over transboundary water resources between states.[20] The Second World Water Assessment is even more forthright about the role of governance, observing that the lack of access to safe and sufficient water to meet global needs is first and foremost a crisis of governance.[21]

Water has also been increasingly seen as a *global* security issue.[22] On the occasion of the Third World Water Forum, UN Secretary-General Kofi Annan reflected that, 'water is likely to become a growing source of tension and fierce competition between nations, if present trends continue, but it can also act as a catalyst for cooperation'.[23] Similarly, Kofi Annan's successor has recently observed that, 'the challenge of securing safe and plentiful water for all is one of the most daunting challenges faced by the world today... Our experiences tell us that environmental stress due to lack of water may lead to conflict and would be greater in poor nations'.[24]

While there has been a growing recognition of the need to address governance and water security issues from the local to the global level, a key challenge that remains is the fragmented legal architecture for transboundary water management.[25] At the global level, states must rely on the general rules and principles of international customary law in their relations with neighbors – rules that may be vague and contested – and which lack a neutral enforcement mechanism. Regionally, only three international legal instruments exist and within just three regions: Europe, Central Asia and Southern Africa.[26] At the basin level, over 400 treaties have been adopted since 1820, but an estimated 158 international watercourses lack cooperative management frameworks.[27] Moreover, a common trend

20 Second World Water Forum, *Ministerial Declaration of the Hague on Water Security in the 21st Century*. (March 22, 2000). Available online at www.waternunc.com/gb/secwwf12.htm (accessed March 14, 2013).

21 World Water Assessment Programme, *Water – A Shared Responsibility. The UN World Water Development Report 2* (UNESCO, 2006), at 3. Available online at http://unesdoc.unesco.org/images/0014/001444/144409e.pdf (accessed March 14, 2013).

22 Wouters, P., Vinogradov, S. and Magsig, B. O. 'Water Security, Hydrosolidarity, and International Law: A River Runs Through It...', *Yearbook of International Environmental Law*, 2008; 19 (1): 97–137.

23 UN, *UN Launches International Year of Freshwater to Galvanize Action on Critical Water Problems* [Press release] (UN Department of Public Information, December 10, 2002). Available online at www.un.org/events/water/launchrelease.pdf (accessed March 14, 2013).

24 Ban Ki-moon, Address as Prepared for Delivery to the Davos (World Economic Forum, January 24, 2011). Available online at www.un.org/apps/news/infocus/sgspeeches/search_full.asp?statID=177 (accessed March 14, 2013).

25 Pegasys Strategy and Development for DfID and WWF-UK, *International Architecture for Transboundary Water Management: Policy Analysis and Recommendations* (WWF-UK, 2010).

26 UN Economic Commission for Europe Convention on the Protection and Use of Transboundary Watercourses and International Lakes (adopted March 17, 1992, entered into force October 6, 1996) 1936 UNTS 269; Directive 2000/60/EC of 23 Oct 2000 of the European Parliament and of the Council establishing a framework for community action in the field of water policy (2000) OJ L327/1 (EU Water Framework Directive); Southern Africa Development Community Revised Protocol on Shared International Watercourses (adopted August 7, 2000, entered into force September 22, 2003) (2001) 40 ILM 321.

27 Giordano and Wolf (note 15), at 7.

has been for states to enter into bilateral arrangements, even in basins that are shared by more than two states.[28] Finally, states are often parties to several disconnected and not harmonized agreements, including when drained by more than one international watercourse or when subject to more than one treaty on the same basin, which makes implementation more difficult.

On the current status of international agreements, UN-Water comments that:

> existing agreements are sometimes not sufficiently effective to promote integrated water resources management due to problems at the national and local levels such as inadequate water management structures and weak capacity in countries to implement the agreements as well as shortcomings in the agreements themselves (for example, inadequate integration of aspects such as the environment, the lack of enforcement mechanisms, limited – sectoral – scope and non-inclusion of important riparian states).[29]

It was the weakness of the legal architecture for transboundary waters that first prompted the UN General Assembly (UNGA) to consider the codification and progressive development of the law of the non-navigational uses of international watercourses. UNGA Resolution 2669 (XXV) notes that,

> despite the great number of bilateral and other regional regulations, as well as the Convention on the Regime of Navigable Waterways of International Concern…and the Convention relating to the Development of Hydraulic Power affecting more than one state,…the use of international rivers and lakes is still based in part on general principles and rules of customary international law.[30]

Accordingly, the work of the UNGA in developing the Convention aimed at addressing a pervasive need to strengthen the legal architecture for transboundary waters at the global level. As seen above, such a need remains true today, especially in the context of an increasingly complex world.

Despite the fact that 103 states[31] voted in favor of the UNWC, this widespread and representative endorsement has not been easily translated into formal

28 Zawahri, N. A. and Mitchell, S. M. 'Fragmented Governance of International Rivers: Negotiating Bilateral Versus Multilateral Treaties', *International Studies Quarterly*, 2011; 55: 835.

29 UN-Water, *Transboundary Waters: Sharing Benefits, Sharing Responsibilities*. Thematic Paper. (UN, 2008). Available online at www.unwater.org/downloads/UNW_TRANSBOUNDARY.pdf (accessed March 14, 2013), at 6.

30 UN General Assembly, Resolution 2669 (XXV) Progressive Development and Codification of the Rules of International Law relating to International Watercourses, 1920th Plenary Meeting (December 8, 1970). Available online at www.un.org/ga/search/view_doc.asp? symbol=A/Res/2669(XXV) (accessed March 14, 2013).

31 UN General Assembly, '99th Plenary Meeting' (May 21, 1977). UN Doc A/51/PV.99, at 8. The official record shows that three additional states intended to vote in favor of the UNWC: Belgium, Nigeria and Fiji.

instruments of ratification.[32] A number of questions follow from the latter observation. Why has the UNWC not entered into force? Does it represent an accurate codification and progressive development of international law in the field? Do states agree with the text of the Convention? If the UNWC entered into force, what impact would it have? Similarly, what would happen if the UNWC did not enter into force? If the UNWC did enter into force, what mechanisms should be in place to ensure its effective implementation? What relevance does the UNWC have within particular regions, river basins and countries? How does the UNWC compare with and potentially complement or contradict other treaty regimes, both related to watercourses, and other environmental agreements, e.g. on climate change or desertification? Is the UNWC still relevant and capable of addressing contemporary challenges such as ensuring equitable and sustainable access to the world's freshwater supplies?

The search for answers to those questions provided the genesis and inspiration for this book. For more than five years, leading experts have been sharing their ideas and opinions on the role and relevance of the UNWC in strengthening international law for transboundary water management. The UNWC Global Initiative, encompassing a wide spectrum of institutional and individual partners, has provided a platform for exchanging experiences and developing tools and knowledge around the Convention. This book brings together some of the key elements of that work, and commits ideas, opinions and experience to paper.

The book is divided into six parts. In addition to this introduction, Part I, in Chapter 2, examines the evolution of international water law, as well as the context and process by which the UNWC was developed and adopted. The same chapter provides an important overview of the text of the UNWC. Part I then explores the ratification process in more detail, by seeking to identify the primary reasons the UNWC has taken so long to enter into force. Chapter 3 looks at the degree to which treaty congestion, lack of leadership and low levels of awareness and capacity have slowed down the ratification process. Chapter 4 considers misconceptions around the Convention's provisions. Chapter 5 complements that analysis by examining the various theoretical accounts as to why states cooperate. Such accounts are then compared to the UNWC context in order to distil some key reasons why states have ratified this global framework instrument.

Part II moves on to consider the potential effects that entry into force of the UNWC could bring. Chapter 6 looks at how the authority and functions of the UNWC might be affected by its entry and non-entry into force. Chapter 7 then offers the context in which the UNWC operates, namely the international architecture for transboundary waters. The section concludes with Chapter 8, which draws upon experience in the implementation of existing multilateral environmental agreements, to ascertain the factors that might influence the effectiveness of the UNWC once it enters into force.

---

32 'Ratification' is used throughout the book in its broadest sense, which thus encompasses acceptance, accession and approval.

Part III examines the role and relevance of the UNWC within specific regions (Chapters 9–12), basins (Chapters 13–15), and countries (Chapters 16–17). The intention of these studies is to examine the particular challenges and opportunities relating to transboundary waters within specific geopolitical landscapes; compare the provisions of the UNWC to the existing legal architecture; and identify the potential value added by the Convention in each case.

Part VI goes on to scrutinize the relationship between the UNWC and other multilateral environmental agreements and global development policies. Such an analysis encompasses consideration of the Climate Change Convention (Chapter 18), the Desertification Convention (Chapter 19), the UNECE Water Convention (Chapter 20), and development and environmental goals adopted under the auspices of the UN (Chapter 21).

Part V examines issues that might become particularly pertinent once the UNWC enters into force. Chapter 22 examines how the implementation of the UNWC, upon entry into force, might be enhanced through institutional design features; Chapter 23 links the Convention to the work conducted by the International Law Commission on transboundary aquifers. Given that the UNWC was adopted in 1997, Chapter 24 examines how customary international law has evolved since the adoption of the UNWC, and if and how the Convention might be reconciled with such developments, focusing on two key areas: environmental impact assessment and the human right to water.

Finally, Part VI looks at the role and relevance of the UNWC within the context of certain emerging challenges and future trends, including climate change adaptation, the concept of benefit sharing, water security and power asymmetry.

Ultimately, the book seeks to take us one step further in understanding how the UNWC, as a global legal instrument of a 'framework' nature, can contribute to the sustainable management of international watercourses and their ecosystems, based on effective and mutually beneficial interstate cooperation, and in the context of the multi-level legal governance of those precious resources.

2 The progressive development of international water law

Stephen C. McCaffrey

This chapter places the UN Watercourses Convention (UNWC) in its historical context, traces its development and provides a general overview of its provisions. Such a background is important to show why the Convention itself and its preparatory work have had widespread influence on interstate relations. Indeed, the Convention's authority derives, to a great extent, from the thorough and inclusive analytical and negotiation process that preceded its adoption. The Convention's entry into force will only reinforce its stature as the most authoritative set of provisions on the law of the non-navigational uses of international watercourses.

The first section of the chapter surveys the evolution of international water law, and describes the existing customary legal framework. The chapter then focuses on the Convention itself, first reviewing the International Law Commission's (ILC's) elaboration of the draft articles on which the Convention is based, then looking at the negotiation of the Convention within the UN, including the positions taken by states on key issues. Finally, the chapter offers a general overview of the UNWC, highlighting provisions which later chapters examine in greater detail.

Evolution of international water law and the existing legal framework

International water law – or, more precisely, the law of the non-navigational uses of international watercourses – developed quite slowly until the middle of the twentieth century.[1] To be sure, the subject was addressed in ancient times[2] and later

---

1 Brown Weiss, E. *The Evolution of International Water Law*. Recueil des Cours: Collected Courses of the Hague Academy of International Law 331. (Martinus Nijhoff Publishers, 2009), at 163–446. See also McCaffrey, S. C. 'The Evolution of the Law of International Watercourses', *Austrian Journal of Public and International Law*, 1993; 45: 87.

2 The first known water agreement concerned non-navigational use of the waters of the Tigris River. It was concluded in approximately 3100BC between the Mesopotamian city-states of Umma and Lagash. This was, in fact, the earliest recorded treaty. It is inscribed on the 'Stela of the Vultures', which is housed in the Louvre. McCaffrey, S. C. *The Law of International Watercourses* (2nd edn, Oxford University Press, 2007), at 59–60.

in diplomatic claims[3] and occasional bilateral treaties, but it lagged far behind navigation in terms of the attention and importance given to each by states in their agreements.[4] In a study of over 2000 international water agreements, Brown Weiss notes that the percentage of such agreements dealing principally with navigation 'peaked in the period 1700–1930', while treaties focused on 'allocation and use issues were most significant as a percentage of total [international water] agreements negotiated during the period 1931–2000'.[5] A prominent exception was the 1919 Treaty of Versailles, which contained provisions on non-navigational uses – including hydropower, irrigation, fishing and water supply.[6] As to treaties dealing chiefly with pollution and the protection of ecosystems, these 'emerged mainly after 1950, although a few agreements, such as the 1909 Boundary Waters Treaty between Canada and the United States, contained at least one significant pollution control provision'.[7]

During the twentieth century, attempts at codification and progressive development of international water law generally paralleled the above trends. One of the very first such efforts was the 1911 Madrid Resolution on International Regulations Regarding the Use of International Watercourses for Purposes other than Navigation, adopted by the Institute of International Law (IIL).[8] The Resolution addresses both contiguous and successive international watercourses. Regarding the latter, where 'a stream traverses successively the territories of two or more States', the Resolution forbids all injurious alterations to the water; and prohibits industrial activities from 'tak[ing] so much water that the constitution, otherwise called the utilizable or essential character of the stream shall, when it reaches the territory downstream, be seriously modified'.

The Madrid Resolution was a half-century ahead of its time. It was followed by the IIL's 1961 Salzburg Resolution, according to which disputes between riparian states are to be settled 'on the basis of equity, taking into consideration the respective needs of the States, as well as any other circumstances relevant to any particular case'[9] – a provision that appears to be consistent with the principle of equitable utilization. Procedural rules also began to appear in this Resolution, which required prior notification of new uses and negotiation of any related disputes. The 1961 Salzburg Resolution further reflected the progress in understanding international watercourses by addressing its rules to the entire hydrographic basin.

3 What has been described as 'the first diplomatic assertion of any rule of international law' concerning non-navigational uses of international watercourses was made by Holland in 1856. Smith, H.A. *The Economic Uses of International Rivers* (P. S. King and Son, 1931), at 137.
4 Brown Weiss (note 1), at 163; McCaffrey (note 2), at 171.
5 Brown Weiss (note 1), at 235. The year 2000 was the cut-off date for Brown Weiss's survey.
6 Treaty of Peace between the Allied and Associated Powers and Germany (signed June 28, 1919) (1919) 225 Consolidated Treaty Series 189.
7 Brown Weiss (note 1), at 235.
8 (1911) 24 *Yearbook of the IIL* 365 (Madrid Resolution).
9 IIL, 'Utilization on Non-Maritime International Waters (Except for Navigation)' (1961) 49(II) *Yearbook of the IIL* 371, at Article 3 (Salzburg Resolution).

The Institute's final resolution of 1979 addresses pollution of international watercourses. The Athens Resolution recognizes the 'common interest' of riparian states 'in a rational and equitable utilization of such resources through the achievement of a reasonable balance between the various interests'.[10] In so doing, the Resolution reflects progress in thinking and state practice concerning the law of international watercourses.

A true milestone in efforts at codification of international water law was achieved by the adoption in 1966 of the Helsinki Rules on the Uses of the Waters of International Rivers by the International Law Association (ILA).[11] The Helsinki Rules take a drainage basin approach to the regulation of the uses of international watercourses, moving well beyond the former focus on rivers and lakes. Their fundamental rule concerning non-navigational uses is equitable utilization. Article IV of the Helsinki Rules provides: '[e]ach basin State is entitled, within its territory, to a reasonable and equitable share in the beneficial uses of the waters of an international drainage basin'. 'Substantial injury' is not prohibited but is one of the factors listed in Article V that is to be taken into consideration in determining what constitutes a reasonable and equitable share within the meaning of Article IV.

The Helsinki Rules also deal with pollution of the waters of an international drainage basin, navigation, timber floating and procedures for the prevention and settlement of disputes. Per Article XXIX, the procedures include the provision of information 'concerning the waters of a drainage basin within [a state's] territory and its use of, and activities with respect to such waters' and the giving of notice by a state, 'regardless of its location in a drainage basin', to potentially affected states concerning proposed measures 'which would alter the regime of the basin in a way which might give rise to a dispute'. These procedures are ahead of their time. Although they are not as detailed, they in fact go beyond the procedural provisions of the UNWC, in particular as to matters on which information is to be exchanged and in their explicit requirement that a state must give prior notification of planned measures regardless of its location in a basin. This latter feature anticipates by several decades the realization that 'harm' can, in effect, flow upstream, in the form of the foreclosure of future upstream uses through the maximization of use in downstream states.[12]

The Helsinki Rules constitute a landmark in the evolution of international water law. They are still referred to by states in negotiations and have had a significant impact on the development of the law. They embody a number of

---

10 IIL, 'The Pollution of Rivers and Lakes and International Law' in (1980) *Yearbook of the IIL* vol. 58(I) 197 (Athens Resolution).

11 ILA, 'Helsinki Rules on the Uses of the Waters of International Rivers Adopted by the International Law Association at the fifty-second conference, held at Helsinki in August 1966' in *Report of the Committee on the Uses of the Waters of International Rivers* (International Law Association, 1967). Available online at http://webworld.unesco.org/water/wwap/pccp/cd/pdf/educational_tools/course_modules/reference_documents/internationalregionconventions/helsinkirules.pdf (accessed March 15, 2013).

12 See Salman, S. M. A. 'Downstream Riparians Can Also Harm Upstream Riparians: The Concept of Foreclosure of Future Uses' *Water International*, 2010; 35: 350.

principles and reflect trends that later found expression in the UNWC. In fact, in 1970, Judge E. J. Manner, the Finnish Chair of the ILA committee that prepared the Rules, proposed in the 6th (Legal) Committee of the UN General Assembly (UNGA), in his capacity as a government delegate, that the ILC begin the study of the law of the non-navigational uses of international watercourses, using the Helsinki Rules as a model.[13] As early as 1959, the UNGA had requested the UN Secretary-General to prepare a report on legal problems relating to the utilization and use of international rivers, in Resolution 1401 (XIV) of November 21, 1959.[14] While the Helsinki Rules were not mentioned in the UNGA Resolution referring the topic to the ILC,[15] they are undeniably an important precursor of the ILC's draft articles on international watercourses and thus of the UNWC.

Also contributing to the development of international water law are the decisions of international courts and tribunals. These decisions have been remarkably few in number, although the pace of filings by states of cases concerning international watercourses seems to be picking up.[16] The International Court of Justice (ICJ) and its predecessor, the Permanent Court of International Justice (PCIJ), have together decided only three cases concerning the non-navigational uses of international watercourses: the *Meuse* case,[17] the *Gabčíkovo* case[18] and the *Pulp Mills* case.[19] The PCIJ also decided a case concerning navigation, the *River Oder* case,[20] which introduces an important concept that was applied to non-navigational uses later, in the *Gabčíkovo* judgment. The *Meuse* case contributed little to the development of the law, while the latter three were of greater importance in this connection. Space constraints accordingly make it advisable to focus on these latter cases.

Gabčíkovo involved a project consisting of a series of dams on the Danube, to be constructed pursuant to a 1977 treaty between Hungary and Czechoslovakia. Hungary ceased working on the project in 1989 and, in May 1992, attempted to

---

13 UN General Assembly 6th Committee, '1225th Meeting: Progressive Development and Codification of the Rules of International Law relating to International Watercourses (A/7991)' (1970) UN Doc A/C.6/SR.1225, at 267.

14 The report, UN Doc A/5409, was later published along with a supplementary report, UN Doc A/CN.4/274, in the 1974 Yearbook of the ILC. *Yearbook of the ILC*, (1974); II(2): 33.

15 The resolution recommended that the ILC 'take up the study of the law of the non-navigational uses of international watercourses with a view to its progressive development and codification'. UNGA, Res 2669 (XXV) 'Progressive Development and Codification of the Rules of International Law Relating to International Watercourses' (December 8, 1970).

16 Four cases concerning international watercourses have been filed with the ICJ in the past 20 years, two of which (*Gabčíkovo* and *Pulp Mills*, both discussed here) involved non-navigational uses.

17 *Diversion of Water from the Meuse (Netherlands v Belgium)*, PCIJ Rep Series A/B No 70. For a discussion of this case, see McCaffrey (note 2), at 206–9.

18 *Case Concerning the Gabčíkovo-Nagymaros Project (Hungary v Slovakia)* (Judgment) [1997] ICJ Rep 7 (*Gabčíkovo*).

19 *Case Concerning Pulp Mills on the River Uruguay (Argentina v Uruguay)* (Judgment) [2010] ICJ Rep 14 (*Pulp Mills*).

20 *Territorial Jurisdiction of the International Commission of the River Oder (Czechoslovakia, Denmark, France, Germany, Great Britain and Sweden/Poland)* PCIJ Rep Series A No 23, at 5–46 (River Oder).

terminate the treaty unilaterally. Czechoslovakia then decided to put the upper part of the project into operation by damming the Danube within its own territory – something that the treaty did not contemplate. The resulting dispute was brought before the ICJ, Slovakia having assumed Czechoslovakia's rights and obligations under the treaty upon the dissolution of the latter state as of January 1, 1993. The Court held that the treaty remained valid and that both countries had breached it.

In the course of its decision, which reinforced the difficulty of terminating treaties unilaterally, the ICJ made a number of important statements concerning international water law and the environment. In particular, with regard to the effect of Czechoslovakia's damming of the Danube, the Court referred to Hungary's 'basic right to an equitable and reasonable sharing of the resources of an international watercourse'.[21] This 'basic right' derived not from the treaty but from customary international law. That the Court characterized the right as 'basic' is a testament to the extent to which equitable and reasonable utilization has become a cornerstone of the law in the field.

In addition, the ICJ applied to non-navigational uses the 'community of interest' concept, first articulated by the PCIJ in the *River Oder* case in the context of navigation rights. According to the PCIJ, the community of interest in an international watercourse

> becomes the basis of a common legal right, the essential features of which are the perfect equality of all riparian States in the use of the whole course of the river and the exclusion of any preferential privilege of any one riparian State in relation to the others.[22]

The ICJ then built on that statement, recognizing that '[m]odern development of international law has strengthened this principle for non-navigational uses of international watercourses as well, as evidenced by the adoption of the [UNWC] by the [UNGA]'.[23] The ICJ thus cited the UNWC as evidence of the contemporary vitality of the principle of community of interest, notwithstanding that the Convention had been concluded only four months earlier. This may well be because of the Convention's provenance: 20 years of work by the ILC followed by negotiations open to all states within the UN, which produced an agreement that conforms closely to the ILC's draft articles and was endorsed by more than 100 governments.

The *Pulp Mills* case was brought by Argentina against Uruguay in relation to two pulp mills authorized by Uruguay, only one of which was ultimately built, on the Uruguay River. The ICJ held that Uruguay had breached its procedural obligations, although not its substantive duties, under the 1975 Statute of the River Uruguay (1975 River Uruguay Statute)[24] – a treaty binding on the two states. As

---

21   *Gabčíkovo* (note 18), at 54.
22   Ibid., at 56, quoting from *River Oder* (note 20), at 27.
23   Ibid.
24   Statute of the River Uruguay (adopted February 26, 1975) 1295 UNTS 339 (1975 River Uruguay Statute).

in the *Gabčíkovo* case, the Court held the parties to their agreement, underscoring that the relevant provisions of the treaty were due diligence obligations, rather than obligations to achieve particular results. The Court noted the importance of monitoring the ecological condition of the river, to ensure that the initial positive indications as to the pulp mill's performance were maintained.

The ICJ also drew upon the principle of equitable and reasonable utilization, as well as the concept of sustainable development, in interpreting the requirements of Article 27 of the 1975 River Uruguay Statute, which provides: 'The right of each Party to use the waters of the river, within its jurisdiction, for domestic, sanitary, industrial and agricultural purposes shall be exercised without prejudice to the application of the procedure laid down in articles 7 to 12 when the use is liable to affect the regime of the river or the quality of its waters'. Referring to the uses mentioned in Article 27, the judgment states:

> [t]he Court wishes to add that such utilization could not be considered to be equitable and reasonable if the interests of the other riparian State in the shared resource and the environmental protection of the latter were not taken into account. Consequently, it is the opinion of the Court that Article 27 embodies this interconnectedness between equitable and reasonable utilization of a shared resource and the balance between economic development and environmental protection that is the essence of sustainable development.[25]

Several aspects of this passage deserve attention. First, since the 1975 River Uruguay Statute does not employ the expression 'equitable and reasonable utilization', the Court must have taken it from the corpus of customary international law governing the use of international watercourses, reflected in the UNWC. This use of the doctrine of equitable utilization is in line with the *Gabčíkovo* judgment and further solidifies the principle as one of the most important and fundamental in the field.

Second, it is significant that the Court found that, for a state's use of an international watercourse to be equitable and reasonable, it must take into account not only the interests of other riparian states but also protection of the environment of the shared resource. This is in line with Article 5(1) of the UNWC, according to which equitable and reasonable utilization requires 'taking into account the interests of the watercourse States concerned, consistent with adequate protection of the watercourse'.

Third and finally, the Court read Article 27 through the lens of contemporary watercourse and environmental law, as also embodied in Article 5(1) of the UNWC, when it said that the provision 'embodies [the] interconnectedness between equitable and reasonable utilization of a shared resource and ... sustainable development'. This interpretation sheds further light on the ICJ's understanding of the content of the principle of equitable utilization.

---

25 *Pulp Mills* (note 19), at 53.

The ICJ also stressed the key role played by the Administrative Commission of the River Uruguay, formed by the 1975 River Uruguay Statute, in the management and protection of the river, recognizing that joint machinery was necessary for the river's optimal and rational utilization. The judgment reflects the Court's view that, when states have established a joint management mechanism, one of them may not opt for procedures outside those specified in their agreement, even if that state believes that the procedures chosen accomplish the same purpose as would have been served by going through the joint body. The *Pulp Mills* case is also discussed in Chapter 24.

The UN Watercourses Convention

The UNWC was negotiated at the UN on the basis of draft articles prepared by the ILC. The ILC's work on the topic began in 1974 and was completed in 1994.[26] During these 20 years, the Commission's work was guided by a succession of five special rapporteurs: Richard D. Kearney (1974–1976), Stephen M. Schwebel (1977–1981), Jens Evensen (1982–1984), Stephen C. McCaffrey (1985–1991) and Robert Rosenstock (1992–1994).[27] The work cannot be said to have been linear, as different approaches were tried, partly because the ideas of some of the special rapporteurs varied from those of others and partly as a result of the regular interaction between the ILC and the UNGA. Such interaction chiefly consisted of the comments on the ILC's annual reports to the UNGA, in the 6th (Legal) Committee. The results of these debates were made available to the ILC and taken into account by its special rapporteurs, as appropriate. Thus, by the time the ILC forwarded the final product of its efforts to the UNGA, it had already taken into account the more strongly-held views of UN member states – at least the ones expressed in 6th Committee debates or in the UNGA itself.

Drafting and voting process

The Convention was negotiated in the 6th Committee, convening for this purpose as a 'Working Group of the Whole', during two sessions in 1996 and 1997.[28] The

---

26 See McCaffrey, S. C. 'The International Law Commission Adopts Draft Articles on International Watercourses', *American Journal of International Law*, 1995; 89: 395–404.

27 For reports of the special rapporteurs, see ILC, 'The Law of Non-navigational Uses of International Watercourses – Analytical Guide' (in ILC, *Analytical Guide to the Work of the International Law Commission 1949–1997*. New York: UN, last update November 2, 2011). Available online at http://untreaty.un.org/ilc/guide/8_3.htm (accessed March 14, 2013. These reports are also available in the ILC's Yearbooks, Volume 2, Part I.

28 For the resolution authorizing the work of the 6th Committee, see UNGA, Res 49/52 'Draft Articles on the Non-Navigational Uses of International Watercourses' (December 9, 1994) UN Doc A/Res/49/52. When the first three-week session provided for in that resolution proved insufficient, a second session, of two weeks, was authorized. UNGA, Res 51/206 'Convention on the Law of the Non-Navigational Uses of International Watercourses' (December 17, 1996) UN Doc A/Res/51/206.

negotiations were open to all UN member states as well as states that are members of UN specialized agencies. Much of the actual negotiating occurred in the Working Group's drafting committee, which was in fact open to all members of the Working Group of the Whole. The working group was chaired by Ambassador Chusei Yamada of Japan, while the drafting committee was chaired by Professor Johan Lammers of The Netherlands.[29] This subsection notes the results of the debates concerning the Convention itself and its most important provisions.

After two sessions of intensive negotiations,[30] the Working Group was not able to reach a consensus on the text of the UNWC as a whole. Accordingly, separate votes were held on Article 3, concerning the relation of the Convention to specific watercourse agreements; Articles 5–7, on equitable and reasonable utilization, factors relating to equitable and reasonable utilization and the obligation to prevent significant harm, respectively; and Article 33, on the settlement of disputes. A compromise proposal by the Chair of the working group on Articles 5–7 was adopted by a vote of 38 votes to 4 (China, France, Tanzania and Turkey), with 22 abstentions.[31] A vote was then taken on the Convention as a whole, the result of which was 42 in favor, 3 against (China, France and Turkey), and 19 abstentions.[32]

In the UNGA, Turkey called for a vote on the draft resolution,[33] something that is contrary to the normal practice of adopting Conventions by consensus. The UNWC was adopted by a vote of 103 to 3 (Burundi, China and Turkey), with 27 abstentions. Subsequent to the vote, Belgium, Nigeria and Fiji informed the Secretariat that they had intended to vote in favor of the Convention, which would bring the total of votes in favor to 106.

Brief review of the UNWC

The present subsection provides a brief overview of the UNWC; a number of its provisions will be the subject of further examination in subsequent chapters.[34] The Convention is divided into seven parts containing 37 articles: I. Introduction (Articles 1–4); II. General Principles (Articles 5–10); III. Planned Measures (Articles

---

29 Professor Lammers' introductory statement is contained in UNGA 6th Committee, 'Summary Record of the 24th Meeting' (October 25, 1996) UN Doc A/C.6/51/SR.24, at 2–9.

30 Ibid.

31 UNGA 6th Committee, 'Summary Record of the First Part of the 62nd Meeting' (April 4, 1997) UN Doc A/C.6/51/SR.62. The compromise proposal by the Chair of the Working Group is found in UN Doc A/C.6/51/NUW/WG/CRP.94.

32 UNGA 6th Committee, 'Summary Record of the Second Part of the 62nd Meeting' (April 4, 1997) UN Doc A/C.6/51/SR.62/Add.1.

33 The draft resolution is contained in UNGA, 'Convention on the Law of the Non-Navigational Uses of International Watercourses' (May 12, 1997) UN Doc A/51/L.72. The record of the 99th plenary meeting of the UNGA, at which the UNWC was adopted, is contained in UNGA, '99th Plenary Meeting' (May 21, 1997) UN Doc A/51/PV.99 (99th Plenary Meeting), which reflects Turkey's request for a vote at 4.

34 On the Convention generally, see Tanzi, A. and Arcari, M. *The UN Convention on the Law of International Watercourses: A Framework for Sharing* (Kluwer Law International, 2001); McCaffrey, S. C. and Sinjela, M. 'The 1997 UN Convention on International Watercourses' *American Journal of International Law*, 1998; 92: 97.

11–19); IV. Protection, Preservation and Management (Articles 20–26); V. Harmful Conditions and Emergency Situations (Articles 27–28); VI. Miscellaneous Provisions (Articles 29–33); and VII. Final Clauses (Articles 34–37). An Annex to the Convention contains provisions on arbitration pursuant to Article 33.

The heart of the UNWC comprises Parts II and III. These are also the parts that gave rise to most controversy during the negotiations in the 6th Committee. Articles 5–6, on equitable and reasonable utilization and the factors relevant for its application, respectively, were not particularly controversial in and of themselves. Most of the discussion concerning Article 5 concerned the efforts of some delegations to update Paragraph 1 to reflect then-modern concepts of environmental protection. Ultimately, the changes of this kind were a) the addition of the words 'and sustainable' after 'optimal';[35] and b) the insertion of the phrase 'taking into account the interests of the watercourse States concerned'.

A major innovation of the Convention is the concept of equitable participation, elucidated in Article 5(2). And when the ink of the UNWC was barely dry, the ICJ quoted Paragraph 2 in its *Gabčíkovo* judgment to reinforce the importance of cooperative participation in the management of international watercourses.[36]

The concept of equitable and reasonable utilization was not particularly controversial itself. However, since some downstream states perceived that concept as benefitting countries upstream, Articles 5–6 had to be balanced against Article 7, dealing with significant transboundary harm. It was the latter article, particularly its Paragraph 2, that proved the most difficult provision of the entire Convention on which to reach agreement. In reaching a compromise formulation, the Working Group made two key changes to Article 7(1): a) the replacement of the ILC's formula, 'exercise due diligence', with 'take all appropriate measures', which appears to be another way of saying the same thing; and b) the addition of an express reference to prevention. Thus, while the ILC's formulation spoke of a due diligence obligation of watercourse states 'to utilize an international watercourse in such a way as not to cause significant harm', the Convention obligates those states to 'take all appropriate measures to prevent the causing of significant harm'.

The final formulation of Article 7(2) has been described as having 'all the hallmarks of a hard-won compromise: it is rather awkward, somewhat ambiguous and probably not entirely satisfying to anyone, but something most delegations could live with'.[37] Hence, the final text of Article 7(2) of the UNWC is an *amended* version of the corresponding provision in the 1994 ILC Draft Articles. Still, the ILC's commentary on the legal effect of its version of Article 7 holds true as the authoritative interpretation of the duty to prevent significant transboundary harm, as finally codified in the Convention: 'a process aimed at avoiding significant harm as far as possible while reaching an equitable result in each concrete case'.[38]

---

35   See Tanzi and Arcari (note 34), at 110–17.
36   *Gabčíkovo* (note 18) at 80. On the concept of equitable participation, as contained in Art 5(2), see Tanzi and Arcari (note 34), at 117–20.
37   McCaffrey and Sinjela (note 34), at 101.
38   ILC, 'Report on its 46th Session' (May 2–July 22, 1994) UN Doc A/49/10 (1994), at 236. Available online at http://untreaty.un.org/ilc/reports/english/A_49_10.pdf (accessed March 14, 2013).

Part III, on planned measures, also gave rise to some controversy. This part contains rather detailed procedures concerning the provision of prior notification to other states of a planned activity that may have a significant transboundary adverse effect. This obligation of prior notification of planned measures was acceptable to most delegations. The states that raised objections in relation to Part III were Ethiopia, Rwanda and Turkey[39] – all upstream states on important international watercourses: the Blue Nile, the White Nile and the Tigris-Euphrates Basins, respectively.

Conclusion

This chapter has attempted to place the UNWC in historical context, by looking at its precursors, its provenance and its main provisions. The chapter has also considered the reception of the Convention by governments, international organisations and international tribunals, points that will be further addressed in subsequent chapters. In conclusion, there is no doubt that both the Convention itself and its preparatory work have had widespread influence on state practice, and continue to do so. As stated earlier, the Convention's entry into force will only reinforce its stature as the most authoritative set of provisions on the law of the non-navigational uses of international watercourses.

---

39 '99th Plenary Meeting' (note 33), at 9 (Ethiopia), 12 (Rwanda) and 4–5 (Turkey).

3 Possible reasons slowing down the ratification process

Joseph W. Dellapenna, Alistair Rieu-Clarke and Flavia Rocha Loures

The UN Watercourses Convention (UNWC) had considerable international support before and during its discussion and adoption at the UN General Assembly (UNGA) in 1997. At that time, there was widespread recognition of the need to codify and progressively develop international water law, to 'ensure the utilization, development, conservation, management and protection of international watercourses and the promotion of the optimal and sustainable utilization thereof for present and future generations'.[1] So, with attention to water resources rising in the global agenda and most transboundary basins still not subject to adequate cooperative mechanisms, why is it taking so long for the Convention to enter into force?

This chapter outlines some of the possible reasons slowing the UNWC's ratification process, looking at factors external to the Convention itself: treaty congestion, the lack of champions for the Convention after its adoption and the low levels of awareness and understanding among government officials and other stakeholders of the Convention's role and relevance.

Of course, there is no terminal date by which the UNWC must enter into force. It will do so whenever it attains the required number of 35 instruments of ratification, acceptance, approval and accession, regardless of how long it may take. As of the time of completing this book, there are 30 contracting states – only five states short of the number required for entry into force.

Treaty congestion

The first factor to consider here is the concept of *treaty congestion*. The UNGA adopted the UNWC in 1997, five years after the UN Rio Conference on Environment and Development (UNCED). Several multilateral environmental

---

1 UNWC, at Preamble, Recital 5.

agreements were negotiated during UNCED and came into force soon after.[2] States worldwide then began to take the initial steps in their implementation. In addition to binding agreements, 'soft-law' instruments emerged from UNCED,[3] further straining the capacity of countries to comply with their treaty obligations or otherwise implement their new commitments.

Such a multiplication of environmental legal and policy instruments in the 1990s may have overwhelmed states with reporting duties and periodical meetings. In Brown Weiss's words, 'treaty congestion leads to overload at the national level in implementing the international agreements. A country needs sufficient political, administrative, and economic capacity to be able to implement agreements effectively'.[4] With other multilateral environmental agreements taking up that capacity, therefore, the ratification process of the UNWC may have been impacted, despite the international community's ample support for its adoption in 1997.

Lack of champions

The lack of champions for the UNWC is another factor that might have affected its ratification process, and which may be related to treaty congestion to some extent as well. Often, after the adoption of a multilateral agreement, key individual governments, UN bodies or regional organizations – with a special role often played by the European Union – push forward the ratification process. In the case of the UNWC, its adoption was succeeded by that of the Kyoto Protocol,[5] later in the same year. The latter's ratification process proved to be a challenge itself: with climate change higher in the global agenda than water resources at the time, the focus quickly shifted away from the UNWC to secure the entry into force of the Kyoto Protocol.

In the last few years, however, this scenario began to change, with a good part

---

2   These include: UN Economic Commission for Europe (UNECE) Convention on the Protection and Use of Transboundary Watercourses and International Lakes (adopted March 17, 1992, entered into force October 6, 1996) 1936 UNTS 269 (UNECE Water Convention); UN Framework Convention on Climate Change (adopted May 9, 1992, entered into force March 21, 1994) 1771 UNTS 107 (UNFCCC); UN Convention on Biological Diversity (adopted June 5, 1992, entered into force December 29, 1993) 1760 UNTS 79; UN Convention to Combat Desertification in Those Countries Experiencing Serious Drought and/or Desertification, Particularly in Africa (adopted June 17, 1994, entered into force December 26, 1996) 1954 UNTS 3.

3   See UNCED (Rio de Janeiro, June 3–14, 1992), 'Rio Declaration on Environment and Development' (January 1, 1993), UN Doc A/CONF.151/26/Rev.1 (Vol. I), at 3; UNCED (Rio de Janeiro, June 3–14, 1992), 'Agenda 21: A Programme for Action for Sustainable Development' UN Doc A/CONF.151/26/Rev.1 (Vol. I), at 9; UNCED (Rio de Janeiro, June 3–14, 1992), 'Non-Legally Binding Authoritative Statement of Principles for a Global Consensus on the Management, Conservation and Sustainable Development of All Types of Forests' (January 1, 1993) UN Doc A/CONF.151/26/Rev.1 (Vol. I), at 480.

4   Brown Weiss, E. 'Symposium: International Environmental Law: Contemporary Issues and the Emergence of a New World Order', *Georgetown Law Journal*, 1993; 81: 675 at 697.

5   Kyoto Protocol to the UNFCCC (adopted December 11, 1997, entered into force February 16, 2005) 2303 UNTS 148.

of the international community finally coming together around the goal of bringing the UNWC into force. For example, the UN Secretary-General's Advisory
Board on Water and Sanitation (UNSGAB) has been raising awareness of the
UNWC and began calling on countries to join it in 2006 as a means to strengthen
transboundary water governance, thereby better enabling progress on access to
water and sanitation.[6] Under the Convention on Biological Diversity (CBD), there
are two recent decisions of the Conference of the Parties to the Convention (CoP)
recognizing the role of the UNWC in the protection and sustainable management
of ecosystems within or dependent upon international watercourses.[7] Finally, for
the past five years, the UN Secretary-General, acting in his capacity as depositary
and ahead of the 2007–2011 UN Treaty Events, reiterated an invitation to all
member states to become a party to the UNWC.[8]

At the regional level, a number of formal and informal calls for ratification have
been issued, including the 2007 Dakar Call for Action on the Ratification of the
UNWC by West African States, the 2008 Mediterranean Civil Society Statement
to the Euro-Mediterranean Ministerial Conference on Water, the Interim Guinea
Current Commission's 2010 OSU Declaration and the 2011 Bangkok Declaration
of African Basin Organizations.[9]

Through those and other efforts, the pace of ratification has finally picked up,
with three additional ratifications in 2011, five in 2012 and one in 2013, bringing

---

6   See UNSGAB, 'Hashimoto Action Plan: Compendium of Actions' (UNDESA March 2006) at 9.
     Available online at www.unsgab.org/docs/HAP_en.pdf (accessed March 15, 2013); UNSGAB,
     'Hashimoto Action Plan II: Strategy and Objectives Through 2012' (UNDESA, January 2010) at 5,
     15. Available online at www.unsgab.org/HAP-II/index.htm (accessed March 15, 2013).

7   CBD CoP-8 (Curitiba, March 20–31, 2006), 'Alien Species that Threaten Ecosystems, Habitats or
     Species (Article 8 [h]): Further Consideration of Gaps and Inconsistencies in the International
     Regulatory Framework' (June 15, 2006) UNEP/CBD/COP/DEC/VIII/27, at 3. Available online
     at www.cbd.int/doc/decisions/cop-08/cop-08-dec-27-en.pdf (accessed March 15, 2013); CBD
     CoP-9 (Bonn, May 19–30, 2008), 'Biological Diversity of Inland Water Ecosystems' (October 9,
     2008) UNEP/CBD/COP/DEC/IX/19, at 1. Available online at www.cbd.int/doc/decisions/cop-
     09/cop-09-dec-19-en.pdf (accessed March 15, 20130.

8   See UN Treaty Collection, 'Treaty Event 2011: Towards Universal Participation and
     Implementation'.   Available   online   at   http://treaties.un.org/pages/TreatyEvents.aspx?
     pathtreaty=Treaty/Focus/Page1_en.xml (accessed March 15, 2013); UN Treaty Collection, 'Past
     Treaty   Events'.   Available   online   at   http://treaties.un.org/Pages/TreatyEvents.aspx?
     pathtreaty=Treaty/PastTreaty/Page1_en.xml (accessed March 15, 2013).

9   West Africa Regional Workshop on the UN Watercourses Convention (Dakar, 20–21 September
     2007), 'Dakar Call for Action on the Ratification of the 1997 UN Convention on the Law of the
     Non-Navigational Uses of International Watercourses' (on file); Euro-Mediterranean Ministerial
     Conference on Water (Barcelona, April 13, 2010), 'Mediterranean Civil Society Statement'.
     Available online at http://www.medaquaministerial2010.net/events/12april/Civil_Society_
     Statement_3.doc (accessed March 15, 2013); 2nd Meeting of the Committee of West and Central
     African Ministers of the Guinea Current Large Marine Ecosystem Project (Accra, July 2, 2010),
     'OSU Declaration'. Available online at http://gclme.iwlearn.org/publications/rapports/2010/
     2010-osu-declaration-english-version/view (accessed March 15, 2013); 1st International
     Environment Forum for Basin Organizations (Bangkok, October 24–25 2011), 'Bangkok
     Declaration of African Basin Organizations' (on file).

the number of contracting states to 30. We view this as an indication that the absence of champions following the adoption of the Convention had indeed been an influential factor on its ratification process.

Low levels of awareness of the UNWC

While support for the UNWC has grown progressively as a result of this recent push by the international community, much work remains with regard to building awareness and understanding of the Convention's scope and substantive and procedural content. In view of the time gap since the adoption of the UNWC, the government officials now in charge of international law and transboundary water management are often not the same as those involved in the Convention's drafting, deliberations and adoption. With so many other treaties to track and implement, such officials have simply not paid enough attention to the UNWC.

Inadequate levels of awareness may also be due to the fact that, in several countries, responsibility for follow-up action on the Convention has been dispersed across several ministries – e.g. irrigation and water, law and justice, environment and foreign affairs – with no particular leading ministry. In such situations, following through with ratification may simply have fallen between the cracks.

For example, a 2008 survey of West African states found that many water-related ministries in the region did not have a solid grasp of the content and relevance of the UNWC, with some representatives not even aware of its existence.[10] More recently, during regional workshops in Central America and among participants from the Mekong Basin countries, the scenario was the same, with many of the officials present claiming no or little knowledge of the Convention.[11]

A related problem is the lack of capacity and funding. Many countries – such as those in West Africa, Central America and Southeast Asia – require technical and financial assistance to enable the consultation and decision-making processes that must precede ratification. With foreign aid often directed to project implementation, certain countries may have found it difficult to devote resources to assessing a global treaty and processing its ratification.

Low levels of understanding of the UNWC's role and relevance

Beyond the factors discussed above, the extent to which states may be willing to be bound by international water law and the way they perceive the value of the UNWC, as a global instrument, may also have been relevant. The Convention is a

---

10 Garane, A. UN Watercourses Convention: Applicability and Relevance in West Africa (UNWC Global Initiative 2008) at 23. Available online at www.internationalwaterlaw.org/bibliography/ WWF/RA_West_Africa.pdf (accessed March 15, 2013).

11 Central America Regional Workshop on the UN Watercourses Convention (Guatemala City, 16 March 2011), 'Final Report' at 23 (on file); Ministry of Water Resources and Meteorology of the Royal Government of Cambodia and others, 'Draft Summary of Notes' (International Water Law and UN Watercourses Convention Regional Awareness Workshop, Siem Reap, Cambodia, 10–11 May 2012) (on file).

global binding treaty *governing* transboundary *waters* and interstate cooperation. These are sensitive and highly politicized issues. The Convention is not alone as such: 12 years were necessary for the UN Convention on the Law of the Sea[12] to enter into force and several major maritime countries still have not ratified it.

Indeed, states' apprehension about loss of sovereignty over shared waters has often been an obstacle to the adoption of stronger, more comprehensive watercourse agreements. In the case of the UNWC, some states, such as China, the Czech Republic, Rwanda and Turkey, would have preferred an express reference, among the Convention's basic principles, to a watercourse state's sovereignty over the parts of an international watercourse located in its territory.[13] Some academics even maintain that China claims absolute territorial sovereignty over watercourses within its boundaries.[14] Such a claim would contradict the basic rules of contemporary international water law, which have long rejected the principle of absolute territorial sovereignty. Instead, it is now generally agreed that the 'management of international watercourses should be determined less by the traditional notion of "restricted sovereignty" than by a positive spirit of cooperation and effective interdependence'.[15]

In the case of China in particular, recent policy statements and bilateral agreements tend to suggest that – while the precise parameters of how China perceives its sovereignty over international watercourses have yet to be clearly and collectively defined – there is a growing outward recognition within the Chinese Government that sovereignty over international watercourses is indeed restricted.[16]

Article 5(2) of the UNWC reflects this balanced understanding, granting states

---

12   UN Convention on the Law of the Sea (adopted 10 December 1982, entered into force 16 November 1994) 1833 UNTS 3.

13   UNGA, 'General Assembly Adopts Convention on Law of Non-Navigational Uses of International Watercourses' (Press Release GA/9248, May 21, 1997). Available online at www.un.org/News/Press/docs/1997/19970521.ga9248.html (accessed March 15, 2013).

14   Freeman, J. D. 'Taming the Mekong: The Possibilities and Pitfalls of a Mekong Basin Joint Energy Development Agreement' *Asian-Pacific Law and Policy Journal*, 2009; 10: 543, at 464–65; Thorson, E. J. 'Sharing Himalayan Meltwater: The Role of Territorial Sovereignty', *Duke Journal of Comparative and International Law*, 2009; 19: 487–514, at 512–13.

15   Green Cross International, National Sovereignty and International Watercourses (GCI, 2000), at 18.

16   See, for instance, Remarks by Song Tao, H. E., Vice-Minister of Foreign Affairs of the People's Republic of China, 'Work Together for Common Development' (1st Summit of the Mekong River Commission, Hua Hin, Thailand, April 2010). Translation available online at www.mrcsummit2010.org/Speech-of-Vice-Minister-Song-Tao.pdf (accessed March 15, 2013). See also Agreement between the Government of the Russian Federation and the Government of the People's Republic of China Concerning Protection, Regulation and Reproduction of Living Water Resources in Frontier Waters of Rivers Amur and Ussury (1994); Agreement between the Government of the Russian Federation and the Government of the People's Republic of China Concerning Guidance of Joint Economic Use of Separate Islands and Surrounding Water Areas in Frontier Rivers (1997); Agreement between the Government of the Republic of Kazakhstan and the Government of the People's Republic of China Concerning Cooperation in Use and Protection of Transboundary Rivers (2001); Agreement between the Government of the Russian Federation and the Government of the People's Republic of China Concerning Rational Use and Protection of Transboundary Waters (2008).

'both the right to utilize the watercourse and the duty to cooperate in the protection and development thereof'. The International Law Commission (ILC) elaborates upon riparian states' right to use the watercourse, explaining that the attribute of sovereignty is correlative with the rights of other states sharing that watercourse. The ILC also notes that, even when states claim absolute territorial sovereignty in disputes over water resources, they end up taking into account the rights of other states when it comes to resolving the dispute.[17] Hence, the Convention's approach to sovereignty reflects customary law.

In addition, states involved in ongoing disputes on specific shared watercourses may feel more comfortable with the greater ambiguity of customary law. Such states, however, fail to realize the eventual need to resolve such disputes on the basis of widely accepted and codified international rules and procedures.

Moreover, some nations may regard a global regulation of international watercourses as not directly relevant to them and, for that reason, have prioritized treaties of immediate national interest. Such cases may include states satisfied with existing regional or watercourse agreements, island states with no transboundary waters or arid states with only transboundary fossil aquifers, to which the Convention does not apply. For example, UN Economic Commission for Europe (UNECE) member states may consider that they do not need to become parties to the UNWC because they have already joined the UNECE Water Convention.[18] Yet, the two instruments are not mutually exclusive and are generally in harmony with each other. The provisions in the UNECE Water Convention are often stricter and more detailed, and arguably focus more on water quality and environmental protection rather than on the water-sharing issues that some consider the heart of the UNWC. On the other hand, states that join the UNWC can, in implementing it, benefit from the wealth of experience and tools adopted under its sister Convention. Hence, the two treaties supplement each other and, in the future, could be implemented in a coordinated, mutually reinforcing manner. For example, the UNECE foreign aid programme for projects affecting international watercourses in the developing world would benefit from an effective global treaty codifying and developing the law in the field.

Related to the point above, certain states have questioned the role of a global Convention on the basis that transboundary waters are *regional* resources, i.e. the exclusive concern of riparian states. In our view, however, that argument overlooks

---

17 ILC, 'Report on the Work of Its 46th Session' (May 2–July 22, 1994) UN Doc A/49/10 (1994), at 98–99. Available online at http://untreaty.un.org/ilc/documentation/english/A_49_10.pdf (accessed March 15, 2013).

18 UNECE Convention on the Protection and Use of Transboundary Watercourses and International Lakes (adopted March 17, 1992, entered into force 6 October 1996) 1936 UNTS 269. The same arguments can also be raised with regard to the Revised SADC Protocol. Southern African Development Community Revised Protocol on Shared International Watercourses (adopted August 7, 2000, entered into force September 22, 2003) (2001) 40 ILM 321.

the importance of the multi-level legal governance of such resources.[19] As a global framework convention, the UNWC was designed to have a supportive and complementary role in relation to regional agreements. In terms of the existing legal architecture, many basins or sub-basins still lack the necessary legal frameworks to govern their management in an equitable and reasonable manner. As a codification and progressive development of international law in the field, the UNWC could play a supplementary role towards addressing such gaps in the law, as well as act as a catalyst for states to develop basin or sub-basin treaties.

Furthermore, the increasing global interdependences associated with water security have placed water as a priority topic in the global policy agenda. Yet, there is no single interstate platform under the auspices of which stakeholders can come together to discuss the global and interregional dimensions of sound water management and the water crisis. There is thus justification in having a global focal point for transboundary water issues, along similar lines to, for example, desertification, wetlands, biodiversity or climate change.[20] Arguably, such a role could be more effectively achieved if the UNWC were supported by the appropriate institutional mechanisms, e.g. secretariat, meeting of the parties and so forth.[21]

Finally, certain officials might consider that the UNWC is too closely tied to customary rules and thus fails to contribute sufficiently to progressive legal development in the field. This view questions the value added by the UNWC's entry into force. As discussed in Chapter 6, however, there is more progressive development in the Convention than meets the eye, especially with regard to two aspects: the UNWC clarifies and solidifies the procedural aspects of dispute prevention and resolution required for the principles of equitable and reasonable use and harm prevention to operate smoothly. Moreover, the UNWC spells out in detail the obligation to protect the ecosystems of international watercourses, including through *joint* action, where necessary. While there is growing evidence of progressive consolidation of a duty around ecosystem protection in international customary law,[22] the precise content and extent, if not the existence, of such a duty remains controversial. There is thus a significant potential gain in bringing the UNWC into force and ensuring its widespread ratification, so as to have that obligation spelled in a global, legally binding and relatively clear treaty.

---

19   See, in this connection, Salman, S. M. A. 'The Future of International Water Law – Regional Approaches to Shared Watercourses?' in Arsanjani, M., Cogan, J. K., Sloane, R. D. and Wiessnre, S. (eds), *Looking to the Future: Essays on International Law in Honor of W. Michael Reisman* (Martinus Nijhoff Publishers, 2011), at 907.

20   See generally Conca, K. 'Transnational Dimensions of Freshwater Ecosystem Governance' in Turton, A., Hattingh, J. H., Maree, G. A., Roux, D. J., Claasen, M. and Strydom, W. F. (eds), *Governance as a Trialogue: Government-Society-Science in Transition* (Springer, 2007), at 101–22.

21   Chapter 22 discusses further the institutional mechanisms that might support the implementation of the UNWC.

22   See, e.g., International Law Association, 'Berlin Rules on Water Resources' in Report of the 71st Conference (Berlin 2004), Articles 22–28.

Conclusion

The factors discussed above pertain to the levels of awareness and capacity around the UNWC, as well as to how states perceive the value of the Convention or relate to its binding nature. Such factors do not seem to reflect any real and direct opposition to the Convention.

The next chapter looks at how states' misconceptions around some of the UNWC's provisions might have been a factor preventing a broader formal endorsement of this global instrument and thus slowing down the ratification process.

4 Misconceptions regarding the interpretation of the UN Watercourses Convention

Salman M. A. Salman

A possible explanation for states' inaction in ratifying or acceding to the UN Watercourses Convention (UNWC) rests on varying and sometimes inaccurate interpretations of the provisions of the Convention. Based on the statements by the different delegations during the Convention's drafting process, as well as presentations by, and informal discussions with, various government officials from different countries, there are at least five misunderstandings and misconceptions about the Convention that stand out.[1] Those misunderstandings often relate to provisions that were contentious during the UNWC's negotiations. This chapter briefly reviews each of those issues.

Relationship between the principle of equitable and reasonable utilization and the obligation to prevent significant harm

The relationship between the principle of equitable and reasonable utilization and the obligation to prevent significant harm was the most important area of contention among states during negotiations. Article 7(1) of the UNWC obliges watercourse states, when utilizing an international watercourse in their territory, to take all appropriate measures to prevent the causing of significant harm to other co-riparians. When significant transboundary harm nevertheless occurs, Article 7(2) of the Convention requires the state causing the harm to 'take all appropriate measures, having due regard for Articles 5 and 6, in consultation with the affected State, to eliminate or mitigate such harm, and where appropriate, to discuss the question of compensation'. Since Articles 5–6 of the Convention deal with equitable and reasonable utilization, Article 7(2) requires giving due regard to that principle when significant harm has nevertheless been caused to another watercourse state.

At the time of the UNWC's adoption, the above formula 'was considered by a number of lower riparians to be sufficiently neutral not to suggest a subordination of the no-harm rule to the principle of equitable and reasonable utilization. A

---

1 Salman, S. M. A. 'The UN Watercourses Convention Ten Years Later: Why Has Its Entry into Force Proven Difficult?' *Water International* 2007; 32 (1): 1–15, at 9.

number of upper riparians thought just the contrary, namely that, that formula was strong enough to support the idea of such subordination.'[2]

Although the compromise facilitated adoption of the UNWC by the UN General Assembly (UNGA), second thoughts about this formula started surfacing soon thereafter. Some upper riparians seem to consider the UNWC as biased in favor of the lower riparians because of its specific and separate mention of the obligation to take all appropriate measures to prevent significant harm. The three countries that voted against the Convention (Burundi, China and Turkey) and some of those that abstained, such as Bolivia, Ethiopia, Mali and Tanzania, are largely upper riparian states. On the other hand, certain downstream states, such as Egypt, Pakistan and Peru, which also abstained, may have come to subscribe to the more widely accepted interpretation,[3] aligned with customary law,[4] that the UNWC subordinates the no-harm rule to equitable and reasonable utilization. Such states may consider that, in so doing, the Convention favors upstream countries.

Those views do not, however, reflect the unified position of either upstream or downstream riparians – there are countries from both groups among the UNWC's contracting states. Even more encouragingly, some states that were initially concerned about the hierarchy between those two norms have since ratified the Convention. For example, France abstained from voting in the UNGA in part because of the subordination of the no-harm rule to the principle of equitable and reasonable utilization. However, in early 2011, France became the 22nd contracting state and is now actively promoting the Convention's ratification and entry into force.

In addition, the subordination of the no-harm rule to the principle of equitable and reasonable utilization does not really favor upstream riparians. Rather, that subordination aims to balance the interests of all riparians. Equitable and reasonable utilization, which has been the guiding principle of international water law since the Helsinki Rules were issued in 1966,[5] duly recognizes and is based on the equality of all riparians in the uses of the shared watercourse. That principle lays down certain objective factors for determining the equitable and reasonable share for each riparian state: those factors include existing water uses but these uses receive no inherent priority over future uses. These approaches have been reconfirmed by the UNWC, in Articles 5–6 and 10.[6]

---

2 See Caflisch, L. 'Regulation of the Uses of International Watercourses' in Salman, S. M. A. and Boisson de Chazournes, L. (eds), *International Watercourses – Enhancing Cooperation and Managing Conflict* (New York: World Bank, 1998), at 13–15.

3 McCaffrey, S. C. 'The Contribution of the UN Convention on the Law of the Non-Navigational Uses of International Watercourses' *International Journal of Global Environmental Issues*, 2001; 1 (3/4): 250–63, at 255.

4 McCaffrey, S. C. and Sinjela, M. 'Current Development: The 1997 UN Convention on International Watercourses' *American Journal International Law*, 1998; 92: 97–107, at 101–2.

5 International Law Association (ILA), 'Helsinki Rules on the Uses of the Waters of International Rivers' in ILA, Report of the 52nd Conference (Helsinki 1966).

6 See Salman, S. M. A. 'The Helsinki Rules, the UN Watercourses Convention and the Berlin Rules: Perspectives on International Water Law' *International Journal of Water Resources Development*, 2007; 23: 625–40.

Furthermore, four months after its adoption, the International Court of Justice (ICJ) endorsed the UNWC and the principle of equitable and reasonable utilization. The ICJ addressed the issue of the suffering of harm by one riparian and attributed that issue to the deprivation of such a riparian of the right to an equitable and reasonable utilization of the shared watercourse, which the court called a 'basic right'. The ICJ stated that

> Czechoslovakia, by unilaterally assuming control of a shared resource, and thereby depriving Hungary of its right to an equitable and reasonable share of the natural resources of the Danube, ... failed to respect the proportionality which is required by international law.[7]

Properly understood, therefore, the UNWC merely codifies existing customary law when clarifying the relationship between equitable and reasonable use and harm prevention – customary law that allows a balancing of the interests of all interested states through the pondering of all relevant factors. Under customary law, that determination must include a concern with sustainability, as well as consideration of the possibility of harm to other states and the need to protect the aquatic environment. Similar results are enshrined in Articles 5–6 and 20–23 of the UNWC.

Notification process

A second and related misconception is the view that the notification process under the UNWC favors downstream riparians and provides them with veto power over projects of upstream riparians. This derives from the misguided belief that harm can only 'travel' downstream; i.e. only upstream riparians can cause harm to downstream riparians. Although erroneous, this belief arises from basic misunderstandings of international water law in general and the Convention in particular.

It is clear that downstream riparians can be harmed by the physical impacts of water quality and quantity changes caused by water uses in upstream riparians. It is much less obvious and generally not understood that the prior use and the claiming of rights by a downstream country can harm an upper riparian through the foreclosure of the latter's future water use.[8] For example, a poor upstream country could, in practice, be precluded from developing the water resources of an international waterway tomorrow if a richer downstream riparian, without consultation or notification, develops it today. This is an important, although not widely understood, principle of international water law, which establishes a clear linkage between equitable and reasonable use and harm prevention.

---

7   *Case Concerning the Gabčíkovo-Nagymaros Project (Hungary v Slovakia)* (Judgment) [1997] ICJ Rep 7 at 53 (*Gabčíkovo*).
8   See Salman, S. M. A. 'Downstream Riparians Can Also Harm Upstream Riparians: The Concept of Foreclosure of Future Uses', *Water International*, 2010; 35: 350–64, at 351.

In this sense, for example, Ethiopia has been protesting most of the projects undertaken by Egypt and Sudan because those projects may foreclose the former's future uses of the Nile waters. In other words, such projects could deprive Ethiopia of its equitable and reasonable share of the basin.[9] The Senegal Water Charter, signed by four West African states, enumerates in Article 4 a number of principles for water allocation: such principles include 'the obligation of each riparian state to inform other riparian states before engaging in any activity or project likely to have an impact on water availability, and/or the possibility to implement future projects'.[10] Jawaharlal Nehru, then prime minister of India, perhaps understood this point best. In an exchange with Pakistan's President Mohammad Ayub Khan, before the birth of Bangladesh, regarding the Ganges River, he noted the practical implications of actions by downstream riparians and of requiring notification only from upstream countries:

> [o]ne more matter to which I must also refer, is the distinction you still seem to make between the rights of upper and lower riparians in paragraph 7 of your letter, which implies that the lower riparian can proceed unilaterally with projects, while the upper riparian should not be free to do so. If this was to be so, it would enable the lower riparian to create, unilaterally, historic rights in its favour and go on inflating them at its discretion thereby completely blocking all development and uses of the upper riparian. We cannot, obviously, accept this point of view.[11]

Accordingly, the World Bank policy for projects on international watercourses, which was established in 1956 and has undergone major revisions and updates since that date, requires notification of all riparians – whether downstream or upstream, or along a boundary river or lake – of any project on an international waterway. A project on the Mekong in Vietnam (the lowest downstream riparian) would require notification of all the other five riparians, just as would a project in China, the upper most riparian in that basin. Similarly, a project in Ethiopia or Egypt would require notifying all the other Nile riparians, whether of the Blue Nile or the White Nile.[12] The practical application of the Bank's policy has clarified this issue, as well as the issue regarding the linkages between the principle of equitable and reasonable utilization and the obligation on the prevention of significant harm.

---

9 See Waterbury, J. *The Nile Basin: National Determinants and Collective Action* (New Haven: Yale University Press, 2002), at 84–5.

10 Charte des Eaux du Fleuve Sénégal [Senegal River Waters Charter] (Mali-Mauritania-Senegal), Organisation pour la Mise en Valeur Du Fleuve Sénégal (OMVS) Resolution 005, adopted by the Conference of Heads of State and Government (May 18, 2002), at Article 4. Available online at http://lafrique.free.fr/traites/omvs_200205.pdf (accessed March 15, 2013) [official text in French].

11 See Crow, B., Lindquist, A. and Wilson, D. *Sharing the Ganges: The Politics and Technology of River Development* (Sage, 1995), at 89.

12 See Salman, S. M. A. *The World Bank Policy for Projects on International Waterways – An Historical and Legal Analysis* (Martinus Nijhoff Publishers, 2009).

The UNWC and, before it, the Helsinki Rules, follow the same approach regarding notification, appropriately reflecting customary law. Neither instrument limits notification to downstream riparians, nor grants any state veto power over the projects of other riparian states. The Convention requires notification in case a project may cause 'significant adverse effects' to other riparians, regardless of whether they are upstream or downstream, and lays down related provisions, including on the eventual failure of the implementing state to notify.[13] In the case of an objection by a riparian state, the UNWC requires the parties to enter into consultations and, if necessary, negotiations, with a view to arriving at an equitable resolution of the situation.[14] The Convention further requires that each state, in good faith, pay reasonable regard to the rights and legitimate interests of its neighbors within the basin.[15]

Hence, the notification process benefits all riparian states, in that it increases states' access to information relating to planned measures. Under the UNWC, the notification process seeks to spur collaboration, encourage the integrated (optimal) development of the shared watercourse and maximize benefits for all parties while minimizing adverse impacts.

Relationship between existing watercourse agreements and the UNWC

A third issue that may have contributed to the reluctance of some states to become parties to the UNWC is the manner in which the Convention deals with existing agreements. The Convention does not affect the rights or obligations of watercourse states arising from agreements that are in force. Nonetheless, the Convention asks the parties to consider, where necessary, harmonizing such agreements with its basic principles. The UNWC also allows watercourse states to enter into agreements that apply and adjust its provisions to the characteristics and uses of a particular international watercourse. When some, but not all, watercourse states to a particular international watercourse are parties to an agreement, nothing in the agreement would affect the rights or obligations under the Convention of watercourse states that have not joined such an agreement.[16]

Certain riparian states that already have agreements in place believe that the UNWC has not fully recognized those agreements; they question the Convention's provision encouraging parties to consider harmonizing such agreements with its principles. Conversely, riparian states that have been left out of existing agreements argue that the Convention should have subjected those agreements to its provisions and should have required consistency between the two. However, the UNWC recognizes both the validity of existing agreements and the right of the riparian

---

13 UNWC, at Articles 12–19.
14 See McCaffrey, S. C. *The Law of International Watercourses* (2nd edn, Oxford: Oxford University Press, 2007), at 471–6.
15 UNWC, at Article 17(2).
16 See UNWC, at Article 3(2)–(3) and (6).

states that are not parties to such agreements in the shared watercourse. The Convention could not simply have provided for the annulment of existing watercourse agreements; such a provision would have been rejected by most members of the UNGA, in addition to resulting in chaos in several basins.[17] A riparian state has the right to enter into an agreement regarding the shared watercourse, but that right is also subject to the rights of other riparians in the uses of the basin. The UNWC has basically reflected this legal and common sense principle. After all, the Convention is a framework treaty that lays down basic principles, which are to be complemented by agreements between the parties, taking into account the characteristics of the specific watercourse.

Dispute settlement provisions

Another area of contention concerns the UNWC's dispute settlement provisions. Some states believe that the dispute settlement provisions of the Convention are too weak because they do not provide sufficiently for binding mechanisms. In contrast, other riparian states view the detailed conflict resolution procedures of the Convention, particularly the fact-finding process, as interfering with their sovereign right to choose among the various means available to solve transboundary water disputes. As with other contentious issues, Article 33 of the Convention seeks a middle ground. Rather than a compulsory dispute settlement procedure, the Convention offers a number of mechanisms, including negotiations; the good offices of, or mediation and conciliation by, a third party; the use of joint watercourse institutions; or submission of the dispute to arbitration or the ICJ.

The UNWC sets forth only one obligatory method: impartial fact-finding. This process comes into effect, upon unilateral request of one of the parties, if agreement has not been reached after six months, 'unless the parties otherwise agree'.[18] At the end of the fact-finding process, states must 'consider in good faith' the report of the fact-finding commission.[19]

The Convention tries to reconcile the divergent views of states: those wishing for strong dispute resolution procedures are given a variety of tools from which to choose, as well as compulsory fact finding if the other options do not yield results. At the same time, states which believe that compulsory dispute resolution methods interfere with their sovereign rights have a menu of options, as well as the right to

---

17   In this context, for example, one has to imagine what would happen in the Indus Basin if the Indus Waters Treaty were to be annulled because Afghanistan and China are not parties. The Treaty was negotiated for almost ten years by India and Pakistan, with active mediation by the World Bank. More than one and a half billion dollars have been invested in implementing the Treaty. See Salman, S. M. A. and Uprety, K. *Conflict and Cooperation on South Asia's Rivers: A Legal Perspective* (World Bank, 2003), at 37–61.

18   UNWC, at Article 33(3).

19   Ibid., at Article 33(8). The World Bank has somewhat similar procedures in case of an objection by one riparian state to a project in another state. See Salman, S. M. A. 'The Baardhere Dam and Water Infrastructure Project in Somalia – Ethiopia's Objection and the World Bank Response' *Hydrological Sciences Journal*, 2011; 56: 630–40.

reject those options. Given that the UNWC is a framework agreement, this structure is a sensible and reasonable compromise.

Role of regional economic integration organizations

The last major area of misunderstanding is the UNWC's expanded definition of the expression 'watercourse state' to encompass 'regional economic integration organizations'.[20] As per Article 2(d), the Convention also applies to:

> an organization constituted by sovereign States of a given region, to which its member States have transferred competence in respect of matters governed by this Convention and which has been duly authorized in accordance with its internal procedures to sign, ratify, accept, approve or accede to it.

Although such entities can become a party to the Convention, this does not 'imply that regional economic integration organizations have the status of states in international law'.[21]

Yet, a number of states have the misimpression that members of regional economic integration organizations could become riparians to watercourses that do not touch their territories simply through their membership in an organization that became a party to the Convention. While the wording in the UNWC may be difficult to follow for a lay person, there is nothing in the language of Article 2(c)–(d) that would allow such an interpretation. The ability of a regional organization to join the Convention does not make its members riparians to all watercourses within the purview of the organization. Rather, this definition recognizes the separate legal personality of such an organization. In fact, the term 'regional economic integration organization' was included specifically to allow the participation of the European Union,[22] as has happened with other conventions.[23]

Conclusion

In sum, the UNWC addresses the issues discussed above in a compromising manner.[24] This has provided the basis for the Convention's wide acceptability and

20 Under Article 2(c) of the UNWC, a 'watercourse state' is 'a State Party...in whose territory part of an international watercourse is situated, or a Party that is a regional economic integration organization, in the territory of one or more of whose Member States part of an international watercourse is situated'.

21 UNGA, 'Convention on the Law of The Non-Navigational Uses of International Watercourses: Report of the 6th Committee Convening as the Working Group of the Whole' (April 11, 1997) UN Doc A/51/869, at 5. Available online at http://www.un.org/law/cod/watere.htm (accessed March 15, 2013).

22 McCaffrey (note 14), at 360.

23 See, e.g., UNECE Convention on the Protection and Use of Transboundary Watercourses and International Lakes (adopted March 17, 1992, entered into force October 6, 1996) 1936 UNTS 269, Article 23.

24 For a more detailed discussion of those issues, see Salman (note 1), at 1–15.

has facilitated its adoption by the UNGA in 1997. However, those compromises have also contributed to some misinterpretations. These misinterpretations may be hindering the Convention's entry into force by leading some states to forsake the treaty without a more careful analysis of the advantages and disadvantages of ratification. Those misunderstandings are clearly related to the issues of capacity and awareness discussed in Chapter 3.

5 Why have states joined the UN Watercourses Convention?

Alistair Rieu-Clarke and Alexander López

Committing to a global, legally binding convention governing international watercourses pushes for a reconceptualization of classic notions of national security, sovereignty and territoriality. Transboundary environmental problems, by their very nature, undermine state sovereignty. While states have therefore come under mounting pressure to manage resources according to international norms, there is also clear evidence of the reluctance of some states to relinquish traditional notions of sovereignty, territoriality and national interests.[1]

However, a record of treaty practice over international watercourses would tend to imply that there has been considerable cooperation. Over 400 basin-specific multilateral and bilateral treaties and a number of regional agreements are in place. However, numerous basins are not covered or are only partially covered by watercourse agreements.[2] UN-Water therefore reports that:

> existing agreements are sometimes not sufficiently effective to promote integrated water resources management due to problems at the national and local levels such as adequate water management structures and weak capacity in countries to implement the agreements as well as shortcomings in the agreements themselves (for example, inadequate integration of aspects such as the environment, the lack of enforcement mechanisms, limited – sectoral – scope and non-inclusion of important riparian States.[3]

Given the above environment, this chapter considers why states have perceived it necessary to join a global, *legally binding* instrument aimed at the governance of international watercourses. The chapter firstly surveys theoretical accounts as to why states cooperate over water generally; secondly, the preparatory documents

1 See, e.g., López, A. 'Environmental Transborder Cooperation in Latin America: Challenges to the Westphalia Order' in Matthew, R. A., Barnett, M. J., McDonald, B. and O'Brien, K. L. (eds) *Global Environmental Change and Human Security* (MIT Press, 2009), at 291–304.
2 Giordano, M. and Wolf, A. T. 'Sharing Waters: Post-Rio International Water Management' *Natural Resources Journal*, 2003; 27: 163–71.
3 UN-Water, *Transboundary Waters: Sharing Benefits, Sharing Responsibilities: Thematic Paper* (New York: UN-Water, 2008). Available online at www.unwater.org/downloads/ UNW_TRANSBOUNDARY.pdf (accessed March 15, 2013).

within the International Law Commission (ILC) and UN General Assembly (UNGA) that led to the decision to adopt a global convention are studied; and thirdly, the official reasons given by states upon their ratification of the UN Watercourses Convention (UNWC) are identified.

Why states cooperate over water – differing worldviews

Power

As applied to transboundary waters, a realist interpretation of state interaction would posit that the most powerful state within a particular basin – the hydro-hegemon[4] – dictates the nature of cooperation.[5] According to a classic realist account, if the most powerful state is upstream it will have little incentive to cooperate, whereas a downstream hegemon would tend to take the lead in fostering cooperative arrangements.[6] In terms of the weaker state, Lowi contends that,

> since the asymmetry of power is not in its favour, it is not in a position to achieve its aims and satisfy its needs in an optimal fashion. . . . In effect, it has little alternative but to accept a modus vivendi dictated by the stronger.[7]

A further argument made by realist scholarship is that 'high' politics of war and diplomacy tend take precedence over 'low' politics of economics and welfare.[8] Two classic examples of such power asymmetries can be seen in the case of China, upstream on the Mekong, and Egypt, downstream on the Nile.[9]

Cascão and Zeitoun offer a more nuanced interpretation of power in transboundary water relations.[10] The latter scholars identify four forms of power:

4   Zeitoun, M. and Warner, J. 'Hydro-Hegemony: A Framework for Analysis of Transboundary Water Conflicts' *Water Policy*, 2008; 8: 435–60.

5   Lowi, M. R. *Water and Power: The Politics of a Scarce Resource in the Jordan River Basin* (Cambridge University Press, 1993), at 10.

6   See Waterbury, J. 'Between Unilateralism and Comprehensive Accords: Modest Steps Toward Cooperation in International River Basins' (1997) 13 *International Journal of Water Resources Development*, 1997; 13: 279–89, at 281; Naff, T. H. 'A Case for Demand-Side Water Management' in Issac, J. and Shuval, H. (eds), *Water and Peace in the Middle East* Studies in Environmental Science Vol. 58 (Elsevier Science, 1994), at 83–92.

7   Lowi (note 5), at 169.

8   Ibid., at 10.

9   See Menniken, T. 'China's Performance in International Resource Politics: Lessons from the Mekong', *Contemporary Southeast Asia: A Journal of International and Strategic Affairs*, 2007; 29: 97–120; Cascão, A. 'Changing Power Relations in the Nile River Basin: Unilateralism vs. Cooperation', *Water Alternatives*, 2009; 2: 245–68.

10 Cascão, A. E. and Zeitoun, M. 'Power, Hegemony and Critical Hydropolitics' in Earle, A., Jägerskog, A. and Öjendal, J. (eds), *Transboundary Water Management: Principles and Practice* (Earthscan, 2010), at 27–42. See also Zeitoun, M. and Mirumachi, N. 'Transboundary Water Interaction I: Reconsidering Conflict and Cooperation', *International Environmental Agreements: Politics, Law and Economics*, 2008; 8: 297–316; Zeitoun, M., Mirumachi, N. and Warner, J. 'Transboundary Water Interaction II: The Influence of "Soft" Power', *International Environmental Agreements: Politics, Law and Economics*, 2010; 11: 159–78.

(i) geographic power – riparian position; (ii) material power – economic, military or technological strength, as well as international political and financial support; (iii) bargaining power – the ability to control the rules of the game and to set agendas; and (iv) ideational power – the capacity to impose and legitimize particular ideas and narratives.[11]

Following this interpretation of power, weaker states within a transboundary watercourse setting are not completely powerless to influence basin dynamics.[12] Using the example of the relations between Iraq, Syria and Turkey on the Tigris and Euphrates River, Daoudy therefore maintains that, 'downstream countries can…mobilize structural factors, such as the codification of new legal "rules" on watercourses to bring upstream riparians to cooperate, and therefore acquire enhanced structural power'.[13] Similarly, Zeitoun and Jägerskog argue that, 'basin bullies can be susceptible to the powers of persuasion, and may be less likely to force an arrangement if they are held accountable to an objective standard, or risk being "named and shamed"'.[14] Following this argument further, it could be maintained that entry into force and widespread support for the UNWC might help strengthen the role of ideational and bargaining power in basin dynamics. International law could therefore be seen as an important tool for 'leveling the playing field' between states, and therefore an incentive for states to ratify the UNWC.[15]

Reciprocity

While realist accounts focus on power as the primary explanatory factor as to why states cooperate over water, liberal institutionalism perceives states as rational egoists that cooperate when they expect to gain more from collective action than unitary action or to manage the risk of costly conflict.[16] The above can be understood in terms of transforming conflicts into win–win options. In the same way, another important motivation is economic reciprocity, which allows the countries involved to generate trade-off scenarios – for example, joint actions such as hydroelectric projects, where cost and benefits can be distributed.

Following game theory, cooperation seeks to 'reduce the shadow of the future'.[17]

11 Cascão and Zeitoun (note 10), at 30.
12 Zeitoun, M. and Jägerskog, A. 'Confronting Power: Strategies to Support Less Powerful States', in Jägerskog, A. and Zeitoun, M. (eds), *Getting Transboundary Water Right: Theory and Practice for Effective Cooperation* (Stockholm International Water Institute, 2009), at 11; Daoudy, M., 'Asymmetric Power: Negotiating Water in the Euphrates and Tigris', *International Negotiation*, 2009; 14: 361–91, at 363.
13 Daoudy, M. 'Hydro-Hegemony and International Water Law: Laying Claims to Water Rights', *Water Policy*, 2008; 10: 89–102.
14 Zeitoun and Jägerskog (note 12), at 12.
15 Ibid.
16 Brochmann, M. and Hensel, P. 'Peaceful Management of International River Claims', *International Negotiation*, 2009; 14: 393–418; Keohane, R. and Ostrom, E. 'Introduction' in Keohane, R. and Ostrom, E. (eds), *Local Commons and Global Interdependence* (Sage Publications, 1994), at 1–26.
17 Bearce, D. W., Floros, K. M. and McKibben, H. E. 'The Shadow of the Future and International Bargaining: The Occurrence of Bargaining in a Three-Phase Cooperation Framework', *Journal of Politics*, 2009; 71: 719–32.

The tensions between national incentives that encourage opportunistic behavior, and the collective good that comes from cooperation, is a common feature of state interaction. If there is continuous interaction among states that share a river basin, that situation could therefore be an incentive to have an instrument that prevents or limits opportunistic behavior and supports cooperation, thereby addressing the tension between national incentives and the common good. Cooperation may thus arise where states perceive that they have a common interest and that the benefits, or the payoffs, of joint actions are greater than those of unilateral actions.

Sadoff and Grey identify four types of benefits that states sharing transboundary waters might enjoy collectively:

> environmental benefits to the river' (e.g. improved water quality, conserved biodiversity); economic benefits from the 'river' (e.g. increased food and energy production); reduction of costs because of the 'river' (e.g. reduced geo-political tensions, enhanced flood management); and benefits beyond the 'river' (catalyzing wider cooperation and economic integration).[18]

Another important benefit that is not explicitly mentioned by Sadoff and Grey is the potential reduction of negative externalities, which would therefore reduce the potential cost and uncertainty generated by the production of externalities that, as a public good, implies costs even to the actors that are not part of one given decision. Externalities are costs (negative externalities) that not only affect the actors that create them. The production of negative externalities could be one reason for a state to join the UNWC because, the larger the externality, *ceteris paribus*, the more likely states will alter their behavior, organizing either to capture the positive, or eliminate the negative, effect of cooperation or non-cooperation.

An important driver of cooperation to emphasize is related to risk and uncertainty – i.e. reducing the costs that stem from the river.[19] Cooperation may be an effective mechanism by which to mitigate the predicted impacts of climate change, including the greater frequency and severity of floods and droughts.[20] Risk of industrial accidents has also been highlighted as an incentive for cooperation.[21] Taking this approach further, others have argued that scarcity – the same factor that leads realist scholars to raise the water wars alarm – may also represent a driver for cooperation.[22] Additional benefits may derive from 'side-payments' or 'issue-linkages'. Falkenmark

18 Sadoff, C.W. and Grey, D. 'Beyond the River: The Benefits of Cooperation on International Rivers', *Water Policy*, 2002; 4: 389–403.

19 Drieschova, A., Fischhendler, I. and Giordano, M. 'The Role of Uncertainties in the Design of International Water Treaties: An Historical Perspective', *Climatic Change*, 2011; 105: 387–408.

20 Rieu-Clarke, A. 'A Survey of International Law Relating to Flood Management: Existing Practices and Future Prospects' *Natural Resources Journal*, 2008; 48: 649–77; Bakker, M. H. N. 'Transboundary River Floods and Institutional Capacity', *Journal of the American Water Resources Association*, 2009; 45: 553–66.

21 Vinogradov, S. 'Regime Building for Transboundary Waters: The Evolution of Regional and Institutional Frameworks in the EECCA Region', *Journal of Water Law*, 2007; 16: 28 at 89.

22 Stinnett, D. M. and Tir, J. 'The Institutionalization of River Treaties' *International Negotiation*, 2009; 14: 229–51, at 249. See also Brochmann and Hensel (note 16), at 393.

and Jägerskog maintain that cooperation can be strengthened by identifying development opportunities which are often outside the '(blue) water sector', such as land/green water, trade, energy and tourism.[23]

In such a situation, 'institutions' or 'regimes' provide an important mechanism by which to reduce the costs of making, monitoring and enforcing rules. As Keohane notes:

> [i]nternational regimes . . . permit governments to attain objectives that would otherwise be unattainable. They do so in part by facilitating intergovernmental agreements. Regimes facilitate agreements by raising the anticipated costs of violating others' property rights, by altering transaction costs through the clustering of issues, and by providing reliable information to members. Regimes are relatively efficient institutions, compared with the alternative of having a myriad of unrelated agreements, since their principles, rules, and institutions create linkages among issues that give actors incentives to reach mutually beneficial agreements.[24]

The 1992 UN Economic Commission for Europe (UNECE) Water Convention[25] can be seen as an example of such a regime at the regional level. Through the framework nature of this legal instrument and the supporting institutional structures, the Convention has strengthened cooperation between transboundary watercourse states across the UNECE region.[26]

Following a liberal institutionalist perspective, states may therefore be motivated to support the UNWC from a recognition of the perceived benefits of cooperation and through a desire to create a global regime by which to foster collective action. In fact, one reason for UNWC support is that a global regime can be useful because it directs attention to the possibility of security cooperation where actors share a set of regional environmental problems or threats without a clear security framework for dealing with such problems. Accordingly, a regime is normally formed and cooperation is more likely if participants perceive that they have common interests and that the benefits – the payoff – of joint actions are greater than those of unilateral action. One of the most important payoffs of regional environmental security regimes is that they can strengthen the governance of transboundary river basins. However, without an effective global governance regime, a 'prisoner's dilemma paradox' may materialize, in which individual rational actor strategies lead to irrational outcomes at the regional level.

---

23 Falkenmark, M. and Jägerskog, A. 'Sustainability of Transnational Water Agreements' in Earle, A., Jägerskog, A. and Öjendal, J. (note 10), at 168.

24 Keohane, R. *After Hegemony: Cooperation and Discord in the World Political Economy* (Princeton University Press, 1984), at 97.

25 UNECE Convention on the Protection and Use of Transboundary Watercourses and International Lakes (adopted March 17, 1992, entered into force October 6, 1996) 1936 UNTS 269.

26 See Vinogradov (note 21).

The power of norms

An alternative worldview emphasizes the power of norms in shaping actor behaviour.[27] Constructivism stresses the notion that shared ideas and normative practices determine the identities and interests of actors.[28] Along similar lines, Jacobs maintains that, 'there have been few attempts to conceptualize and understand the multi-layered interplay of the principled content of cooperation between actors, scale and the hard and soft law that binds them together'.[29] Jacobs' own work emphasizes the importance of accounting for 'local configuration, domestic policy, political identities and social and cultural institutions'.[30] Such an approach speaks to the criticism that most research tends to focus predominantly on states as actors.[31] A constructivist approach to analyzing transboundary water relations therefore emphasizes the need to account for a range of actors and networks. The importance of epistemic communities is recognized by Jägerskog, who notes that 'members of an epistemic community might... have a decisive influence on the construction of policy in an area and, since the communities' nature is international, it might also lead to a general convergence of policies at the international level'.[32] An example of such an epistemic community can be seen by the work of the ILC and the UNGA, during the development, negotiation and adoption of the UNWC.[33] Such activities have helped to foster shared understanding among states, at a global level, as to which laws do, or should, apply to the non-navigational uses of international watercourses.[34]

Following a constructivist interpretation of why states might support the UNWC, it could be argued that they do so because of a desire to foster and

27 Nagtzaam, G. *The Making of International Environmental Treaties: Neoliberal and Constructivist Analyses of Normative Evolution* (Edward Elgar, 2009), at 50–79.

28 See, e.g., Wendt, A. 'Anarchy Is What States Make of It: The Social Construction of Power Politics', *International Organization*, 1992; 46: 391–425; Finnemore, M. and Sikkink, K. 'International Norm Dynamics and Political Change', *International Organization*, 1998; 52: 887–917; Ruggie, J. G. 'What Makes the World Hang Together? Neo-Utilitarianism and the Social Constructivist Challenge', *International Organization*, 1998; 52: 855–85; Brunnée, J. and Toope, S. J. 'International Law and Constructivism: Elements of an International Theory of International Law', *Columbia Journal of Transnational Law*, 2000; 39: 19–74.

29 Jacobs, I. *Norms and Transboundary Co-operation in Africa: The Cases of the Orange-Senqu and Nile Rivers* (PhD thesis, University of St Andrews, 2009).

30 Ibid., at 233–35.

31 Furlong, K. 'Hidden Theories, Troubled Waters: International Relations, the "Territorial Trap", and the Southern African Development Community's Transboundary Waters', *Political Geography*, 2006; 25: 438–58.

32 Jägerskog, A. 'Water Regimes: A Way to Institutionalize Water Co-operation in Shared River Basins' (UN Educational, Scientific and Cultural Organization and Green Cross International Conference on Water for Peace, 2002), at 165. See also Haas, P. M. 'Introduction: Epistemic Communities and International Policy Coordination', *International Organization*, 1992; 46: 1–35.

33 Salman, S. M. A. 'The Helsinki Rules, the UN Watercourses Convention and the Berlin Rules: Perspectives on International Water Law', *Water Resources Development*, 207; 23: 625–40.

34 See Brunnée, J. and Toope, S. J. 'The Changing Nile Basin Regime: Does Law Matter?', *Harvard Journal of International Law*, 2002; 43: 105–59.

strengthen shared understanding around the global norms that do, or should, regulate international watercourses.

Why go global? – perspectives from the UNGA and ILC

The fact that the legal architecture for transboundary watercourses is made of a fragmented system of bilateral and multilateral arrangements was a major reason why states opted to develop the UNWC in the first place. UNGA Resolution 2669 (XXV) of 1970 stressed that, 'despite the great number of bilateral treaties and other regional regulations . . . [t]he use of international rivers and lakes is still based in part on general principles and rules of customary law'.[35]

A key concern of the ILC during its work was how to strike a balance between global arrangements and regional and basin-specific agreements. In recognizing the uniqueness of international watercourses, one of the ILC's Special Rapporteurs, Jens Evensen, cautioned that:

> [i]n drawing up a draft convention on this topic, it seems essential to recog-
> nize and accept the common features of international watercourses, but also to
> accept the limitations to the venture of drawing up an international instru-
> ment on international watercourses on account of the unique features of each
> watercourse. Consequently the Special Rapporteur agrees that specific water-
> course agreements pertaining to a special watercourse of parts thereof, to the
> watercourses of a region or to special activities in or uses of watercourses, may
> frequently be required for the satisfactory administration and management of
> international watercourses. Nevertheless, such concrete approaches to specific
> watercourses or specific problems do not make a general framework agree-
> ment on the topic superfluous. A framework convention should accept the
> necessity and validity of such specific watercourse agreements, whether
> concluded prior or subsequent to the adoption of a general convention on the
> non-navigational uses of international watercourses.[36]

Following this sentiment, some states claimed that the value of the Convention arises within three situations: firstly, where no governing regime for transboundary water exists; secondly, where not all basin states are party to an existing agreement; and thirdly, where an agreement only partially covers matters addressed by the

---

35 UNGA, Res 2669 (XXV) 'Progressive Development and Codification of the Rules of International Law Relating to International Watercourses' (December 8, 1970). Available online at www.un.org/ga/search/view_doc.asp?symbol=A/Res/2669(XXV) (accessed March 16, 2013).

36 Evensen, J. (Special Rapporteur), *Second Report on the Law of the Non-Navigational Uses of International Watercourses. Topic: Law of the Non-Navigational Uses of International Watercourses* (April 24, 1984), UN Doc A/CN.4/381 and Corr.1 and Corr.2, Extract from the Yearbook of the ILC, 1984; II(1): 101–27, at 104. Available online at http://untreaty.un.org/ilc/documentation/english/a_cn4_381.pdf (accessed March 16, 2013).

rules.[37] A clear example was offered by Poland, which maintained that three key UNECE Conventions 'to a greater extent' cover the matters specified by the draft Articles.[38]

As noted by the Nordic countries,

> the framework agreement approach, adopted by the Commission in drafting the articles provides a good basis for further negotiations. It leaves the specific rules to be applied to individual watercourses to be set in agreements between the states concerned, as has been the current practice.[39]

Ultimately, this approach was widely supported by the ILC and 6th Committee of the UNGA.[40]

However, one issue that requires careful consideration is why states perceived there to be a need to adopt a convention on the subject, rather than merely using the ILC draft articles as a guide to state practice. This point was raised by McCaffrey, who observes that, regardless of entry into force,

> the history of state conduct over the past decade in relation to the international watercourses strongly suggests that states will continue to base their water-related negotiations on the Convention and that it will continue to have a stabilizing influence on the relations between countries sharing freshwater resources.[41]

Upon ratification, Hungary went on to point out the role that the UNWC could play, as 'a model for adaptation', where transboundary water agreements were not yet in place.[42] Greece also recognized the need to establish minimum standards at the global level for the protection and sustainable management of water where no or only partial agreements exist and observed that entry into force of the UNWC will form the basis for the conclusion of other regional treaties and strengthen

---

37 ILC, *The Law of the Non-Navigational Uses of International Watercourses: Comments and Observations Received from Governments. Topic: Law of the Non-Navigational Uses of International Watercourses.* Extract from the Yearbook of the ILC, 1993, Vol. II(1) (March 3, April 15, May 18 and June 14, 1993) UN Doc A/CN.4/447 and Add.1–3, at 147. Available online at http://untreaty.un.org/ilc/documentation/english/a_cn4_447.pdf (accessed March 16, 2013).

38 Ibid., at 165.

39 Ibid., at 164.

40 ILC, 'Draft Articles on the Law of the Non-Navigational Uses of International Watercourses and Commentaries thereto and Resolution on Transboundary Confined Groundwater', in *ILC, Report of the International Law Commission on the work of its forty-sixth session, 2 May–22 July 1994, Official Records of the General Assembly, Forty-ninth session, Supplement No. 10* (May 2–July 22, 1994) UN Doc A/49/10 (1994). Extract from the Yearbook of the ILC, 1994, Vol. II(2). Available online at http://untreaty.un.org/ilc/documentation/english/A_49_10.pdf (accessed March 16, 2013).

41 McCaffrey, S. C. 'The 1997 UN Watercourses Convention: Retrospect and Prospect', *Pacific McGeorge Global Business and Development Law Journal*, 2008; 21: 165–74.

42 See 'Some Considerations on the Ratification of the New York Convention by Hungary' (State Security for Water, World Water Forum, Istanbul March 16–22, 2009) (on file).

existing ones.[43] Additionally, Finland noted the importance of entry into force in ensuring that the UNWC provides a valuable point of reference for many regional and bilateral arrangements.[44] While it might therefore be suggested that the UNWC may shape state practice at the basin level regardless of entry into force, it would appear from the above state sentiments that there is a perception that entry into force will provide/would provide the instrument with greater authority and influence in shaping state practice.

States have also recognized the role that entry into force can play in raising the profile of water issues at the global level. France thus maintains that the Convention is a step that 'may seem small but is no less symbolic for global water governance'.[45] Along similar lines, France referred to its desire to 'set an example' in the run-up to the 6th World Water Forum in Marseille in March 2011.[46] Similarly, Burkina Faso noted that ratification of the Convention can raise the awareness of the urgency and impact of climate change, and demonstrate a commitment to finding solutions;[47] and Portugal notes the need for ratification as 'a timely matter' given the global need to promote the peaceful management of transboundary waters.[48]

States, when becoming parties to the UNWC, have also put forward various other incidental reasons. For instance, Hungary referred to the UNWC as providing a platform by which to 'share her Danubian and bilateral transboundary cooperation experience with interested parties' and thus noted the UNWC's ability to develop 'mutual understanding'.[49] From a domestic perspective, France recognized the role that the UNWC could play in making 'the issue of water and access to it a priority for its foreign policy'.[50] A number of European Union States have also justified their membership, at least in part, by noting that there would be no additional burden in terms of reforming their domestic legislation.[51]

43 Greek Minister of Finance and others, 'Explanatory Memorandum on the Draft Law on Ratification of the Convention on the Law of the Non-Navigational Uses of International Watercourses to the Hellenic Parliament' [Greek] (June 7, 2010). Available online at http://assets.panda.org/downloads/unwatercoursesconv_gr_justification.pdf (accessed March 16, 2013).

44 Assemblée Nationale, 'Report on Behalf of the Committee on Foreign Affairs on the Bill No. 2009 Authorizing Membership in the 1997 UN Convention on the Law of the Non-Navigational Uses of International Watercourses' [French] (April 6, 2010). Available online at www.assemblee-nationale.fr/13/rapports/r2433.asp (accessed March 16, 2013).

45 Ibid.

46 Ibid.

47 Burkina Faso, 'Explanatory Memorandum on the Draft Law on Ratification of the Convention on the Law of the Non-Navigational Uses of International Watercourses' [French] (on file).

48 Ministry of Foreign Affairs of Portugal, 'Multilateral Affairs Director-General Approves the Convention on the Law of the Non-Navigational Uses of International Watercourses'. Available online at http://awsassets.panda.org/downloads/nota_justificativa_clean.pdf (accessed March 16, 2013).

49 See 'Some Considerations on the Ratification of the New York Convention by Hungary' (note 42).

50 Assemblée Nationale (note 44).

51 See, e.g., Deutscher Bundestag, 'Draft Law on the Convention on the Law of the Non-navigational Uses of International Watercourses' (Drucksache 16/738, February 21, 2006). Available online at http://assets.panda.org/downloads/ratification_justification_germany.pdf [German] (accessed March 16, 2013).

Conclusion

This chapter has demonstrated that there are several theoretical accounts seeking to explain why states do, or do not, join international watercourse agreements. Additionally, the chapter has identified a range of 'official' reasons put forward by states joining the UNWC. A question that remains to be addressed is the degree to which the theoretical accounts match the practice.

From a classic realist perspective, it would appear that states have very little incentive to become parties to a global convention – or to spend much time negotiating the text of an international agreement, for that matter. Such an account might explain why those states that have traditionally been considered 'hydro-hegemons' (for example, China, Egypt or Turkey) have not become party to the Convention but fails to explain why other 'hydro-hegemons', such as South Africa, have endorsed the UNWC. The more nuanced interpretation of power put forward in this chapter would appear to offer a more accurate account to explain why states like Syria or Iraq have become party to the Convention – namely as a means by which to increase their bargaining power within a particular trans-boundary watercourse context. Such motivations are not likely to be found in the official statements of states when ratifying the UNWC but they are no less important when understanding why states formally adopt agreements.

It could be maintained that some of the official statements by governments on ratifying the UNWC echo constructivist interpretations of how behavior is influenced by shared ideas and normative practices, while also reflecting notions of 'ideational power'. Such an interpretation can be seen in Hungary's reference to 'mutual understanding' or France's desire to 'set an example'.

Finally, following a liberal institutionalist account, it could also be claimed that the mere fact that deliberations within the ILC and the UNGA took place at all represents a collective desire on the part of many states in the international community to foster cooperation over a common goal – the equitable and sustainable management of the world's international watercourses. Such a desire is also reflected in some of the official government statements upon ratification of the UNWC.

It may therefore be concluded that no one theory fully explains why states have joined the UNWC, nor can one single reason be identified in the official statements of the governments upon their ratification. While a suite of reasons may be behind each state's ratification of the UNWC, it would appear that three key underlying themes are prevalent, namely a recognition of the weakness of the current legal architecture and of the role that a global framework can play in complementing the existing architecture, and a belief that the process of state ratification can lead to the UNWC playing a more authoritative and persuasive role.

Part 2

Entry into force and widespread endorsement

Potential effects on international law and state practice

Part 2

Entry into force and
commencement

6   The authority and function of the UN Watercourses Convention

Flavia Rocha Loures, Alistair Rieu-Clarke, Joseph W. Dellapenna and Johan Lammers

Several scholars have studied the impact of the UN Watercourses Convention (UNWC) on state practice since its adoption. Based on those studies, some experts believe that the Convention could remain influential even if it never entered into force. Much less investigated are the potential additional effects from the Convention's entry into force and widespread ratification.

This chapter explores how entry into force might affect the UNWC's authority and its related capacity to support effective water cooperation. We note that, since the Convention's adoption, states have continued to rely largely on customary law in the absence of watercourse agreements. Moreover, states have often spent time debating the merits of resorting to the UNWC, have failed to interpret its provisions appropriately for a lack of understanding or even bypassed the Convention altogether, either because they did not know of its existence or they insufficiently appreciated its authority. We conclude that entry into force is vital to consolidate and strengthen the Convention's legacy in codifying and contributing to the progressive and orderly development of international water law. An effective and widely ratified UNWC would serve as a stronger tool for enabling sustainable and equitable transboundary water management.

We structure our analysis by looking at the Convention as an authoritative codification, clarification and framework for law development. We then consider what, in our view, are the Convention's key functions, with regard to the drafting of water treaties; the interpretation and application of existing watercourse agreements; the governance of international watercourses in the absence of applicable agreements; the development of global treaty law to govern emerging issues; and support for the implementation of water-related multilateral environmental conventions. We also show how, once in force, the UNWC is expected to have stronger effects on non-parties than it has had generally on states since its adoption.

An authoritative framework for law codification, clarification and development

The UNWC is widely recognized as the most authoritative source of the international law governing the non-navigational uses of international watercourses.[1] The Convention represents the culmination of over 50 years of work on the topic under the auspices of the United Nations (UN) that included thorough discussions among renowned international jurists and exhaustive interstate negotiations.

There are several building blocks for the authoritative status of the UNWC. The text was drafted by the UN International Law Commission (ILC) through the leadership of five eminent international lawyers. In the course of the drafting and negotiation procedures, the ILC received and considered comments from a large number of countries, including on whether the proposed articles accurately reflected the status of international law in the field; delegates had many opportunities to make oral and written statements and the process was open for participation by all UN member states.

The UN General Assembly (UNGA) approved the UNWC in May 1997, under the sponsorship of 38 states and with 106 votes in favor, 26 abstentions and only three votes against.[2] This majority included key basin countries, such as Brazil and Zambia, as well as major donors in the water sector, including Japan and the UK. As McCaffrey explains,

> that only three states could bring themselves to vote against . . . suggests a sense among the overwhelming majority of delegations that the rules embodied in the Convention are generally acceptable and, on the whole, reflect a reasonable balance between the interests of upstream and downstream states.[3]

In fact, 'the level of endorsement makes it one of the most successful international instruments recently adopted'.[4]

Among states that had originally abstained or were absent, Benin, France, Iraq and Uzbekistan have since acceded to the Convention. Furthermore,

> at least half of the absent states were island countries with no apparent interest in transboundary water resources. Most of the other states in this group could not participate in the final deliberations and voting because of unrelated circumstances ranging from military conflicts to internal political unrest.[5]

---

1   See, e.g. McCaffrey, S. 'The Contribution of the UN Convention on the Law of the Non-Navigational Uses of International Watercourses', *International Journal of Global Environmental Issues*, 2001; 1 (3/4): 250–63.

2   There were 103 recorded votes in favor but three more states informed subsequently that they had intended to cast positive votes. UNGA, 51st Session, 99th Plenary Meeting at 2, 7–8, UN Doc A/51/PV.99 (May 21, 1997) (UNWC Voting Records).

3   McCaffrey, S. C. *The Law of International Watercourses* (2nd edn, Oxford University Press, 2007), at 375.

4   Wouters, P. 'The Legal Response to International Water Scarcity and Water Conflicts: The UN Watercourses Convention and Beyond', *German Yearbook of International Law*, 1999; 42: 293–336, at 315.

5   Ibid.

Even the three countries that voted against the UNWC have never formally refuted the fundamental principles contained in its text. As McCaffrey postulates, 'Burundi's vote may have more to do with Egypt's historical concern with activities in the upper basin than with the hydrogeographic reality'.[6] In turn, China and Turkey felt that territorial sovereignty should have been explicitly included in the text of the UNWC.[7] Those two countries were further concerned over the compulsory nature of the third-party fact-finding provisions, although both accepted their fundamental obligation to settle disputes peacefully.[8] Turkey also expressed concern over the level of detail contained in what was meant to be a *framework* convention.[9] Yet, subsequent practices by China and Turkey have been generally in line with the Convention. For example, since 1997, China has concluded several watercourse agreements, which incorporate some of the Convention's fundamental principles;[10] and, if Turkey joined the European Union, the country would be subject to the more stringent provisions of the applicable European legal instruments.[11]

The Convention remains the first and only set of rules on the non-navigational uses of international watercourses endorsed by the international community in a formal vote at the UNGA. As Salman highlights, 'it is widely agreed that [the Convention] reflects and embodies the basic principles of international water law'.[12] It is because of this authority that the UNWC has influenced state practice, even pending entry into force.

The question we examine next is how progressively increasing levels of support for the Convention – as expressed through entry into force, followed by widespread ratification and implementation – would strengthen its authority as a framework for law codification, clarification and progressive development.

Codification and clarification of customary law

Customary law is the only global binding legal framework in the field of international water law. Yet, customary norms are often unclear, vague and contested, and may thus aggravate power imbalances:

6 McCaffrey (note 3), at 375.
7 UNGA, Verbatim records of plenary meeting No 99, 21 May 1997, UN Doc A/51/PV.99. Available online at http://www.un.org/ga/search/view_doc.asp?symbol=A/51/PV.99 (accessed March 19, 2013) at 4 6.
8 Ibid.
9 Ibid., at 5.
10 See, e.g. Agreement between Kazakhstan and China on Cooperation in the Use and Protection of Transboundary Rivers (adopted 12 September 2001) [Russian]. Available online at http://iea.uoregon.edu/pages/view_treaty.php?t=2001-UseProtectionTransboundary Rivers.EN.txt&par=view_treaty_html (accessed March 19, 2013).
11 See Rieu-Clarke, A. S., Wouters, P. and Loures, F. *The Role and Relevance of the UN Convention on the Law of the Non-Navigational Uses of International Watercourses to the EU and its Member States* (University of Dundee Centre for Water Law, Policy and Science, 2008). Available online at http://www.internationalwaterlaw.org/bibliography/WWF/RA_European_Union.pdf (accessed March 19, 2013).
12 Salman, S. M. A. 'The UN Watercourses Convention Ten Years Later: Why Has its Entry into Force Proven Difficult?', *Water International*, 2007; 32 (1): 1–15, at 13.

Customary international law is more complex and uncertain than international law found in treaties or conventions... The process of identifying the norms of customary international law, even when successful... often leaves gaps and ambiguities.... Customary international law... frequently is ill-defined and uncertain. Identifying when a practice has crystallized as customary law and the precise content of such customary law has been difficult, requiring research into the proffered reasons for a practice in what often are obscure and inconclusive sources.... Relying upon an informal legal system alone to legitimate and limit claims to use shared water resources is inherently unstable.[13]

Furthermore, customary law lacks a neutral enforcement mechanism and, for this reason, 'has proven unable by itself to solve the problem of managing transboundary water resources... [T]he settlement of [water] disputes... has nearly always required negotiation of a treaty regime'.[14]

Reliance on customary international law alone is thus insufficient to address the challenge of transboundary water management. The UNWC offers a universal common ground arrived at after exhaustive discussions and which makes all watercourse states aware of their minimum rights and duties. Codification of the applicable law in the UNWC is, in this sense, crucial to providing legal clarity and stability and thus preventing unnecessary disputes and balancing power.

A recognition of the Convention as a codification of customary international law is expressed in the decision of the International Court of Justice (ICJ) in the Gabčíkovo-Nagymaros case.[15] In McCaffrey's words, 'the most fundamental obligations contained in the [UNWC]... reflect customary norms'.[16] This is true, particularly for equitable and reasonable use, harm prevention and notification before major planned measures, as well as, arguably, any collateral obligations that directly derive from those three basic principles, such as information exchange.[17] Given that such rules reflect customary law, they are binding upon all states, including non-parties, and even pending entry into force.

The UNWC, however, does more than simply codify existing customary norms in their basic formulation. The Convention clarifies and details the content, scope and extent of such rules and principles, how they relate to one another and some of the specific obligations they entail. The Convention sheds light on aspects of customary law that, otherwise, would remain unclear or uncertain in the absence

---

13 Dellapenna, J. W. 'International Water Law in a Climate of Disruption', *Michigan State Journal of International Law*, 2008; 17: 43–95, at 64, 66, 72.

14 Dellapenna, J. W. 'The Nile as a Legal and Political Structure in The Scarcity of Water: Emerging Legal and Policy Responses' in Brans, E. H. P., de Haan, E., Nollkaemper, A. and Rinzema, J. (eds), *The Scarcity of Water: Emerging Legal and Policy Responses* (Kluwer Law International, 1997), at 121, 123.

15 *Case concerning Gabčíkovo-Nagymaros Project (Hungary v Slovakia)*, 1997 ICJ 7 (25 September), Para 85 (Gabčíkovo-Nagymaros case).

16 McCaffrey (note 1), at 259.

17 Ibid., at 260.

of their codification. In so doing, the Convention facilitates the application, interpretation and implementation of the basic, minimum standards governing the
non-navigational uses of international watercourses'.[18]

For example, in Articles 9 and 31, the UNWC codifies a data-sharing obligation, clarifying the nature of the information to be exchanged and of the exchange
process itself, as well as the extent of such a duty, other correlated obligations and
the applicable exception. The same applies to Articles 11–19 of the UNWC, on
planned measures, which address a void in the normative content of the duty to
notify and its correlated obligations. While the duty to notify itself is recognized as
part of customary law, the UNWC goes further by determining the specific
requirements and timelines for notification, information exchange, consultations
and negotiations regarding the utilization of international watercourses.

Given the importance of codifying and clarifying the law in the field and the
UNWC's authoritative role in this regard, what are the potential effects on such an
exercise from entry and non-entry into force?

The entry into force, widespread endorsement and effective implementation of
the UNWC will reinforce and progressively strengthen its legal and political
authority as a framework codifying and clarifying international water law.[19] In this
sense, it is stated in the Convention's Preamble that the successful codification and
progressive development of international water law 'would assist in promoting and
implementing the purposes and principles set forth in Articles 1 and 2 of the
Charter of the [UN]'. In our view, the objective of codifying and progressively
contributing to the development of the law in this field can only be considered as
having been successfully achieved when the Convention enters into force and
progressively receives widespread endorsement from a representative number of
states involved in transboundary water issues.

In other words, growing levels of support for the UNWC would consolidate this
instrument as a widely accepted code of rules, necessary for providing some legal and
political stability to the relations among watercourse states. After all, the exercise initiated at the UNGA was driven by the recognized need for a global convention
codifying and progressively developing international water law. At the time, UN
member states and experts agreed on the need for the UNWC, given that most
international watercourses lacked appropriate cooperative arrangements and, therefore, watercourse states still largely relied on general norms of customary law. As we
discuss below, this remains the case today pending the Convention's entry into force.

Furthermore, the UNWC's implementation process would foster interstate
discussions and potentially case law elucidating its provisions that remain ambiguous or insufficiently developed. For example, a question that commonly arises in
connection with Article 7(2) relates to burden of proof. According to the ILC, once
the harmed state demonstrates the occurrence of injury and establishes the link
with an activity within another state's territory, 'the burden of proof for establishing that a particular use is equitable and reasonable lies with the state whose use of

18   Ibid., at 261.
19   McCaffrey (note 3), at 376.

the watercourse is causing significant harm'.[20] Arguably, the interpretation of this aspect of Article 7(2) can only be clarified in more precise terms through case law and state practice in the Convention's implementation upon entry into force.

Failure to support the UNWC could have the opposite effect: theoretically, it 'may give rise to doubts whether at the time of its conclusion it did already embody existing rules or principles of customary international law or else may affect [their] continued validity'.[21] In this regard, Tanzi recalls that,

> when a codification convention fails to obtain broad support from states in terms of ratifications, it can run counter to the basic purpose of clarifying and developing the basic general rules applicable in a given area, leaving customary law in a worse state than it was before codification was attempted.[22]

As a result, non-entry into force could strengthen the negotiating position of states tactically resorting to extreme theories to plead their case and exert pressure on weaker co-riparians. As Tanzi explains, 'one of the factors that allowed riparian states to flirt with the notion of an absolute and unlimited sovereignty over transboundary rivers was the underdeveloped and unclear state of international water law in the early 1990s'.[23] At this stage, 'this potential effect has not taken place... There is... no practice of states justifying such doubts or pointing to such a negative development'.[24] But it seems fair to assume that non-entry into force would prevent the Convention from becoming a globally agreed common denominator that encourages equitable solutions and levels the playing field among all watercourse states – whether they are weak or strong, rich or poor, capable or inexperienced negotiators.

In particular, if entry into force were unduly delayed, arguably, Article 18 of the Vienna Convention[25] would no longer apply. In such a scenario, signatory states or other countries that have somehow given their consent to be bound by the Convention would eventually no longer be under the obligation to refrain from acts which would defeat its object and purpose.[26]

---

20   ILC, *Report of the International Law Commission on the Work of its Forty-sixth Session, 2 May–22 July 1994, Official Records of the General Assembly, Forty-ninth Session, Supplement No. 10. Topic: Multiple Topics.* Extract from the International Law Commission 1994, Vol. II(2) UN Doc A/49/10 (1994), at 104. Available online at http://untreaty.un.org/ilc/documentation/english/A_49_10.pdf (accessed March 19, 2013).

21   Lammers, J. G. *Potential Effects from the Non-Entry into Force of the UN Watercourses Convention,* Presentation at the Seminar 'The UN Watercourses Convention: Legacy, Prospects and Value for the Realization of International Policy Goals', World Water Week (Stockholm, August 17–23, 2008).

22   Tanzi, A. and Arcari, M. *The UN Convention on the Law of International Watercourses: A Framework for Sharing* (Kluwer Law International, 2001) at 29.

23   Ibid., at 12.

24   Lammers (note 21).

25   Vienna Convention on the Law of Treaties (adopted May 23, 1969, entered into force January 27, 1980) 1155 UNTS 331.

26   Lammers (note 21).

Progressive development of customary law

The UNWC not only codifies but also aims to contribute to the progressive development of customary law. This aspect of the Convention's authority is of major relevance, given that, if progressive norms became recognized as customary law, they would bind even non-parties.[27]

For example, assuming that a general duty to protect ecosystems is part of customary law, the UNWC contributes to the development of that duty in a certain direction. Article 20 transcends existing custom by requiring watercourse states to protect and preserve the ecosystems of international watercourses for their own sake. Unlike other environmental obligations established by the Convention, Article 20 does not refer to significant transboundary harm. In this sense, activities posing a threat to the integrity of aquatic ecosystems, but not necessarily causing significant injury to co-riparians, could arguably fall under Article 20. The goal is to maintain the 'continued viability [of aquatic ecosystems] as life-support systems, thus providing an essential basis for sustainable development'.[28] The provision goes even further, establishing that the duty in question must be implemented through individual or, where appropriate, *joint* actions.

Even pending entry into force, 'the provisions of the Convention that do not reflect current law are likely to give rise to expectations of behavior on the part of riparian states that may, over time, ripen into international obligations'.[29] The ICJ corroborates the value of codification through the work of the ILC, as in the case of the UNWC, in developing customary international law:

> A convention adopted as part of the combined process of codification and progressive development of international law may well constitute, or come to constitute the decisive evidence of generally accepted new rules of international law... The convention may serve as an authoritative guide for the practice of states faced with the relevant new legal problems, and its provisions thus become the *nucleus around which a set of generally recognized legal rules may crystallize*.[30]

---

27 The only exception to this statement would be where a state, pursuant to the persistent objector rule, has clearly and consistently objected to the existence of a rule or principle prior to its widespread recognition as a customary norm. See Fisheries Case (UK v. Norway), December 18, 1951, ICJ Reports 1951, 116, at 131.

28 1994 ILC Report (note 20), at 119, 122.

29 McCaffrey, S. C. and Sinjela, M. 'Current Development: The 1997 UN Convention on International Watercourses', *American Journal of International Law*, 1998; 92: 97–107, at 106. See also Tanzi, A. *The Relationship between the 1992 UNECE Convention on the Protection and Use of Transboundary Watercourses and International Lakes and the 1997 UN Convention on the Law of the Non-Navigational Uses of International Watercourses*. (UNECE Task Force on Legal and Administrative Aspects, 2000), at 51, 54.

30 North Sea Continental Shelf (FRG v. Denmark; FRG v Netherlands), 1969 ICJ 3, 244 (20 February).

Entry into force will consolidate and strengthen the Convention's role as a framework for the crystallization of emerging norms in the field, such as the obligation to protect ecosystems of international watercourses. With entry into force, state practice in the implementation of such norms is likely to become more intense and easily detectable, including among non-parties, accelerating the process under which the law evolves.[31] As experts have noted,

> if a widespread and representative number of states agree to be bound by a treaty and apply the provision of the treaty in their state practice, the rules originally found in the treaty may, sometimes even in a short period of time, come to reflect customary international law and therefore indirectly bind the states not party to the convention.[32]

Non-entry into force within a reasonable time could have the opposite effect; i.e. it could prevent or slow down the progressive development of customary law.

The functions of the UN Watercourses Convention

In this section, we show how a widely accepted and effective UNWC, in performing its various functions, can 'provide watercourse states with firm common ground as a basis for negotiations – which is what watercourse negotiations lack most at the present time'.[33] McCaffrey identifies three key functions the UNWC is to perform:

> It provides a starting point for the negotiation of agreements relating to specific watercourses, and, in the absence of any applicable agreement, sets basic parameters governing the conduct of states riparian to those watercourses. Even where there is an applicable agreement, the convention may play an important role in the interpretation of that agreement.[34]

In addition to these functions, the UNWC is expected to play a role as a framework for the development of global treaty law in the field and in the application and implementation of water-related multilateral environmental conventions.

Inspiring negotiations on watercourse agreements

The UNWC offers a common denominator for informing the process of drafting and negotiating watercourse agreements, taking into account the guidance and

---

31   See McCaffrey and Sinjela (note 29) at 106, n.58, citing Anthony D'Amato, *The Concept of Custom in International Law*, ch. 8 (1971).

32   Rieu-Clarke, A. 'Entry into Force of the 1997 UN Watercourses Convention: Barriers, Benefits and Prospects', *Water*, 2007; 21: 12–16, at 14. Available online at www.iwapublishing.com/pdf/ Water21%20Dec07p12to16.pdf (accessed March 19, 2013).

33   1994 ILC Report (note 20), at 93.

34   McCaffrey (note 1), at 261.

parameters in its Article 3. The UNWC does so through its cohesive set of minimum flexible standards – or 'core elements' – which facilitate and inform negotiations on watercourse agreements. Such standards can be applied and adjusted to the special characteristics of each watercourse and to the needs of the basin states concerned. Among such 'core elements' are, for example, the principles of equitable and reasonable use and participation and harm prevention, the duty to protect the ecosystems of international watercourses, the procedural rules on planned measures and the dispute settlement mechanisms.

The scope of this function may relate to: a) the adoption of *new* legal instruments, where none exists, or to supplement existing treaties; and b) the revision of existing treaties considered outdated, weak or inequitable. Revision may entail the strengthening, readjustment, expansion in scope or updating of the original treaty to reflect emerging challenges, or the progressive development of international water law.

This has been and will continue to be a function of the utmost importance. Only 40 percent of the world's transboundary watersheds are subject to a cooperation regime and, among existing treaties, 80 percent are bilateral, even in cases where there are more riparian states in the basin.[35] Among other flaws, many water treaties: a) simply define borders or apply to specific joint projects; b) neither deal with pollution nor provide for ecosystem protection or integrated river basin management[36] – crucial approaches to managing freshwaters; c) favor the most powerful riparian within the watercourse; or d) disregard the needs of local communities. Specifically, only half of the existing watercourse agreements contain monitoring provisions and 80 percent lack or have inadequate enforcement mechanisms.[37]

Hence, 'states negotiating future agreements will [and should] resort to [the Convention's] provisions at least as a point of departure'.[38] Since its adoption, the Convention has played this role in many places: for example, the 2000 SADC Protocol[39] repeats most of the UNWC provisions, with small adjustments here and there; the Preamble of the Senegal River Water Charter[40] expressly refers to the Convention; other agreements 'contain express reference to the basic principles...

---

35   UNEP, *Challenges to International Waters: Regional Assessments in a Global Perspective* (UNEP, 2006) at 35. Available online at www.unep.org/dewa/giwa/publications/finalreport/giwa_final_report.pdf (accessed March 19, 2013).

36   Hamner, J. and Wolf, A. 'Patterns in International Water Resources Treaties: The Transboundary Freshwater Dispute Database', *Colorado Journal of International Environmental Law and Policy, 1997 Yearbook,* 1998.

37   UNDP, *Protecting International Waters/Sustaining Livelihoods* (New York: UNDP Global Environment Facility, 2004), at 8.

38   McCaffrey (note 3), at 377.

39   Revised Protocol on Shared Watercourse Systems in the Southern African Development Community Region, 40 ILM 321 (adopted August 7, 2000, entered into force September 22, 2003).

40   Charte des Eaux du Fleuve Sénégal (adopted May 28, 2002, not in force). Available online at www.lexana.org/traites/omvs_200205.pdf (accessed March 19, 2013).

consolidated in the text of the [Convention]'[41], including the Lake Victoria Protocol[42] and the Zambezi Commission Agreement.[43]

Considering the extent to which the UNWC has influenced watercourse negotiations since its adoption, how would entry into force affect the Convention's ability to perform such a key function? Given that the Convention is not in force, its potential to support cooperation at the basin level remains under-explored. Once in force and widely accepted, the UNWC is likely to have a much wider and more useful influence on the process of treaty-making among co-riparians. An effective UNWC would address two problems: a) insufficient awareness of the Convention among key stakeholders; and b) lack of acceptance by a considerable number of riparian states, through the act of accession, of the Convention's provisions as a starting point for negotiations.

In certain places, government officials are not even aware that the Convention exists or, at best, are not familiar with its authority, content or functions.[44] This happens because the Convention is not in force, and there is no official and effective pressure for states to devote time and resources to understanding and applying its provisions. States are already overwhelmed with multiple other global and regional environmental treaties. Why would they pay attention to a treaty that has not yet entered into force?

Furthermore, widespread ratification would consolidate and reinforce the Convention's status as a generally accepted framework to inform negotiations among watercourse states. In particular, parties to the Convention would benefit from having a commonly agreed basis as a starting point for dialogue and, ideally, some shared understanding over the key rules and principles of international water law. Currently, as already noted, states must rely on customary law and the numerous existing water agreements at various levels as diffuse, sometimes conflicting sources for informing treaty negotiations. States also spend excessive time debating the merits of resorting to the UNWC. They fail to interpret its provisions appropriately for a lack of understanding or even bypass the Convention altogether, because they do not know of its existence or they insufficiently appreciate its authority.

By addressing the problems above, an effective UNWC would more efficiently facilitate interstate negotiations and thus potentially fuel the process of treaty-making. As Rieu-Clarke explains,

> while some states have relied on the convention to formulate regional, basin-specific or bilateral agreements, the number of these agreements remains quite

---

41   Tanzi (note 22), at 28.

42   Protocol for Sustainable Development of Lake Victoria Basin (adopted November 29, 2003). Available online at http://bd.stp.gov.ml/padelia/pdf/CHARTEDESEAUXDUFLEUVE SENEGAL.pdf (accessed March 19, 2013).

43   Agreement on the Establishment of the Zambezi Watercourse Commission (adopted July 13, 2004). Available online at www.icp-confluence-sadc.org/documents/agreement-establishment-zambezi-watercourse-commission-2004 (accessed March 19, 2013).

44   Reports of regional workshops in West Africa, Central America (on file with authors).

low in comparison to the number of international watercourses. Entry into force of the convention would therefore hopefully increase awareness of the need to strengthen existing agreements and provide the impetus to establish new agreements where they do not exist.[45]

In that regard, Article 3(5) of the UNWC establishes that, when a co-riparian considers that 'adjustment and application of the provisions of the convention is required because of the characteristics and uses of a particular international water-course,' a party must consult and negotiate in good faith for the purpose of concluding an agreement. This provision, along with the Convention's recommendation to review existing agreements according to its standards, is likely to trigger and frame interstate dialogue and interaction based on the Convention's rules and principles. It is likely that, in some cases, such dialogue will lead to the drafting and adoption of new or revised watercourse agreements.

In addition to stimulating treaty-making, entry into force would enable the Convention to create peer pressure on states for them to strive to ensure consistency among watercourse agreements. As highlighted during negotiations, the UNWC serves 'as an important general framework for orderly international processes to deal with international water issues'.[46] In practice, however, the Convention has not played such a coordinating role in an efficient manner. An effective UNWC would ultimately provide 'a certain degree of stability to the process of creating [such] agreements'.[47] This is especially important where states share multiple river basins and thus may be parties to several treaties, as is the case in West Africa. From a broader perspective, consistency with customary legal norms promotes the rule of law in international affairs, including as a key tool to level the playing field among co-riparian states.

Finally, many agreements, even among those postdating the UNWC's adoption, fail to incorporate key elements of international water law, as reflected in the Convention.[48] At least to some extent, the inadequacy of such agreements may be associated with the lack of an effective, widely endorsed global framework setting the legal stage for negotiations. Entry into force would contribute to bettering the levels of knowledge and understanding of the UNWC and thus of international water law in general. Ultimately, this would enable the development of better agreements, well grounded on customary law. Furthermore, a widely endorsed UNWC would offer a universal template for assessing the adequacy and bringing to light the strengths and weaknesses of existing treaties.

---

45 Rieu-Clarke (note 32), at 14.
46 Nordic countries' statement, A/C.6/58/SR.20, at 6, January 6, 2005, 6th Committee, Summary Record of the 20th Meeting, 58 UNGA), Meeting of November 3, 2003.
47 Schroeder-Wildberg, E. *The 1997 International Watercourses Convention: Background and Negotiations.* Working Paper on Management in Environmental Planning 004/2002 (Technical University of Berlin, 2002), at 7. Available online at http://hydroaid.tinext.net/FTP/Data_Research/ E.%20Schroeder-Wildberg-The%201997%20Int%20Watercourse%20Convention.pdf (accessed March 19, 2013).
48 See Part III for numerous examples of such agreements.

Aiding parties in the application and interpretation of watercourse agreements

Another function played by the UNWC has to do with supporting the application and interpretation of watercourse agreements. This is especially important in the case of disputed provisions with ambiguous or obscure language, as well as outdated treaties calling for a progressive reading and application consistent with the modern developments in international law incorporated into the Convention. After all, a watercourse agreement,

> is not something standing alone, but is supported by, limited by and tested against a set of general international law standards, the content of which and the validity of which are not determined by the agreement in question. The conventional law of any drainage basin can be effectively applied only with the aid of principles and rules drawn from the larger international legal system.[49]

In this sense, 'while the intent of the parties at the time of a treaty's conclusion obviously cannot be disregarded, developments in the law may be relevant to the treaty's interpretation'.[50]

Since its adoption, the UNWC has guided lawyers and diplomats in framing negotiations and pleading their case, and has provided an analytical framework for arbitral or judicial decision-making. For example, in 1997 – the year of the Convention's adoption – the ICJ justified its decision on the Gabčíkovo-Nagymaros case on the basis of the Convention's principles.[51] According to Tanzi, this ICJ Decision 'enhanced the normative value of the general principles embodied in the Convention... as authoritative parameters for the interpretation and application of existing special watercourse treaties'.[52]

The UNWC has delivered on this function even pending entry into force because of its authority. As discussed earlier, entry into force would enhance the UNWC's authority and drive awareness raising and knowledge development around its content, relevance and functions. These factors combined would ultimately increase the likelihood of states and judges invoking the Convention's provisions in support of the interpretation and application of existing treaties.

Governing interstate relations in the absence of applicable agreements

As discussed above, the UNWC's most fundamental principles are recognized as part of customary law. Such principles are binding on all states, whether they have

---

49 Tanzi (note 22), at 27, citing R. D. Hayton, 'The Formation of the Customary Rules of the Drainage Basin Law' in Garretson, A. H., Hayton, R. D. and Olmstead, C. J. (eds) *The Law of International Drainage Basins* (Oceana, 1967), at 834–36.

50 Ibid., quoting *Legal Consequences for States of the Continued Presence of South Africa in Namibia (South West Africa) notwithstanding Security Council Resolution 276 (1970)*, 1971 ICJ Rep 16, 31, Para 53 (Advisory Opinion of June 21).

51 Gabčíkovo-Nagymaros Case (note 15), paras 85, 147. See Chapter 2 for a more detailed analysis of this case.

52 Tanzi (note 22), at 28.

acceded to the Convention or not, and even pending entry into force. Once in force, however, in the absence of applicable watercourse agreements, the UNWC as a whole will serve as a solid, binding, legal basis for determining the rights and duties of watercourse states that are parties to it – rights and duties that will be enforceable through dispute prevention and settlement mechanisms. As Wouters explains, the Convention operates as 'a flexible rule governing legal entitlement, accompanied by the requirement of preventive behavior and complemented by a comprehensive set of relatively detailed procedural rules'.[53]

The Convention's substantive rules would offer a firm common ground, formally endorsed through the accession process, to govern relations among parties. This is central to dispute prevention. In the absence of relevant watercourse agreements, the Convention elucidates the applicable law, thereby setting out the substantive rules of the game needed to inform fruitful negotiations.

Procedurally, the principle of cooperation in Article 8 requires states to discuss and cooperatively promote the equitable and reasonable use of their shared waters to their mutual benefit. In addition, the Convention incorporates rules under which parties must consult and negotiate with each other in good faith, such as in the case of planned measures. As a general rule, 'a request by one watercourse state to enter into consultations may not be ignored by other watercourse states'.[54] The Convention's well-developed body of procedural rules is vital for its ability to govern interstate relations directly in the absence of watercourse agreements. As McCaffrey and Sinjela explain,

> the facts and circumstances of each case, rather than any *a priori* rule, will ultimately be the key determinants of the rights and obligations of the parties. Difficult cases, which are bound to proliferate in the future, will be solved by cooperation and compromise, not rigid insistence on rules of law'.[55]

A firmly grounded, widely accepted and effective UNWC would offer greater juridical stability than customary law. Such a regulatory role will be especially relevant in situations where co-riparian states become parties to the Convention, but fail to negotiate watercourse agreements; or during prolonged negotiations and pending the entry into force of specific treaties. This function remains relevant today, with 60 percent of the world's transboundary watersheds not covered by agreements.[56]

In performing this function, an effective UNWC would: govern interstate relations in the complete absence of a watercourse agreement;[57] and serve as a common framework for the entire watershed, where only partial watercourse agreements were in place and all riparians became parties to the Convention. These

53   Wouters (note 4), at 315.
54   1994 ILC Report (note 20), at 102.
55   McCaffrey and Sinjela (note 29), at 101.
56   UNEP (note 35), at 35.
57   McCaffrey (note 1), at 261.

two aspects are important, given the lack of clarity that characterizes customary law. Rather than relying on general norms of custom, the Convention would have all co-riparians accept in a framework, and yet binding, treaty the nature and extent of their basic rights and duties. In addition, the UNWC would apply directly among parties to supplement an existing agreement in situations not governed by such an agreement, but within the scope of the Convention's provisions. As Tanzi explains, watercourse agreements:

> hardly ever cover every aspect of the uses of the international basin... that may be of legal relevance... In order to deal with such spectrum of situations, riparian states would necessarily have to resort to the general principles and rules as terms of reference for their dealing.[58]

In the absence of an effective UNWC, customary law remains the main source that states may invoke to address legal gaps and failings in existing water treaties.

Enabling progressive treaty making and implementation at the global level

The UNWC offers a foundation for the progressive development of global treaty law on emerging transboundary water issues. Once in force, the Convention can serve as the universal, widely accepted and authoritative basis for the codification of the evolving rules in the field.

The UNWC has already served as the primary basis for the development of the Draft Articles on Transboundary Aquifers.[59] However, only an effective UNWC can offer an adequate template for the progressive and coherent development of treaty law at the global level.[60] Of course, this would require parties to the Convention eventually to agree on the necessary procedures for the adoption of future protocols, amendments and guidelines.

Furthermore, the future implementation of the UNWC will facilitate the detection of relevant areas in need of better or new international regulation, contributing to the process of law development.[61] One of these areas is, for example, public participation in transboundary water negotiations – a topic not sufficiently covered by the Convention, which only codifies, in Article 32, the non-discrimination principle.

Finally, the UNWC has the potential to act as an official world forum, under the auspices of the UN, for global dialogue on the international management of shared

---

58   Tanzi (note 22), at 26.
59   ILC, 'Draft Articles on the Law of Transboundary Aquifers, with Commentaries' in ILC, *Report of the International Law Commission, Sixtieth Session (5 May–6 June and 7 July–8 August 2008)*, General Assembly Official Records Sixty-third Session Supplement No. 10. UN Doc A/63/10 (UN, 2008) at 19–23. Available online at http://untreaty.un.org/ilc/reports/2008/2008report.htm (accessed March 20, 2013).
60   See chapter 23, discussing the potential adoption of a protocol to the UNWC to govern groundwater resources of relevance to international law.

water resources, including policy coordination and the sharing of knowledge, information, experience and best practices. The value of this potential role can be illustrated by the many ways in which the UNECE Water Convention has contributed to the implementation and development of international water law and policy. In particular, the UNECE Water Convention has supported the implementation and strengthening of water agreements in Europe and neighboring regions.

As Rieu-Clarke explains, 'while the [UNECE Water Convention] benefits from an institutional structure...there is no reason why parties to the [UNWC] – armed with a "common language" – could not conduct such activities through informal networks'.[62] Arguably, however, the UNWC's ability to enable progressive treaty-making and implementation at the global level could be enhanced if the appropriate mechanisms were agreed upon among parties. Again here, only entry into force would justify triggering formal discussions among parties on the benefits and options regarding the establishment of institutional mechanisms to support the UNWC's implementation.

Building on synergies with other multilateral environmental conventions

The UNWC offers a legal foundation for international water cooperation in support of the implementation of other multilateral environmental conventions. By codifying and developing international water law, the Convention can add value to the regulatory framework and help to advance the goals under environmental conventions that depend on sound cooperation between riparian states.[63] An effective UNWC would enable parties to formally engage with implementing bodies under related environmental agreements, allowing them to build on synergies, avoid duplication and coordinate implementation.

For example, the Convention on Biodiversity (CBD)[64] relies on the ecosystem approach as its fundamental principle. In the case of transboundary watersheds, international cooperation becomes an indispensable tool for enabling the application of the ecosystem approach at the basin level. In the future, cooperative efforts among co-riparian states under the UNWC, which are also CBD parties, may develop into joint initiatives to protect and preserve the ecosystems of international

---

61   Compare UNWC with The Berlin Rules on Water Resources, in ILA, *Report of the Seventy-First Meeting of the International Law Association*, 334 (2004). Available online at www.ila-hq.org/download.cfm/docid/B6F3AD1C-11B5-45A3-89534097AD1FEE95 (accessed March 20, 2013).

62   Rieu-Clarke (note 32), at 16.

63   See Chapters 18–19, looking at the potential value added by the UNWC to support the implementation of the Conventions on Climate Change and Desertification, respectively.

64   UN Convention on Biological Diversity (CBD) (adopted June 5, 1992, entered into force December 29, 1993). Available online at www.cbd.int/doc/legal/cbd-en.pdf (accessed March 20, 2013). See Brels, S., Coates, D. and Loures, F. *Transboundary Water Resources Management: The Role of International Watercourse Agreements in the Implementation of the CBD*, CBD Technical Series No 40 (Secretariat of the CBD, 2008), for a detailed analysis of the complementarities between the UNWC and the CBD.

watercourses. Recognizing this, CBD Decision VIII/27, on invasive species, explicitly urges parties to ratify and implement the UNWC.[65] This decision was later reiterated by CBD parties at their 9th Conference, with regard to the programme of work on inland waters.[66]

In addition, the UNWC will inform and reinforce the implementation of Article 5 of the Ramsar Convention.[67] This provision requires parties to consult with each other regarding transboundary wetlands. Further guidance on this provision has been adopted by parties,[68] but the Ramsar Convention itself lacks provisions aimed specifically at fostering cooperation among watercourse states. Hence, riparian states that are parties to the Ramsar Convention would be better equipped to pursue their objectives related to transboundary wetland conservation and sustainable use within the UNWC's legal framework.

Legal and political effects on non-parties

As discussed above, the UNWC has had an important influence on state practice since its adoption, as an authoritative instrument evidentiary of customary law. Such an influence is likely to continue and grow progressively upon entry into force in relation to non-parties, in tandem with the levels of endorsement of the Convention. Entry into force and widespread implementation would enhance the Convention's legal and political authority and thus its persuasive force on non-parties. Therefore, albeit not directly binding on non-parties, the Convention is likely to have a growing impact even on basins where not all states are parties to it.

This persuasive force is typical of a codification instrument and has crucial importance in the field of international watercourses: 'given the frequently highly controversial differences between the parties negotiating over uses of international

65 CBD CoP-8 (Curitiba, March 20–31, 2006), *Decision Adopted by the Conference of the Parties to the Convention on Biological Diversity at its Eighth Meeting. VIII/27. Alien Species that Threaten Ecosystems, Habitats or Species (Article 8 (h)): Further Consideration of Gaps and Inconsistencies in the International Regulatory Framework* (UNEP, June 15, 2006) UNEP/CBD/COP/DEC/VIII/27, at 3. Available online at www.cbd.int/doc/decisions/cop-08/cop-08-dec-27-en.pdf (accessed March 20, 201).
66 CBD CoP-9 (Bonn, May 19–30, 2008), *Decision Adopted by the Conference of the Parties to the Convention on Biological Diversity at its Ninth Meeting. IX/19.* Biological Diversity of Inland Water Ecosystems (UNEP, October 9, 2008) UNEP/CBD/COP/DEC/IX/19, at 1. Available online at www.cbd.int/doc/decisions/cop-09/cop-09-dec-19-en.pdf (accessed March 20, 2013).
67 Convention on Wetlands of International Importance especially as Waterfowl Habitat (adopted February 2, 1971, entered into force December 21, 1975) 996 UNTS 245 (Ramsar Convention). Available online at www.ramsar.org/cda/en/ramsar-documents-cops-1971-final-act-of-the/main/ramsar/1-31-58-136^20803_4000_0__ (accessed March 20, 2013).
68 Ramsar Convention CoP-7 (San José, Costa Rica, May 10–18, 1999) *Guidelines for International Cooperation under the Ramsar Convention,* Resolution VII.19. Available online at www.ramsar.org/pdf/res/key_res_vii.19e.pdf (accessed March 20, 2013); and *Resolution VII.18: Guidelines for Integrating Wetland Conservation and Wise Use into River Basin Management,* Available online at www.ramsar.org/cda/en/ramsar-documents-resol-resolution-vii-18/main/ramsar/1-31-107%5E20586_4000_0__ (accessed March 20, 2013).

watercourses, the compromises reached at the bargaining table often produce ambiguous provisions'.[69] This is why it is so important to have complementary, mutually reinforcing legal frameworks at various levels, with the UNWC reflecting globally agreed basic legal standards.

Once in force, the Convention will have an impact on non-parties in the interpretation and application of watercourse agreements and customary law, as well as through interstate negotiations and decision making. In this sense, an effective UNWC,

> will have significant bearing upon controversies between states, one or more of which is not a party to the Convention. In addition, the Convention may be of value in interpreting other general or specific agreements... that are binding on the parties to a controversy, whether or not the Convention is itself binding on those parties.[70]

At the same time, as discussed above, widespread support for the Convention could accelerate the process for consolidating additional provisions as part of international custom. This process would progressively build the Convention's binding force on states that are not parties to it.[71] Arguably, this impact on non-parties could weaken over time and could cease completely in the remote case of no additional contracting states and non-entry into force. Such a scenario could leave unclear the status of those provisions today recognized as part of customary law and eventually preclude such principles from remaining binding on non-parties. Furthermore, as Lammers explains:

> Non-ratification and non-entry into force may... have a negative effect on the completion or further development of an ongoing process of emerging customary international law at the time of the conclusion of the Convention or may prevent progressively developed elements of law from becoming emerging or eventually established customary international law.[72]

Therefore, non-entry into force could preclude emerging customary international law proposed within the text of the Convention from evolving into established custom and thus from becoming binding on non-parties.

Conclusion

More than a decade since the adoption of the UNWC, the challenges in international watercourses remain and, in many places, have gotten worse. On the positive

69   Tanzi (note 22), at 26.
70   McCaffrey, S. C. International Water Law for the 21st Century: The Contribution of the U.N. Convention, *Journal of Contemporary Water Research and Education*, 2001; 118 (1): 11–19, at 17.
71   See Vienna Convention (note 25) Article 38.
72   Lammers (note 21).

side, awareness of the Convention has increased, and the international community is now engaged in its ratification process. The value of the Convention resides in its legal and political authority, which is essential for the UNWC to deliver its functions, as examined above. As Salman explains, 'the adoption of the [UNWC] marked an historic moment in the evolution of international water law'.[73] For this reason, the Convention has influenced and will continue to influence state practice, even pending entry into force.

However, there are specific advantages in having an effective and widely endorsed UNWC. As the next step to the work initiated at the UNGA, the international community should now secure the entry into force of the Convention. States, international organizations and the UN should then remain engaged to progressively strengthen the Convention's authority and thus its ability to perform the intended functions examined above. As such, the Convention would service as a solid basis for the sustainable development, management and protection of international watercourses and their ecosystems.

73   Salman (note 12), at 13.

7 Impacts on the international architecture for transboundary waters

Alistair Rieu-Clarke and Guy Pegram

Legal architecture

UN General Assembly Resolution 2669 (XXV) pointed out that, 'despite the great number of bilateral treaties and other regional regulations... the use of international rivers and lakes is still based in part on general principles and rules of customary law'.[1] In 1970, it was therefore recognized that there was a need at the global level to supplement a fragmented system of bilateral legal instruments relating to international watercourses.

The *Atlas of International Freshwater Agreements* identifies 400 water agreements adopted since 1820.[2] In terms of adoption rates, Conca, Wu and Mei observe that a few agreements per year were adopted throughout the 1980s, then there was a significant increase in treaty adoption activity following the 1992 UN Conference on Environment and Development, followed by a drop-off in the number of agreements adopted towards the turn of the century.[3]

Despite these legal developments, the international legal architecture regulating international watercourses remains fragmented.[4] The majority of basin-specific agreements cover multilateral rivers, even though 67 percent of the world's international rivers are bilateral.[5] Additionally, there has been a common trend to adopt bilateral agreements within multilateral river basins.[6] Such a fragmented system led

1 UNGA, Resolution 2669 (XXV) 'Progressive Development and Codification of the Rules of International Law Relating to International Watercourses' (December 8, 1970), at 127. Available online at www.un.org/ga/search/view_doc.asp?symbol=A/Res/2669(XXV) (accessed March 20, 2013).

2 UN Environmental Programme and Food and Agriculture Organization of the UN, *Atlas of International Freshwater Agreements* (UNEP 2002). Available online at www.transboundarywaters.orst.edu/publications/atlas (accessed 20 March, 2013).

3 Conca, K., Wu, F. and Mei, C. 'Global Regime Formation or Complex Institution Building? The Principled Content of International River Agreements', *International Studies Quarterly*, 2006; 50: 263–85, at 270–71.

4 Zawahri, N. A. and Mitchell, S. M. 'Fragmented Governance of International Rivers: Negotiating Bilateral Versus Multilateral Treaties', *International Studies Quarterly*, 2011; 55: 835–58.

5 Ibid.

6 Song, J. and Whittington, D. 'Why Have Some Countries on International Rivers Been Successful Negotiating Treaties? A Global Perspective', *Water Resources Research*, 2004; 40: 1–18, at 3. doi: 10.1029/2003WR002536.

Wolf and Giordano to observe that 158 of the world's 263 international basins lack any type of cooperative framework; and, of the 105 basins covered by agreements, approximately two-thirds do not include all basin states.[7] While such statistics do not account for global and regional treaty regimes, or rules and principles of customary international law, they do demonstrate that governance frameworks at the basin level are often lacking.

However, it should be noted that the legal architecture within different regions varies. Throughout Africa, there are 59 transboundary river basins, which make up 62 percent of the continent's land surface.[8] Of these transboundary river basins, 16 are covered by basin-wide agreements, three are partially covered by agreements and 40 have no basin-specific agreements in place.[9] Additionally, Angola, Botswana, Congo, Lesotho, Malawi, Mauritius, Mozambique, Namibia, Seychelles, South Africa, Swaziland, Tanzania, Zambia and Zimbabwe have ratified the Southern African Development Community (SADC) Revised Protocol on Shared International Watercourses, which is applicable to the 15 international watercourses across the Southern African region and embodies much of the content of the UN Watercourses Convention (UNWC).[10]

Asia is home to 57 transboundary river basins, which account for 39 percent of the continent's land surface.[11] Ten river basins, constituting 3,270,600 km^2 of the land mass, are covered by basin-wide agreements.[12] 15 river basins, representing 12,584,400 km^2, are partially covered by basin agreements, and 32 river basins, representing 1,933,060 km^2, are not covered by any basin agreement.

Across Europe, there are 64 transboundary river basins covering 54 percent of the continent's land surface;[13] 35 rivers are covered by basin-wide agreements, whereas ten are partially covered and 19 have no basin-specific agreements in place.[14] However, most European states are also obligated to implement two

---

7   Giordano, M. and Wolf, A. T. 'The World's International Freshwater Agreements: Historical Developments and Future Opportunities' in UNEP and FAO (note 2), at 7. Wolf and his colleagues have since revised the number of international basins upward, to 276. See, e.g. de Stefano, L., de Silva, L., Edwards, P. and Wolf, A. T. *Updating the International Water Events Database (Revised)*, (UNESCO, 2009), at 2. Available online at http://unesdoc.unesco.org/images/0018/001818/181890E.pdf (accessed March 20, 2013); De Stefano, L., Duncan, J., Dinar, S., Stahl, K., Strzepek, K. and Wolf, A. T. *Mapping the Resilience of International River Basins to Future Climate Change-Induced Water Variability* (International Bank for Reconstruction and Development and World Bank, 2010), at 5.

8   Wolf, A. T. 'International Rivers of the World', *International Journal of Water Resources Development*, 1999; 15: 387–428, at 395–99. For more in-depth analysis of the legal architecture for Africa, see Chapters 9–10, 12 and 16 of this book.

9   UNEP and FAO (note 2), at 27–50.

10 SADC Revised Protocol on Shared International Watercourses (adopted August 7, 2000, entered into force September 22, 2003) (2001) 40 ILM 321 (Revised SADC Protocol). More detailed analysis of the Revised SADC Protocol is provided in Chapter 10.

11 Wolf (note 8), at 399–403. More detailed analysis of the Aral Sea Basin and the Mekong is provided in Chapters 13 and 15.

12 UNEP and FAO (note 2), at 51–76.

13 Wolf (note 8), at 404–8.

14 UNEP and FAO (note 2), at 77–132.

relatively stringent regional agreements, namely the EU Water Framework Directive and the UN Economic Commission for Europe Water Convention.[15] These two regional instruments include commitments that go beyond the requirements of the UNWC.[16]

In North America, there are 41 transboundary river basins that cover 35 percent of the continent's land surface.[17] There are 28 basin-wide agreements, and a further four river basins are partially covered by agreements.[18] Only nine river basins therefore have no basin-specific agreements in place, representing 76,000 km^2.[19]

Last but not least, South America is home to 38 transboundary river basins, which make up 60 percent of the continent's land surface.[20] Of these river basins, 23 are covered by basin-wide agreements, whereas 15 basins are not subject to any basin agreements.[21]

An analysis of the legal architecture would not be complete without recognition that, in addition to the UNWC, there are numerous global conventions that, at least in part, relate to transboundary watercourses. For instance, the Ramsar Convention was adopted in 1971, and currently has 158 contracting parties, who are obliged to promote the wise use of wetlands within their territory.[22] In relation to international watercourses, the Ramsar Convention stipulates that,

> contracting parties shall consult with each other about implementing obligations arising from the Convention especially in the case of a wetland extending over the territories of more than one contracting party or where a water system is shared by contracting parties.[23]

Around 30 percent of Ramsar sites are located in international river basins. Similarly, the Convention on Biological Diversity, ratified by 191 parties, aims to promote the sustainable use of the world's biodiversity.[24] In relation to international watercourses, states are obliged to notify, exchange information and enter into consultations on activities in one state's jurisdiction or control that are likely to

---

15 See, generally, Rieu-Clarke, A. 'Major Trends in Conflict and Cooperation' in UNEP, University of Dundee and Oregon State University, *Hydropolitical Vulnerability and Resilience along International Waters* (UNEP 2009), at 43–64.

16 Rieu-Clarke, A. 'The Role and Relevance of the UN Convention on the Law of the Non-navigational Uses of International Watercourses to the EU and its Member States', *British Yearbook of International Law*, 2008; 78: 389–428.

17 Wolf (note 8), at 408–10.

18 UNEP and FAO (note 2), at 133–62.

19 Wolf (note 8), at 410–16.

20 Ibid.

21 UNEP and FAO (note 2), at 163–70.

22 Convention on Wetlands of International Importance especially as Waterfowl Habitat (adopted February 2, 1971, entered into force December 21, 1975) 996 UNTS 245 (Ramsar Convention).

23 Ibid., at Article 5.

24 UN Convention on Biological Diversity (adopted June 5, 1992, entered into force December 29, 1993) 1760 UNTS 79.

significantly adversely affect the biodiversity of other states.[25] Pursuant to the Climate Change Convention, ratified by 192 contracting parties, parties are obliged to 'develop and elaborate appropriate and integrated plans for coastal zone management, water resources and agriculture, and for the protection and rehabilitation of areas, particularly in Africa, affected by drought and desertification, as well as floods'.[26] Drought and desertification is also covered in the Desertification Convention, ratified by 193 contracting parties.[27] In relation to international watercourses, contracting parties are obliged to develop 'long-term *integrated* strategies that focus simultaneously, in affected areas, on improved productivity of land, and the rehabilitation, conservation and sustainable management of land and water resources'.[28] Burgeoning agricultural production due to greater demand for food within an ever increasing globalized world will also mean that trade and investment regimes, such as the World Trade Organization,[29] will have a growing impact on legal arrangements concerning international watercourses.[30]

Institutional architecture

A wealth of institutions at the subnational, national, transboundary basin and global levels claim an interest in the management of international watercourses.

At the transboundary basin level, institutions conduct a range of water-related functions related to planning, implementation and monitoring. Figure 7.1 provides a mapping of institutions operating at a transboundary level, together with their linkages with institutions both at a national and subnational level, and global and regional levels.

In terms of geographic distribution, Dombrowsky calculates that, of the 276 international river basins in the world, there are 62 with international river basin organizations, of which 26 are bipartite and 36 are multipartite, whereas only seven basin organizations are basin-wide.[31] Furthermore, the Transboundary Freshwater Dispute Database identifies 38 river basin organizations in Africa, 25 in Asia, 58 in Europe, 43 in North America, and 36 in South America.[32]

25 Ibid., at Article 14.
26 UN Framework Convention on Climate Change (adopted May 9, 1992, entered into force March 21, 1994) 1771 UNTS 107, Article 4(1)(e). For a more detailed analysis, see Chapter 18.
27 UN Convention to Combat Desertification in Those Countries Experiencing Serious Drought and/or Desertification, Particularly in Africa (adopted June 17, 1994, entered into force December 26, 1996) 1954 UNTS 3. See also Chapter 19 of this collection.
28 Ibid., at Article 2(2) [emphasis added].
29 Marrakesh Agreement Establishing the World Trade Organization (adopted April 15, 1994, entered into force January 1, 1995) 1867 UNTS 3.
30 Brown Weiss, E., Boisson de Charzones, L. and Bernasconi-Osterwalder, N. *Fresh Water and International Economic Law* (Oxford University Press, 2005).
31 Dombrowsky, I. *Conflict, Cooperative and Institutions in International Water Management: An Economic Analysis* (Edward Elgar, 2007), at 98; de Stefano, de Silva, Edwards, and Wolf, (note 7), at 2.
32 Oregon State University, 'International River Basin Organizations (RBO) Data' (Oregon State University and others). Available online at www.transboundarywaters.orst.edu/research/RBO (accessed March 20, 2013).

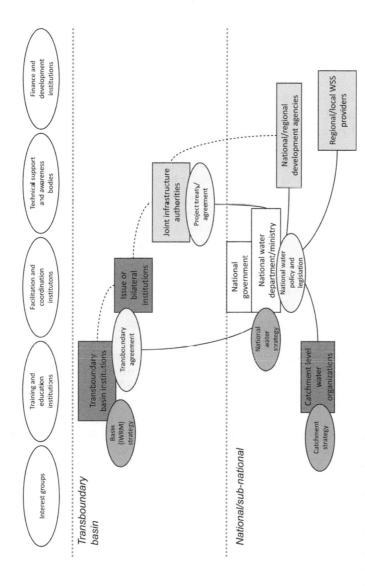

Figure 7.1 Institutional map of transboundary water management and national linkages

Source: (reproduced with permission from *International Architecture for Transboundary Water Management – Policy Analysis and Recommendations* (WWF, Department for International Development, Centre for Water Law, Policy and Science, Pegasys Strategy and Development, 2010)

Institutions operating at the transboundary level conduct a range of functions. For instance, joint infrastructure authorities are typically responsible for the development, financing or operation of water infrastructure on behalf of two or more countries. Such arrangements are usually enabled by a treaty ratified by the relevant states, and, in some circumstances, these institutions may take on some management responsibilities, such as in the case of the Zambezi River Authority.[33] A further type of institution tends to be issue-based and often bilateral. These types of arrangements are usually established by a legal agreement between the parties. Such institutions tend to engage in water issues of common concern, including water sharing, water quality and flooding.[34] Additionally, issue-based institutions tend to be effective negotiating forums between parties. Another form of transboundary basin institution, typically established by a legal agreement among all riparian states within a basin, has the function to advise the parties on a range of transboundary water management issues and priorities.[35]

In conceptualizing transboundary institutions, it is important to bear in mind that the function of institutions may evolve over time, thus reflecting different needs and circumstances within a particular basin. Figure 7.2 presents the different forms that transboundary institutions may take, linking these to the legal instruments that are in place between two or more of the riparian states, in terms of the basin institutional establishment or the basin water management.

In terms of institutional development, there are a number of qualitatively different tiers ranging from firstly, no body facilitating negotiation between states; secondly, a formal basin committee to negotiate or administer a basin agreement, but no permanent secretariat of organization; and, thirdly, a river basin organization established with a commission and permanent secretariat, as a legal entity. A further important distinction exists between transboundary institutions that advise states, and those that have powers to make and implement decisions in terms of basin water strategy or assignments for monitoring and management.

As activities related to international watercourses move from the transboundary level to the regional and global levels, functions move away from an implementation-based focus towards created an enabling environment for such implementation to take place. Five generic areas related to the global and regional levels can be identified. Firstly, institutions play a role in the financing of water infrastructure and water management or governance initiatives. While such activities are commonly also part of the responsibility of national governments, resource limitations and existing legal constraints on basin organizations require most financing from global and regional institutions, such as the World Bank, Global Environment Facility, cooperating partners and increasingly regional development

---

33   Ibid.
34   See, for example, Convention for the Establishment of the Lake Victoria Fisheries Organization (adopted June 30, 1994, entered into force May 24, 1996) 1930 UNTS 128.
35   See, e.g. Convention on the Cooperation for the Protection and Sustainable Use of the Danube River (adopted June 29, 1994, entered into force October 22, 1998). Available online at www.icpdr.org/icpdr-pages/drpc.htm (accessed March 20, 2013).

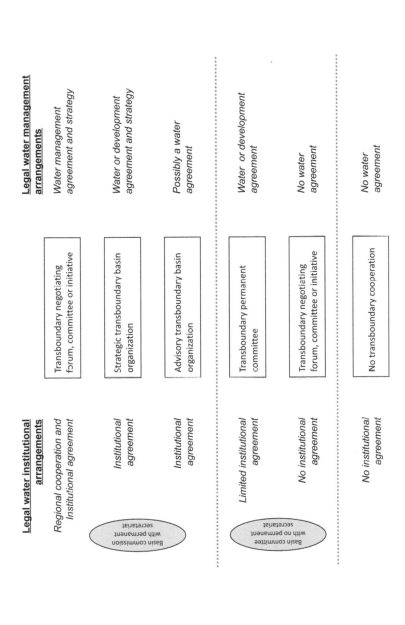

Figure 7.2 Evolution of transboundary basin institutions

Source: (reproduced with permission from *International Architecture for Transboundary Water Management – Policy Analysis and Recommendations* (WWF; Department for International Development, Centre for Water Law, Policy and Science, Pegasys Strategy and Development, 2010)

banks, for example the African Development Bank and the Asian Development Bank. Most of the transboundary financing goes through national governments, rather than basin institutions, although this may be changing with the development of basin trust funds.

Secondly, global and regional institutions play a role in policy and technical support and awareness promotion. At the global level institutions such as UN-Water, World Water Council, Global Water Partnership, the International Network of Basin Organizations, World Bank and cooperating partners play such a role. At the regional level, examples including the UN Economic and Social Commission for Asia and the Pacific, SADC, the Economic Community of West African States, regional water partnerships, African Network of Basin Organizations, Network of Asian River Basin Organizations, and so forth. Many of the latter organizations focus on transboundary organizations as part of the support to integrated water resources management.

Thirdly, training, education and research-oriented institutions at the global and regional levels have a long tradition, such as UNESCO, or the International Association of Hydrological Sciences. Significant resources have also been allocated to their regional counterparts, and these are gradually strengthening. Continued emphasis must be placed on institutional capacity building at a national level to 'level the playing field' and strengthen basin management, but in the context of local requirements, opportunities and constraints (rather than a universal model or approach).

Fourthly, global and regional institutions play a role in terms of facilitation, coordination and conflict resolution related to the negotiation and implementation of transboundary water agreements. At the global level, the World Bank has played this role as 'honest broker', while some regional institutions (such as SADC) have a clear mandate in this regard. This is an area that requires greater attention and institutional development for the successful implementation of transboundary management.

Fifthly, interest-based representation from civil society or private sector institutions has begun to emerge at the global level. The environmental civil society organizations, such as WWF and the International Union for the Conservation of Nature, are in the vanguard, followed by social civil society and water-related non-governmental organizations that are more focused on national and local issues related to water supply and sanitation. Corporate bodies, such as the World Economic Form and CEO Mandate, are also beginning to explore business risks around waters.

An overview of the global and regional institutions influencing transboundary cooperation is provided for in Figure 7.3.

Conclusion

Chapter 6 demonstrated that entry into force and widespread support for the UNWC would provide significant benefits to the existing legal architecture relating to transboundary waters. Most notably, the UNWC would 'inspire' the

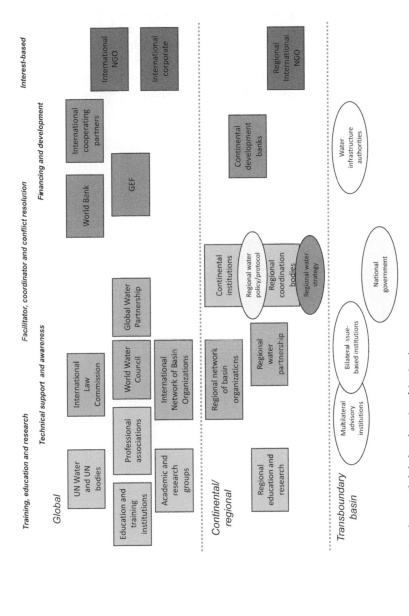

Figure 7.3 Institutional map of global and regional institutions

Source: (reproduced with permission from *International Architecture for Transboundary Water Management – Policy Analysis and Recommendations* (WWF, Department for International Development, Centre for Water Law, Policy and Science, Pegasys Strategy and Development, 2010)

negotiations of new watercourse agreements and the revising of existing treaties, as well as providing a governance framework in the absence of watercourse-specific agreements. The previous chapter also demonstrated that the UNWC could play an important role in aiding parties in the application and interpretation of existing watercourses agreements. This chapter has demonstrated that there is a significant need to perform such functions, owing to the highly fragmented nature of legal agreements for transboundary waters within different regions. Additionally, the chapter has shown that there is a wide range of institutions that support trans-boundary water cooperation at various levels. While a global legal instrument cannot supplement these institutions, entry into force and widespread support for the UNWC could provide a focal point by which to support transboundary water cooperate at the global level, and better harmonize existing efforts.

8 Factors that could limit the effectiveness of the UN Watercourses Convention upon its entry into force

Alistair Rieu-Clarke and Alexander López

Whilst International agreements relating to natural resources date back to the 19th century,[1] most agreements were adopted following the UN Conference on the Human Environment in 1972.[2] Since 1972, there has been an extensive exercise in treaty making covering a wide range of global natural resource issues.[3] However, by the time the Rio Conference took place in 1992, questions over the extent to which these agreements had actually been implemented reached the global stage. Signals that more needed to be done already started to appear in Agenda 21, which called upon states, 'to ensure the effective, full and prompt implementation of legally binding instruments'.[4] Academics and policy analysts – primarily from the disciplines of political science, law and, to a certain degree, economics – responded with vigor to the call to focus on the implementation and effectiveness of agreements.[5] In the past few decades, numerous groups have studied the implementation, compliance and effectiveness within the context of these agreements. Yet, Young cautions that 'our knowledge of institutional dynamics is comparatively underdeveloped...This is particularly true regarding multilateral environmental agreements and other forms of international and transnational cooperation pertaining to environmental issues'.[6]

---

1 Sands, P. *Principles of International Environmental Law* (Cambridge University Press, 2003), at 25–69.
2 Ibid.
3 Brown Weiss, E. 'International Environmental Law: Contemporary Issues and the Emergence of a New World Order', *Georgetown Law Journal*, 1992; 81: 675–710. Brown Weiss argues that, 'ironically, the success that countries have had in negotiating a large number of new international environmental agreements has led to an important and potentially negative side effect: treaty congestion. This affects the international community as a whole, particularly international institutions, as well as individual governments that may want to participate in the negotiation and implementation of agreements but have scarce professional resources'. Ibid., at 697.
4 UN Conference on Environment and Development (Rio de Janeiro, Jun3 3–14, 1992), 'Agenda 21: A Programme for Action for Sustainable Development' (January 1, 1993) UN Doc A/CONF.151/26/Rev.1 (Vol. I), at para. 39.3(e) (Agenda 21).
5 Although studies related to the effectiveness of international law predate the Rio Conference. See, e.g. Henkin, L. *How Nations Behave* (Columbia University Press, 1968).
6 Young, O. *Institutional Dynamics: Emergent Patterns in International Environmental Law* (MIT Press, 2010), at 3.

The purpose of this chapter is to understand what factors might limit the effectiveness of the UN Watercourses Convention (UNWC). Such an approach will be undertaken with an acceptance that knowledge and understanding around what makes international agreements effective remains 'work in progress'. Additionally, the chapter offers an indication of what factors may influence the effectiveness of the Convention. Given that the Convention has not yet entered into force, any empirical analysis of its impact would be premature. The chapter therefore adopts a prospective approach to the analysis of the UNWC's effectiveness.

Defining effectiveness

Prior to examining effectiveness factors, it is important to be clear on what is meant by 'effectiveness'. Chambers describes effectiveness as, 'a term that is randomly used in legal discussion but rarely defined consistently in the world of public international law'.[7] Similarly, Kütting comments that, 'effectiveness means distinctly different things to different communities'.[8] Bodansky identifies three different meanings of effectiveness: legal, behavioral and problem-solving.[9] According to Bodansky, legal effectiveness relates to 'whether outcomes conform to what a legal rule requires'; behavioral effectiveness considers whether a particular instrument has resulted in an actor behaving differently; and problem-solving effectiveness focuses on whether the instrument 'solves the problem it addresses'.[10] Young and Levy add further dimensions to the concept of effectiveness.[11] They consider that, 'an economic definition of effectiveness would incorporate the legal definition and add an efficiency criterion'; and a normative approach would 'think about effectiveness in terms of normative principles, such as fairness or justice, stewardship, participation and so forth'.[12]

Alongside the concept of effectiveness, the concepts of implementation and compliance should also be considered. Implementation can be defined as 'the process of putting international commitments into practice'.[13] According to the UN Environment Programme's (UNEP) Guidelines on Multilateral Environmental Agreement Compliance and Enforcement, implementation implies enacting and promulgating relevant laws, regulations, policies and other measures

---

7   Chambers, B. W. *Interlinkages and the Effectiveness of Multilateral Environmental Agreements* (UN University Press, 2008), at 97.

8   Kütting, G. *Environment, Society and International Relations: Towards More Effective International Environmental Agreements* (Routledge, 2000).

9   Bodansky, D. M. *The Art and Craft of International Environmental Law* (Harvard University Press, 2009), at 253.

10 Ibid.

11 Young, O. R. and Levy, M. A. 'The Effectiveness of International Environmental Regimes' in Young, O. R. (ed.) *The Effectiveness of International Environmental Regimes: Causal Connections and Behavioural Mechanisms* (MIT Press, 1999), at 1.

12 Ibid., at 3–6.

13 Raustiala, K. and Slaughter, A. M. 'International Law, International Relations and Compliance' in Carlsnaes W, Risse T and Simmons BA (eds), *Handbook of International Relations* (Sage Publishing, 2002), at 539.

and initiatives necessary for parties to meet their obligations under an international agreement.[14] Implementation measures can be taken both at the international and domestic levels and rarely follow a single pattern.[15]

While implementation therefore focuses on the mechanism by which instruments are put into effect, 'compliance' refers to, 'a state of conformity or identity between an actor's behaviour and specified rule'.[16] Jacobson and Weiss observe three dimensions of non-compliance with treaty provisions: non-compliance can occur when states fail to meet either their substantive or procedural obligations or to adhere to 'the spirit of the treaty'.[17] This definition differs, however, from one offered by Mitchell, who understands 'compliance' as 'behaviour that conforms to a treaty's explicit rules'.[18] Implementation is a critical precursor for compliance, but the latter can occur without the former if, for instance, treaty commitments reflect existing practice.[19]

It is also important to point out that even full implementation of and compliance with treaty provisions, or even the underlying spirit of the instrument, may not be sufficient to solve the underlying problem or to reach agreed policy goals. Conversely, implementation and compliance may not be necessary for obtaining problem-solving or goal-reaching effectiveness.[20]

The above analysis provides a context in which to analyze the UNWC. In light of the various interpretations of 'effectiveness', this chapter focuses on the ability of the Convention to achieve its key objective. In so doing, we examine whether the design of the treaty might limit the effectiveness of the instrument. However, given that this framework agreement is broad in scope – namely, that it seeks to 'ensure the utilization, development, conservation, management and protection of international watercourses and the promotion of the optimal and sustainable utilization thereof for present and future generations' (Preamble) – it could be argued that the Convention fully covers 'the problem'. More specifically, therefore, the analysis will consider what factors might limit the Convention's effectiveness in solving the problem of ensuring that international watercourses are governed in an equitable and sustainable manner. Unfortunately, as noted above, an examination of the extent to which the UNWC has *actually* achieved its goal, is well beyond the scope

---

14 UNEP, Guidelines on Compliance with and Enforcement of Multilateral Environmental Agreements (UNEP Division of Environmental Law and Conventions 2002), at 2.

15 Victor, D. G. Raustiala, K. and Skolnikoff, E. B. (eds) *The Implementation and Effectiveness of International Environmental Commitments: Theory and Practice* (MIT Press, 1998), at 3.

16 Raustiala, K. 'Compliance and Effectiveness in International Regulatory Cooperation', *Case Western Reserve Journal of International Law*, 2000; 32: 387–439.

17 Jacobson, H. K. and Brown Weiss, E. 'A Framework for Analysis' in Brown Weiss, E. and Jacobson, H. K. (eds), *Engaging Countries: Strengthening Compliance with International Environmental Accords* (MIT Press, 1998). Brown Weiss and Jacobson observe that 'preambles or initial articles in treaties place these specific obligations in a broad normative framework, which we refer to as the spirit of the treaty'. Ibid., at 4.

18 Mitchell, R. 'Compliance Theory: An Overview' in Cameron, J., Werksman, J. and Roderick, R. (eds) *Improving Compliance with International Environmental Law* (Earthscan, 1996), at 5.

19 Raustiala (note 16), at 392.

20 Ibid., at 394.

of this paper and premature. Rather, the focus of the paper is to consider whether the Convention has the requisite ingredients – based on insights from the analysis of multilateral environmental agreements – by which it can be implemented, complied with and ultimately achieving its overall goal.

What makes a treaty effective?

A considerable amount of literature has sought to examine the factors that are likely to influence the effectiveness of global conventions. While most of this literature has tended not to consider international watercourses, a review shows that certain generic insights might be relevant within the context of the UNWC. The purpose of this section is to attempt to identify a number of common trends, or key messages, that emerge from the literature.

Before examining specific factors, it is worth noting that, regardless of the approach adopted, studies have tended to show that, to a lesser or greater extent, regimes and treaties make a difference. Ulfstein, Marauhn and Zimmerman, for instance, maintain that, 'treaties remain the most important instrument for regulating international affairs and the intercourse between states'.[21] Similarly, Young resoundingly argues 'without hesitation that regimes do matter in international society, so that there is nothing to be gained from perpetuating the debate between neo-institutionalists and neo realists about "the false promise of institution"'.[22] However, Breitmier and others caution that, 'clear pathways to effectiveness are complex, often involving a number of factors that interact with one another'.[23]

Rule determinacy

A number of studies emphasize the importance of rule determinacy in relation to multilateral environmental agreement implementation, compliance and effectiveness. Franck defines 'textual determinacy' as 'the ability of the text to convey a clear message, to appear transparent in the sense that one can see through the language to the meaning'.[24] Fuller's 'principles of legality' also emphasizes the need for rules to be 'clear and intelligible'.[25]

21   Ulfstein, G., Marauhn, T. and Zimmermann, A. 'Introduction' in Ulfstein G (ed.) *Making Treaties Work: Human Rights, Environment and Arms Control* (Cambridge University Press, 2007), at 3.
22   Young, O. R. 'Regime Effectiveness: Taking Stock' in Young, O. R. (note 11), at 249. See also Miles, E. L. and Underdal, A. 'Epilogue' in Miles, E. L., Underdal, A., Andresen, S., Wettestad, J., Skjearseth, J. B. and Carlin, E. M. (eds), *Environmental Regime Effectiveness: Confronting Theory with Evidence* (The MIT Press 2002), at 467–74.
23   Breitmeier, H., Underdal, A. and Young, O. R. 'The Effectiveness of International Environmental Regimes: Comparing and Contrasting Findings of Quantitative Research', *International Studies Review*, 2011; 13 (4): 579–605, at 586.
24   Franck, T. M. 'Legitimacy in the International System', *American Journal of International Law*, 1998; 82: 705–59, at 721.
25   Fuller, L. *The Morality of Law* (New Haven, 1964), at 63.

Similarly, Sand cautions that, 'a lack of precise objectives is a major difficulty in measuring achievement'.[26] Beyerlin, Stoll and Wolfrum also stress the importance of considering compliance when drafting treaties, so as to ensure that rights and obligations are articulated as 'clearly and definitely as possible'.[27] Such precision not only relates to the effectiveness of multilateral environmental agreements, but also the ability to measure a particular regime, with more precise norms being easier to assess.[28] However, precision is not always possible; as Franck reminds us, complex problems can require complex rules.[29]

Process features, including reporting, participation and compliance

A further key aspect highlighted in most studies relates to process features, and the systems that are in place for the interpretation, implementation and development of the primary rules of a regime. Underdal highlights the role regimes can play in providing collective learning forums.[30] Through the exchange of information, reporting requirements, technical and scientific programmes and so forth, regimes have served as a vehicle for developing a better understanding of problems and their necessary responses.[31] Such a role is considered by some analysts as funda-mental. Miles and Underdal, for example, warn that, 'regimes engaging beyond standard-setting – in particular, functions such as planning and implementation – tend to be more effective than those that do not'.[32] Similarly, Barrett concludes that, 'international regimes, like all institutions, need to be nurtured. If neglected, they can wither and die'.[33] An appropriate institutional framework is considered to constitute a vital aspect in ensuring that a regime is constantly capable of adapting to changing circumstances. The role of secretariats, meetings of the parties, work-ing groups and so forth is therefore considered crucial to the effectiveness of many multilateral environmental agreements.[34]

26　Sand, P. (ed.) *The Effectiveness of International Environmental Agreements: A Survey of Existing Legal Instruments* (Cambridge University Press, 1992), at 9.
27　Beyerlin, U., Stoll, P. T. and Wolfrum, R. 'Conclusions Drawn From the Conference on Ensuring Compliance with MEAs' in Beyerlin, U., Stoll, P. T. and Wolfrum, R. (eds) *Ensuring Compliance with Multilateral Environmental Agreements: A Dialogue Between Practitioners and Academia* (Martinus Nijhoff Publishers, 2006), at 360.
28　Jacobson, H. K. and Brown Weiss, E. 'Assessing the Record and Designing Strategies to Engage Countries' in Brown Weiss and Jacobson (note 17), at 524.
29　Franck (note 24), at 724. Franck recognizes that '[i]ssues that cannot be reduced to simple binary categories invite regulation by more complex rule texts which, while avoiding the problem of reductio ad absurdum, suffer the costs of elasticity'. Ibid.
30　Underdal, A. 'Conclusions: Patterns of Regime Effectivness' in Miles, *et al.* (note 22), at 440.
31　Ibid. See also, Miles and Underdal, 'Epilogue', in Miles, *et al.* (note 22), at 467–68; Young (note 22), at 254; Sand (note 26), at 14.
32　Miles and Underdal (note 22), at 467.
33　Barrett, S. *Why Cooperate? The Incentive to Supply Global Public Goods* (Oxford University Press, 2007), at 192. Young describes this phenomenon as 'arrested development', namely 'regimes get off to a promising start but then run into barriers or obstacles that block further development'. Young (note 6), at 10–11.
34　Miles and Underdal (note 22), at 472.

Numerous studies stress the central role that reporting plays in ensuring that the regime is 'nurtured'.[35] Beyerlin and others identify a number of functions that reporting can provide, including: (i) the proper assessment of facts and the exchange of information; (ii) a means to monitor implementation of treaty obligations; (iii) 'a dialogue between the regime body assessing reports and the reporting Member State(s) which may considerably facilitate further implementation and compliance'; and (iv) a persuasive mechanism by which to remedy cases of non-compliance.[36] Brunneé also recognizes the role of reporting in enhancing transparency and trust as to a party's performance.[37] However, it has been recognized that the benefits of reporting are largely dependent on the quality and reliability of the reports and the underlying system.[38] Sands observes that, 'there are wide differences in the quality of national performance reports as a means to monitor compliance'.[39]

Another aspect that is crucial when it comes to the effectiveness of regimes in general is stakeholder participation. The perception that rules have emerged from a process that stakeholders consider to be legitimate increases the likelihood that the appropriate actors will adhere to their commitment.[40] One of the weaknesses that could have implications on effectiveness is the missing link between the macro and micro scales, meaning the link between the global, national and subnational levels. That can be seen in the stakeholder context, where there is almost no room for non-state actors. In developing countries, border areas are normally underdeveloped, and participation of civil society is fundamental; multilateral conventions must take this into consideration and push for the involvement of these local actors. For the above reason, the call for *bottom-up* procedures with regard to the concrete design of implementation processes seems justified.

Closely linked to participation's role in increasing the legitimacy of a regime is the recognition that non-state actors, e.g. international and national non-governmental organizations, (multi-)national corporations and individuals, can provide relevant information concerning the activities related to an agreement, or offer an opinion on the claims of state parties.[41] Compliance mechanisms ultimately place reporting requirements and the assessment of those reports in the hands of state parties, either through a compliance committee, meeting of the parties or a combination of the two. Greater participation of non-state parties within the process of compiling and assessing reports is seen as a useful mechanism for increasing transparency and legitimacy.[42]

The way in which non-compliance is managed also appears to have an

---

35   Jacobson and Brown Weiss (note 28), at 525–8.
36   Beyerlin, Stoll and Wolfrum (note 27), at 363.
37   Brunnée, J. 'Compliance Control' in Ulfstein (note 21), at 374.
38   Ibid.
39   Sands (note 1), at 13.
40   Franck (note 24), at 706; Jacobson and Brown Weiss (note 28), at 521; Breitmeier, H., Young, O. R. and Zürn, M. (eds) *Analyzing International Environmental Regimes: From Case Study to Database* (MIT Press, 2006), at 235.
41   Jacobson and Brown Weiss (note 28), at 527.
42   Beyerlin, Stoll and Wolfrum (note 27), at 364.

important bearing on multilateral environmental agreement effectiveness.[43] Sanctions are often seen as the least effective option for ensuring compliance within the context of international environmental problems.[44] Based on the assumption that lack of capacity to comply is often the root cause of non-compliance, positive incentives are seen as an important tool for addressing cases of non-compliance.[45] Such incentives can include: 'special funds for financial or technical assistance, training programmes and materials, access to technology, or bilateral or multilateral assistance outside the framework of the convention, from governments, multilateral development banks, or, in some cases, the private sector'.[46]

In certain circumstances, 'negative incentives' may also be appropriate; such measures include 'formal cautions, public naming and shaming of the non-complying Party, and the imposition of other sanctions, including the suspension of certain treaty rights or privileges'.[47]

External factors

While the above discussion shows that internal design features of international agreements are important within the context of implementation and effectiveness, a range of additional 'external' factors may prove influential – or even pivotal.

Perhaps the most influential external factors are country-specific. As Hathaway points out, 'much of international law is obeyed primarily because domestic institutions create mechanisms for ensuring that a state abides by its international legal commitments whether or not particular governmental actors wish to do so'.[48] Formal mechanisms may include national legislation, and executive or administrative mandates and structures.[49] The functioning of such mechanisms will be highly contingent on administrative capacity.[50] Jacobson and Brown Weiss identify a range of 'sub-factors' related to administrative capacity, including knowledge, educated and trained personnel, adequate financial support and an appropriate legal mandate.[51] Similarly, Lindemann observes that, where there are insufficient financial and administrative resources to plan and administer water projects, and inadequate technical capacities for data generation and project implementation, treaty effectiveness is likely to suffer.[52]

---

43 See, generally, Chayes, A. and Chayes, A. *The New Sovereignty: Compliance with International Regulatory Agreements* (Harvard University Press, 1995).
44 Brunnée (note 37), at 374.
45 Jacobson and Brown Weiss (note 28), at 546.
46 Ibid.
47 Beyerlin, Stoll and Wolfrum (note 27), at 36.
48 Hathaway, O. A. Between Power and Principle: An Integrated Theory of International Law , *University of Chicago Law Review*, 2005; 72: 469–536, at 497.
49 Bodansky (note 9), at 212–16.
50 Jacobson and Brown Weiss (note 28), at 532.
51 Ibid.
52 Lindemann, S. 'Explaining Success and Failure in International River Basin Management: Lessons from Southern Africa' (Global Environmental Change, Globalization and International Security: New Challenges for the 21st Century, University of Bonn, Germany, October 9–13, 2005).

Differing political systems may also influence treaty implementation at the national level, such as the degree to which democratic features are in place, or the level political stability within a country. Jacobson and Brown Weiss justify such factors based on that fact that:

> [d]emocratic governments are normally more transparent than authoritarian governments, so interested citizens can more easily monitor what their governments are doing to implement and comply with accords. In democratic governments, it is possible for citizens to bring pressure to bear for improved implementation and compliance. Also, nongovernmental organizations generally have more freedom to operate under democratic governments. In addition, fully independent courts can be used by nongovernmental organizations and citizens to force governmental action.[53]

Country-specific factors are not limited to the formal mechanisms of government. As noted above, non-governmental organizations can play an important role in ensuring that treaty commitments are implemented, and are effective. Additionally, non-governmental actors 'do not work alone'.[54] Koh identifies an important role played by,

> government officials who...act as allies and sponsors for norms...Once engaged, these governmental norm sponsors work inside bureaucracies and governmental structures to promote the same changes inside organized government that nongovernmental norm entrepreneurs are urging from the outside.[55]

It is also worth pointing out that national actors, be they governmental or non-governmental, tend to collaborate with international actors. Jacobson and Brown Weiss thus recognize the importance of non-governmental organizations at the international level, which 'have become an instrument of universalizing concern'.[56] Lindemann also stresses the important role that international actors can play in providing financial resources, offering technical and cognitive expertise and acting as '"independent" mediators'.[57] 'Transnational issue networks', made up of national and international actors, may therefore constitute an important means by which organizations mobilize their efforts and resources around a particular issue.[58]

In addition to the above-mentioned national dimension, the regional or international context plays a critical role, especially when it comes to international

---

53 Jacobson and Brown Weiss (note 28), at 533.
54 Koh, H. 'The 1998 Frankel Lecture: Bringing International Law Home', *Houston Law Review*, 1998; 35: 623–81, at 647.
55 Ibid., at 648.
56 Jacobson and Brown Weiss (note 28), at 529.
57 Lindemann (note 52), at 9.
58 Koh (note 54), at 649.

regimes such as the UNWC. This is what is called in system theory the state of the suprasystem. In the context of the Convention, this means the state of bilateral relations between neighbor states. Thus, it is important to account for the existence or non-existence of conflicts between states. If bilateral relations are strained by historical or current foreign policy conflicts, the trust needed for the implementation and further development of the water regime may be lacking. Put another way, trust and confidence between the parties constitutes an important influence on communication, joint management, data exchange and so forth, which in turn has direct implications on the effectiveness of any legal arrangement.

Problem structure

Last but not least, the nature of the problem may influence implementation, compliance and effectiveness as well. Victor, Raustiala and Skolnikoff highlight three aspects of a problem that are significant: the ratio of costs to benefits, the distribution of those costs and benefits and 'strategic' considerations such as international economic competitiveness.[59] Underdal comments that, 'some problems are substantively more intricate or complicated than others, implying that more intellectual capacity and energy are needed to arrive at an accurate description and diagnosis and to develop good solutions'.[60] Therefore, the 'malignancy of a problem'[61] is not, by itself, a key factor, but rather the combination of complexity and weak problem-solving capacity. Along similar lines, Bodansky maintains that,

> problems are usually easier to address when a strong scientific consensus exists, when the costs of addressing the problem are low and do not affect a country's competitiveness relative to other countries, when relatively few countries are involved (whose behaviour is easy to monitor), and when countries' interests are aligned.[62]

Hence, a strong correlation exists between the nature of the problem, rule determinacy, process features and external factors.

Does the UNWC have the right ingredients to be effective?

Taking the above-mentioned factors, as summarized in Table 8.1, into account, the purpose of this section is to consider whether the UNWC has the necessary elements to ensure that it is implemented and complied with, and is ultimately effective. As noted in the introduction, no definitive answer can be provided at this stage, but some insights might be gained.

59 Victor, Raustiala and Skolnikoff (note 15), at 67.
60 Ibid.
61 Breitmeier, Underdal and Young (n 23), at 5.
62 Bodansky (note 9), at 263.

Table 8.1 Factors influencing the effectiveness of multilateral environmental agreements

Factor		
Internal	Rule determinacy	Textual clarity
		Precise objectives
	Process features	Reporting
		Exchange of information
		Institutional framework (secretariats, meeting of the parties, working groups, etc.)
		Participation (state and non-state actors)
		Non-compliance strategies (technical assistance, training, finance)
External		National legislative and administrative mechanisms
		Administrative capacity (knowledge, trained personnel, financial support, legal mandate)
		Political system (transparency, civil society participation)
		International networks
Problem structure	Intellectual	Complexity
		Knowledge and understanding
	Political	Positive and negative externalities

Rule determinacy

As noted previously, rule determinacy can constitute a significant impediment to the implementation and effectiveness of any legal arrangement. In that regard, questions over the interpretation of certain provisions of the UNWC might be raised. Indeed, Chapter 4 of this collection identifies and discusses a number of misconceptions relating to the Convention's text. However, while some misconceptions might be resolved through a careful analysis of the UNWC's text and recourse to the *travaux préparatoires*,[63] numerous provisions remain subject to differing interpretations. For example, Article 5 obliges states to utilize an international watercourse in an 'equitable and reasonable' manner. What constitutes 'equitable' or 'reasonable' may differ from case to case. Similarly, under Article 7, states must take 'all appropriate measures' to prevent 'significant harm'. Again, what constitutes 'appropriate' measures may differ from state to state, and the Convention does not define what is meant by 'significant', or what constitutes 'harm'. Other questions could be raised over open-textured standards such as 'good faith' (Articles 3(5),

---

63 For the preparatory documents leading to the adoption of the UNWC, see McCaffrey, S. C. Convention on the Law of the non-Navigational Uses of International Watercourses, New York, May 21, 1977. Audiovisual Library of International Law. Available online at http://untreaty.un.org/cod/avl/ha/clnuiw/clnuiw.html (accessed March 20, 2013).

4(2), 8(1), 17(2), 31 and 33(8)), 'special regard' (Article 10), 'significant adverse affect' (Article 12), 'timely notification' (Article 12), 'best efforts' (Articles 9(2) and (3) and 26), 'where appropriate' (Articles 4, 7(2), 9(3), 20, 21(2), 23, 25(1), 27 and 28(3) and (4)), 'adequate protection' (Articles 5(1) and 8(1)), 'sustainable development' (Article 24(2)(a)), 'optimal and sustainable' (Article 5(1)), and so forth.

Greater clarity within treaty provisions is likely to be more conducive to compliance for two main reasons. Firstly, the subjects of the applicable rules will be able to clearly ascertain what they must do, must not do, or are entitled to do. Secondly, it will be easier for states and other actors to identify actions that either comply with or fail to comply with those provisions. Why then would states draft texts of treaties that leave provisions open to interpretation?

In this regard, Franck observes that, 'issues that cannot be reduced to simply binary categories invite regulation by more complex rule texts ... which suffer the costs of elasticity'.[64] This is certainly the case within the context of international watercourses. For instance, the notion of equity has evolved as the only suitable standard to accommodate all likely factors and circumstances that must be taken into account when reconciling competing interests.[65] While a rule that prioritized agricultural uses over industrial needs, for example, would provide greater clarity than the rule of equity, such a rule would not satisfy the interests of all international watercourses and their states.

Elastic rules may, therefore, be a necessary evil when dealing with complex problems such as international watercourses. However, such open-textured rules do not necessarily lack determinacy. What will be important is that effective mechanisms are in place, so that 'ambiguity can be resolved case by case'.[66] Interpretative processes are thus crucial in the case of the UNWC, as will be discussed in the next section.

Process features, including reporting, participation and compliance

A key strength of the UNWC is its focus on process-orientated rules. Various provisions of the Convention set out a process by which states can, and in many instances must, work collectively in addressing common issues within a particular international watercourse.

In terms of existing and new agreements, Articles 3–4 require watercourse states to consult with each other on the adjustment and application of the provisions of the UNWC, and where necessary, the conclusion of specific watercourse agreements.

---

64 Franck (note 24), at 706.
65 See Fuentes, X. 'The Criteria for the Equitable Utilization of International Rivers', *British Yearbook of International Law*, 1997; 67: 337; Fuentes, X. 'Sustainable Development and the Equitable Utilization of International Watercourses', *British Yearbook of International Law*, 1999; 69: 119; Lipper. J, 'Equitable Utilization' in Garretson, A. H., Hayton, R. D. and Olmstead, C. J. (eds) *The Law of International Drainage Basins* (Oceana Publications, 1967), at 15–88.
66 Franck (note 24), at 724.

With respect to the substantive norms of the UNWC, Article 5(2) obliges states to *participate* in the use, development and protection of an international watercourse, which is further elaborated in Article 8, and the duty to cooperate. In accordance with Article 8, under which states are obligated to cooperate 'on the basis of sovereign equality, territorial integrity, mutual benefit and good faith in order to attain optimal utilization and adequate protection of an international watercourse'.

Additionally, Article 8(2) requires that states 'consider the establishment of joint mechanisms or commissions, as deemed necessary by them, to facilitate cooperation on relevant measures and procedures'. Article 24, which obliges watercourse states to 'enter into consultations concerning the management of an international watercourse, which may include the establishment of a joint management mechanism', reinforces Article 8(2). There is therefore no definitive obligation to establish joint mechanisms. Yet, given the indivisibility of international watercourses, their numerous multi-sectoral uses and the strong pressures placed on them, joint mechanisms will likely be the most appropriate tool by which watercourse states can comply with the substantive and procedural commitments under the Convention.[67]

Additionally, a number of procedural rules are provided for within the UNWC. For instance, Article 9 requires states to share data and information on the condition of the international watercourse on a regular basis; Part III of the Convention sets out detailed procedures for the notification and consultation over planned measures; and Article 28 obliges states to notify each other of emergency situations, and cooperate to prevent, mitigate and eliminate harmful effects of any emergency. The need to cooperate over the protection, preservation and management of international watercourses is emphasized throughout Parts IV and V of the UNWC, whereby states must jointly, where appropriate, protect and preserve aquatic ecosystems; prevent, reduce and control pollution, including consulting on mutually agreeable pollution measures and methods; take measures to protect and preserve the marine environment; regulate flows, including the construction and maintenance or defrayal of costs of regulation works; consult where necessary on the safe operation and maintenance of installations; adopt measures to prevent or mitigate harmful conditions; and develop contingency plans for responding to emergencies.

In terms of treaty interpretation, as discussed in the previous section, the UNWC provides detailed provisions by which to address any disputes relating to the interpretation and application of the Convention. Most notably, under Article 33, states are obliged to settle their disputes by any number of peaceful means, including negotiation, good offices, conciliation, arbitration or adjudication. One of the most significant process-oriented mechanisms within the Convention is the requirement that States must submit their dispute to 'third party fact-finding' if, within six months, they are unable to resolve their dispute by other means. Detailed procedures for the establishment of the third party fact-finding commission are also

---

67 McCaffrey, S. C. *The Law of International Watercourses* (2nd edn, Oxford University Press, 2007), at 167.

provided Pursuant to Article 33(8), the commission is to 'submit a report to the parties, setting forth its findings and the reasons therefor and such recommendations as it deems appropriate for an equitable solution of the dispute'. The parties must then consider the findings of the commission in good faith.

From the perspective of reporting, the UNWC provides little within its text. Conversely, recent treaty practice related to international watercourses offers numerous examples of reporting requirements. For example, the 2000 Revised Southern African Development Community (SADC) Protocol requires watercourse institutions to 'provide on a regular basis or as required by the Water Sector Co-ordinating Unit, all information necessary to assess progress on the implementation of the provisions of [the] Protocol, including the development of their respective agreements'.[68] Under the EU Water Framework Directive, EU member states must send copies of river basin management plans and all subsequent updates to the European Commission, and to any other member states concerned. Member states must also submit, within three years of publication of each river basin management plan or update, an interim report describing progress in the implementation of the planned programme of measures set out pursuant to the Directive.[69] Under the UN Economic Commission for Europe (UNECE) Water Convention, riparian parties are obligated to carry out 'joint or coordinated assessments of the conditions of transboundary waters and the effectiveness of measures taken for the prevention, control and reduction of transboundary impact'.[70] Article 17 of the latter Convention also establishes a Meeting of the Parties, with the tasks to, inter alia, continuously review implementation. States are then obligated to '[e]xchange information regarding experience gained in concluding and implementing bilateral and multilateral agreements or other arrangements'.[71]

At the basin level, reporting requirements can also be seen in treaty practice. For instance, the 2003 Convention on the Sustainable Management of Lake Tanganyika provides that, '[e]ach Contracting State shall report periodically... on measures that it has taken to implement this Convention and on the effectiveness of these measures in meeting the objective of this Convention'.[72] Similarly, the 2003 Protocol for Sustainable Development of Lake Victoria Basin requires that each state 'periodically... report on measures, which it has taken for the implementation of the

68 SADC Revised Protocol on Shared International Watercourses (adopted August 7, 2000, entered into force September 22, 2003) (2001) 40 ILM 321, Art 5(3)(c).
69 Directive 2000/60/EC of October 23, 2000 of the European Parliament and of the Council establishing a framework for community action in the field of water policy [2000] OJ L327/1 (EU Water Framework Directive).
70 UNECE Convention on the Protection and Use of Transboundary Watercourses and International Lakes (adopted March 17, 1992, entered into force October 6, 1996) 1936 UNTS 269, Article 11 (UNECE Water Convention).
71 Ibid., at Article 17(2)(b).
72 Convention on the Sustainable Management of Lake Tanganyika (adopted June 12, 2003, entered into force September 2005), Article 22. Available online at www.ltbp.org/FTP/LAKECONV.pdf (accessed March 21, 2013).

provision of this Protocol and their effectiveness in meeting the objectives of this Protocol'.[73]

During the work of the ILC preceding the adoption of the UNWC, McCaffrey proposed a reporting requirement in his 6th Report as Special Rapporteur.[74] Drawing on examples from state practice largely related to multilateral environmental agreements, McCaffrey proposed a draft article that would require the establishment of a 'conference of the parties', which would meet no later than two years after entry into force, and hold regular meetings at least every two years.[75] Among other activities, such meetings would allow the parties to 'review...implementation', and 'receive and consider any reports presented by any Party or by any panel, commission or other body'.[76] Ultimately, however, such an institutional framework was not incorporated into the text of the UNWC, which might prove unfortunate in terms of providing a mechanism by which to evaluate the future implementation and effectiveness of the Convention.

The text of the UNWC is also weak in terms of stakeholder participation. The only explicit reference to non-state actors appears in Article 32, which provides that:

> [u]nless the watercourse States concerned have agreed otherwise for the protection of the interests of persons, natural or juridical, who have suffered or are under a serious threat of suffering significant transboundary harm as a result of activities related to an international watercourse, a watercourse State shall not discriminate on the basis of nationality or residence or place where the injury occurred, in granting to such persons, in accordance with its legal system, access to judicial or other procedures, or a right to claim compensation or other relief in respect of significant harm caused by such activities carried on its territory.

Such a limited role for non-state actors can be contrasted with more recent treaty practice. For example, the 2003 Lake Victoria Basin Protocol requires that states 'create an environment conducive for stakeholders' views to influence governmental decisions on project formulation and implementation'.[77] In addition, the Protocol requires states to:

---

73 Protocol for Sustainable Development of Lake Victoria Basin (adopted November 29, 2003), Article 45. Available online at www.internationalwaterlaw.org/documents/regionaldocs/Lake_ Victoria_Basin_2003.pdf (accessed March 21, 2013) (2003 Lake Victoria Basin Protocol).

74 McCaffrey, S. C. (Special Rapporteur), Sixth Report on the Law of the Non-Navigational Uses of International Watercourses. Topic: Law of Non-Navigational Uses of International Watercourses. (February 23 and June 7, 1990), Extract from the *Yearbook of the International Law Commission* (1990) II(1) 41, at 64. UN Doc A/CN.4/427. Available online at http://untreaty.un.org/ilc/ documentation/english/a_cn4_427.pdf (accessed March 21, 2013).

75 Ibid.

76 Ibid.

77 2003 Lake Victoria Basin Protocol (note 73), at Article 22.

promote and encourage awareness of the importance of, and the measures required for, the sustainable development of the Basin; and... co-operate, as appropriate, with other States and international organisations in developing educational and public awareness programmes, with respect to conservation and sustainable use of resources of the Basin... [and] promote community involvement and mainstreaming of gender concerns at all levels of socio-economic development.[78]

At the regional level, the UNECE Water Convention requires riparian parties to 'ensure that information on the conditions of transboundary waters, measures taken or planned to be taken to prevent, control and reduce transboundary impact, and the effectiveness of those measures, is made available to the public'.[79] The 2000 EU Water Framework Directive goes further, by stipulating that 'Member States shall encourage the active involvement of all interested parties in the implementation of this Directive, in particular in the production, review and updating of the river basin management plans'.[80]

As noted above, a final process feature that is considered important in terms of treaty effectiveness relates to compliance. Compliance mechanisms are a common feature in more recent watercourse agreements. For instance, the 1998 Agreement between Spain and Portugal on Shared Basins stipulates that 'the Parties shall, jointly or individually, adopt the technical, legal, administrative, and other necessary measures in order to... promote actions to review compliance with the Convention... [and] promote actions to strengthen the effectiveness of the Convention'.[81]

In turn, the Zambezi Watercourse Commission Agreement stipulates that,

> [i]n the event of any Member State failing to fulfil its obligations..., such Member State shall forthwith, and in any event no later than thirty (30) days after such failure, send written communication to the Secretariat explaining the failure and setting forth the reasons therefore, including any measures taken to remedy the failure.[82]

Following the written communication described above, the agreement goes on to stipulate that, 'the Secretariat shall immediately enter into consultations with such Member State with a view to providing such assistance as may be necessary to

---

78 Ibid., at Articles 21 and 23.
79 UNECE Water Convention (note 70), at Article 16(1).
80 EU Water Framework Directive (note 69), at Article 14(1).
81 Convention on the Co-operation for the Protection and Sustainable Use of the Waters of Luso-Spanish River Basins (adopted November 20, 1998, entered into force January 31, 2000), Article 10(l)–(m). [Spanish] Available online at http://faolex.fao.org/docs/pdf/bi-80627.pdf (accessed March 21, 2013).
82 Agreement on the Establishment of the Zambezi Watercourse Commission (adopted July 13, 2004), Article 20(1). Available online at www.icp-confluence-sadc.org/sites/default/files/ZAMCOM_AGREEMENT_2004.pdf (accessed March 21, 2013).

procure the fulfilment of the obligations in question'.[83] A 'mechanism to facilitate and support implementation and compliance' has also been adopted under the UNECE Water Convention.[84]

External factors

As noted above, the problem structure and external factors will also have an important impact on the effectiveness of the UNWC. A comprehensive analysis of country-specific factors that may influence the effectiveness of the Convention is beyond the scope of this chapter. Such an analysis requires in-depth consideration of national legal and administrative structures, the role of non-state actors, administrative capacity and so forth. However, a number of chapters in this collection provide some generic insights.[85]

Problem structure

International watercourses represent a classic case of an 'international common pool resource'.[86] Benvenisti describes the problem with common pool resources as follows:

> Because different states enjoy access to transboundary natural resources, they face a collective action problem. Each state is interested in getting more out of the resource with minimal costs, and these interests conflict with those of the other users. This conflict can lead the parties to a race to the bottom.[87]

Similarly, Lindeman sees the problem of international watercourses as one of 'transboundary externalities', whereby,

> [n]egative externalities arise when the upstream country imposes costs on the downstream country without compensating it for the inflicted harm (e.g. in the case of water abstraction or pollution upstream). Positive externalities, on

83   Ibid., at Article 20(2).
84   UNECE Legal Board, 'Possible Drafting Language for a Mechanism to Facilitate and Support Implementation and Compliance' UN Doc ECE/MP.WAT/AC.4/2011/6. Available online at http://live.unece.org/fileadmin/DAM/env/water/meetings/legal_board/2011/ece.mp.wat.ac.4.2 011.6._final.pdf (accessed March 21, 2013). The Implementation Committee for the UNECE Water Convention was approved at the Sixth Meeting of the Parties in Rome, Italy, November 2012. For more information see UNECE, 'Sixth Session of the Meeting of the Parties to the Water Convention'. Available online at http://www.unece.org/env/water/mop6.html (accessed May 31, 2013).
85   See, in particular, Part III of this collection, which examines the role and relevance of the UNWC in specific regions, basins and countries, as well as Chapters 5 and 28, which deal with political aspects of transboundary waters.
86   Benvenisti, E. *Sharing Transboundary Resources: International Law and Optimal Resource Use* (Cambridge University Press, 2002), at 2.
87   Ibid., at 31. See Hardin G, 'The Tragedy of the Commons' (1968) 162 Science 1243 at 1244.

the other hand, are less frequent and exist when one riparian country produces a public good without receiving full compensation for its efforts (e.g. the provision for flood control upstream).[88]

As mentioned above by Victor, a crucial factor within the context of international watercourses will be the ratio of costs to benefits, and the distribution of those costs and benefits. Collective action theory maintains that benefits can be maximized by a shift away from an emphasis on individual entitlements, towards stronger internal interaction.[89] Within the context of international watercourses, Chapter 5 of this collection has demonstrated that the range of benefits can be significantly increased through collective action; however, such action comes at a cost. The previous section highlights the need to address the knowledge gap when dealing with complex problems. Hence, addressing the challenge of a common pool resource, such as transboundary waters, requires strong process-based mechanisms that not only fill the knowledge gap, but do so in a legitimate manner. While the precise nature of such mechanisms may differ from one watercourse to another, 'process-based' instruments are likely to be more effective than agreements primarily concerned with defining individual entitlements. Additionally, as maintained by Benvenisti, 'global efforts must be made to reduce the cost of endogenous interaction between the resource users'.[90]

Last but not least, it is important to stress that, in an anarchic international system, which path states choose to go down will be contingent on political will. The political aspects of international watercourses are therefore given further attention in Chapter 28.

Conclusion

This chapter has sought to identify the factors that are commonly considered as important when considering whether a treaty is effective. A clear message that can be derived from the chapter is that effectiveness will be contingent on three key areas; namely, factors relating to the internal aspects or design features, external influences, and the problem structure. What is also clear is that each of these three areas must be considered collectively. For instance, an agreement that is considered 'effective' on paper may face problems if the factors conducive to implementation and compliance are not taken into account or the agreement does not fully address the targeted problem.

While this chapter has not been able to conduct a comprehensive, empirical analysis of the effectiveness of the UNWC, a number of useful insights can be gained from the analysis. In terms of rule determinacy, it is clear that the UNWC contains a number of open-textured norms that might be applied differently in a

---

88   Lindemann (note 52), at 5.
89   Olson, M. *The Logic of Collective Action* (Harvard University Press, 1965).
90   Benvenisti (note 87), at 42.

range of contexts. Such a finding demonstrates that the UNWC is sensitive to the complex nature of international watercourse challenges.

Additionally, the finding suggests a need for strong process-based features by which to ensure that such norms are interpreted and applied in a legitimate and fair manner. The analysis of process features of the UNWC demonstrates that the instrument contains a significant number of process-based norms, which potentially provide an effective system by which to interpret and implement its key substantive norms. However, in certain areas, such as reporting, participation and compliance review, contemporary treaty practice at a regional, basin, and bilateral, level has gone further than the UNWC. A question that can be raised is, therefore, whether the international community would be prepared to invest additional resources in support of these three key process-based features. This issue is further considered in Chapter 22.

In terms of external factors, it has not been possible to conduct specific empirical analysis related to UNWC, given that it has not yet entered into force. However, the chapter has shown that, once in force, the effectiveness of the Convention will be largely contingent on states being able to account for, and, where necessary, address, factors such as national legal and administrative structures, administrative capacity, the role of non-state actors and so forth.

Finally, the nature of the problem must also be taken into account. The analysis above shows that there is a strong political aspect to the problem, which is discussed further in Chapter 28. For the purpose of this chapter, it is important to note 'a common pool resource' nature of international watercourses, which calls for collective action among states, and a reduction of 'transaction costs'. Again, the need to 'align'[91] internal and external features, along with the problem structure, becomes imperative.

---

91   Young (note 6), at 13–16.

Part 3

The potential role and relevance of the UN Watercourses Convention in specific regions, basins and countries

9 West Africa

Amidou Garane and Teslim Abdul-Kareem

Close to one-fifth of the entire continent, the westernmost region of Africa is bordered by the Sahara Desert in the north, by the Atlantic Ocean in the west towards the south, and eastwards by the Benue Trough extending towards the Chad Basin. There are 16 countries in West Africa: Benin, Burkina Faso, Cape Verde, Côte d'Ivoire, The Gambia, Ghana, Guinea, Guinea-Bissau, Liberia, Mali, Mauritania, Niger, Nigeria, Senegal, Sierra Leone, and Togo.[1] The region is characterized by many cultures and tribes, with some extending over two or more countries. Although West African countries are generally categorized as developing, industrialization is fast-growing, while agricultural farming and fishing are still important means for people's livelihoods. Additionally, West Africa is endowed with significant natural resources and interconnected waters. Figure 9.1 presents the riparian states in the region, with some of their waters as they link to each other.

As can be seen in Figure 9.1, the major river systems include the Bandama River in Côte d'Ivoire, the Gambia River in Gambia, the Niger River running from Guinea to Nigeria, the Senegal River running across Senegal, Mauritania and Mali, and the Benue River in Nigeria, as well as some prominent lakes. Most of these systems flow into the Atlantic Ocean. Riparian states in West Africa can be classified as upstream, midstream or downstream depending on the river and/or basin in question, and many of them are highly dependent on transboundary waters for their needs. In general, there are about 60 transboundary watersheds in Africa, covering more than 60 percent of the continent's total area;[2] and, even though there are more than two dozen transboundary rivers in West Africa, the major basins are Niger, Senegal, Gambia, Chad, Volta and Koliba-Korubal. The assessment

---

1 They all belong to the Economic Community of West African States (ECOWAS) except for Mauritania, which decided to leave the organization in 1999. See Goodridge, R. B., Jr. *Economic Integration of West African Nations: A Synthesis for Sustainable Development* (International MBA Thesis, National Chengchi University, 2006), at 33–46. Available online at http://nccur.lib.nccu.edu.tw/handle/140.119/33930 (accessed March 21, 2013); Economic Community of West African States, 'ECOWAS Member States'. Available online at www.ecowas.int (accessed March 21, 2013).

2 Ferraro, P. J. *Regional Review of Payments for Watershed Services: Sub-Saharan Africa* (Sustainable Agriculture and Natural Resource Management Collaborative Research Support Programme, 2007) Virginia Tech Working Paper No. 08-07, at 10. Available online at www.oired.vt.edu/sanremcrsp/documents/research-themes/pes/Sept.2007.PESAfrica.pdf (accessed March 21, 2013).

WEST AFRICA

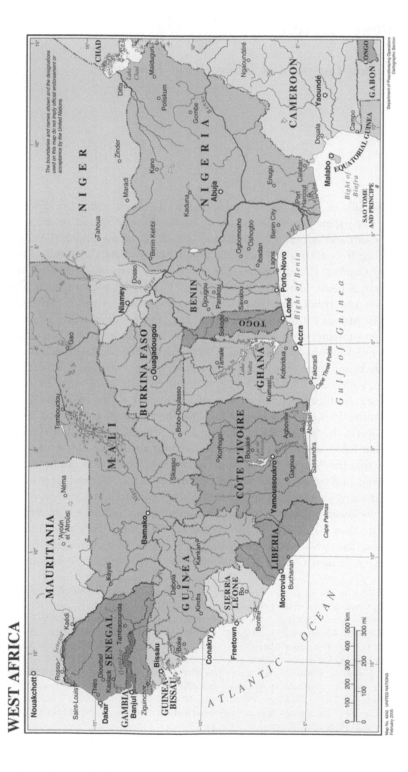

Figure 9.1 West Africa and an overview of its watercourses

Source: Map No 4242, UN Cartographic Section, 2005, reproduced with permission

carried out in this chapter focuses on these basins; furthermore, a comparative analysis is presented, with relevant watercourse agreements (where they exist) and consideration is given to the present situation as regards the adoption of the UN Watercourses Convention (UNWC) by African countries.

Transboundary water cooperation in West Africa

In a bid to ensure cooperation for the control, management and protection of the international watercourses in West Africa, agreements have been reached and transborder institutions have also been established. However, as will be explained, many of these international agreements are still inadequate in a number of respects. Although the UNWC has yet to enter into force, it can still play a great role in augmenting the agreements in West Africa.

The Niger River Basin

Through an international course from the Fouta Djallon highlands (close to the Sierra Leonean border) in Guinea, with a meander in Mali and a fair sail across southwest Niger, to the delta in Nigeria, runs Africa's third longest river – the Niger River. The entire basin spans nine countries,[3] which are characterized by diverse cultures, varying human needs, unique landforms and weather conditions.

Agreements signed on the Niger Basin date as far back as the colonial era, with the first one concerning the division of territory among France, Germany and Great Britain.[4] In 1960, the conditions for shared development of the Niger River Basin were defined by a working session between Nigeria and Mali, and in the same year, by the meeting of the 16th Session of the Commission for Technical Cooperation in Sub-Saharan Africa in Mamou, Guinea.[5] Following that meeting, many other agreements were reached. A major achievement in the cooperation of states of the Niger River Basin was the creation of the Niger Basin Authority (NBA) in 1980,[6] which built on a 1964 agreement establishing the River Niger Commission. More recently, the basin states adopted the Revised Convention pertaining to the Creation of the Niger Basin Authority of 1987 (Revised NBA Convention)[7] and the 2008 Niger Basin Water Charter (Niger Basin Water

---

3 Benin, Burkina Faso, Cameroon, Chad, Côte d'Ivoire, Guinea, Mali, Niger and Nigeria. Algeria and Sierra Leone also share in the Niger Basin, but are not parties to many of the basin's conventions.
4 Anderson, I., Dione, O., Jarosewich-Holder, M., Olivry, J.-C. and Golitzen, K. G. *The Niger River Basin: A Vision for Sustainable Management* (World Bank, 2005), at 8.
5 Ibid.
6 Convention Creating the Niger Basin Authority with Protocol Relating to the Development Fund of the Niger Basin (adopted November 21, 1980, entered into force December 3, 1982) 1346 UNTS 207.
7 Revised Convention pertaining to the Creation of the Niger Basin Authority (adopted October 29, 1987) [French]. Available online at www.fao.org/docrep/W7414B/w7414b0e.htm (accessed March 21, 2013).

Charter).[8] Table 9.1 presents some of the main agreements reached in the Niger River Basin.

Table 9.1 Some notable treaties on the Niger River Basin

S/N	Date	Signatories	Treaty name
1	July 12, 1988	Republic of Mali and Republic of Niger	Protocol of the agreement between the Republic of Niger and the Republic of Mali relative to cooperation in the utilization of resources in water of the Niger River
2	October 29, 1987	Benin, Burkina Faso, Cameroon, Chad, Côte d'Ivoire, Guinea, Mali, Niger and Nigeria	Revised convention pertaining to the creation of the Niger Basin Authority, signed at N'Djamena
3	October 27, 1987	Algeria, Benin, Cameroon, Chad, Guinea, Cote d'Ivoire, Mali, Niger, Nigeria and Burkina Faso	Revised financial procedures of the Niger Basin Authority, done at N'Djamena
4	November 21, 1980	Benin, Cameroon, Chad, Cote d'Ivoire, Guinea, Mali, Niger, Nigeria and Upper Volta	Convention creating the Niger Basin Authority and protocol
5	November 25, 1964	Benin, Burkina Faso, Cameroon, Chad, Côte d'Ivoire, Guinea, Mali, Niger and Nigeria	Agreement concerning the River Niger Commission and the navigation and transport on the River Niger
6	October 26, 1963	Cameroon, Chad, Côte d'Ivoire, Dahomey, Guinea, Mali, Niger, Nigeria and Upper Volta	Act regarding navigation and economic cooperation between the states of the Niger Basin
7	April 20, 1921	France and Great Britain, among others	Convention of Barcelona
8	February 26, 1885[a]	Austria-Hungary, Belgium, Denmark, France, Germany, Great Britain, Italy, Netherlands, Norway, Portugal, Russia, Spain, Sweden, Turkey and United States of America	General act of the conference of Berlin … respecting: 1) freedom of trade in the basin of the Congo; 2) the slave trade; 3) neutrality of the territories in the basin of the Congo; 4) navigation of the Congo; 5) navigation of the Niger; and 6) rules for future occupation of the coast of the African continent

Notes: [a] Treaties include Congo and Niger Basins.
Source: Adapted from UN Environment Programme (UNEP) and others, *Atlas of International Freshwater Agreements* (UNEP, 2002).

8   Niger Basin Water Charter, adopted by the 8th Summit of the Niger Basin Authority Heads of State and Government (April 30, 2008) [French]. Available online at www.abn.ne/attachments/article/39/Charte%20du%20Bassin%20du%20Niger%20version%20finale%20francais_30-04-2008.pdf (accessed March 21, 2013).

The Revised NBA Convention contains some vital principles which are important in ensuring cooperation among states, but they do not reflect substantive and procedural rules in any great depth. Instead, the Revised NBA Convention is more focused on setting up institutional mechanisms like the permanent organs of the Authority and their responsibilities (Article 5), financial provisions (Articles 10–13), and so on. There are no provisions incorporating the principle of equitable and reasonable use and there is no provision dealing with the proper management of water distribution. There is a short clause on making information available under Article 4, where states are obliged to inform the Authority of all projects and work that is to be embarked upon, but this provision is not as detailed as the UNWC's Article 12 on 'notification concerning planned measures with possible adverse effects' or Articles 9 and 11 on information exchange.

Article 4 of the Revised NBA Convention also directs member states to 'to refrain from carrying on the portion of the River, its tributaries and sub-tributaries within their territorial jurisdiction, all work likely to pollute or negatively alter the biological characteristics of the fauna and flora'. Such a clause can still be supplemented by Article 21 of the UNWC related to pollution, which is a specific application of the general principles on equitable utilization in Article 5,[9] and the obligation not to cause significant harm, in Article 7 of the UNWC.

When there is a dispute between states, they must – according to the Revised NBA Convention – settle their differences peacefully via direct negotiations unless the dispute is brought to the Authority by any party, after which the decision made is final. The relevance of the UNWC in terms of dispute settlement procedures is important to note. In the situation where no agreement can be reached, the Revised NBA Convention does not employ dispute settlement mechanisms including good offices, mediation or conciliation by a third party, settlement by arbitration or involvement of the International Court of Justice (ICJ) and submission to an impartial fact-finding body, as is contained under Article 33 of the UNWC.

Attempts were made to address some of the gaps contained in the Revised NBA Convention with the Paris Declaration,[10] when it was adopted by member countries in 2004, but beside the fact that the Declaration itself is lacking in some key areas, its directives only further establish the need to complement the Revised NBA Convention with the UNWC.[11] Another related Niger River Basin agreement is the bilateral agreement between the Federal Republic of Nigeria and the

---

9 Article 5 of the UNWC directs states to participate in protecting an international watercourse in an equitable and reasonable manner.

10 NBA, 'Conférence sur le Bassin du Niger: Déclaration de Paris: Principe de Gestion et de Bonne Gouvernance pour un Développement Durable et Partagé du Bassin du Niger' (Paris, April 26, 2004). [French] Available online at www.office-du-niger.org.ml/internet/index.php?option=com_docman&task=doc_download&gid=24&Itemid=50 (accessed 10 Apr 2012).

11 Garane, A. *UN Watercourses Convention: Applicability and Relevance in West Africa* (UNWC Global Initiative 2008), at 23. Available online at www.internationalwaterlaw.org/bibliography/WWF/RA_West_Africa.pdf (accessed 21 March, 2013).

Republic of Niger.[12] Although it incorporates some relevant and important principles, these principles do not apply to the entire Niger Basin.

One of the eight decisions reached at the Eighth Summit of the NBA Heads of State and Government held in Niamey, Niger, was the adoption of the Niger Basin Water Charter.[13] This occurred on 30 April 2008, making it the most recent and all-encompassing regulatory framework in the Niger Basin. The Niger Basin Water Charter considers the 1980 Convention Creating the NBA and the 1987 Revised NBA Convention, and it was also based, in particular, on the conclusions of the UN Conference on Environment and Development, such as the Rio Declaration on the Environment and Development[14] and Agenda 21.[15]

In the Charter, the pollution of an international watercourse was included in the definitions under 'General Provisions' referring to any change harmful to the composition of the quality of the water of an international watercourse (Article 1(25)), thus linking pollution provisions to human behavior and the concept of harm expressed in the UNWC. Transboundary pollution, which is referred to in Article 1(26) as pollution caused by activities of a state thereby generating harmful effects for the environment of one or several other states, was also included.

One of the aims of the Water Charter, expressed in Article 2, is to 'determine the rules related to the protection and preservation of the environment in accordance with the sustainable development objectives'; per Article 4, the existence of a minimum environmental flow to be preserved is to be taken into account under equitable and reasonable participation and use. Article 7 is primarily concerned with prevention; therefore, the principle of preventive and corrective action of harm caused to the environment is taken into account.

Furthermore, a careful study of the definition of pollution in the Water Charter and UNWC reveals that the Water Charter has actually adopted some concepts on the codification of its environmental provisions from the UNWC. Despite such recognition, the Niger Basin Water Charter would benefit from including specific requirements for states to act jointly, where appropriate, so as to effectively control, prevent and reduce pollution.[16] Additionally, a stronger central role for the principle of equitable use and reasonable utilization would be advantageous.

---

12 Agreement between Niger and Nigeria concerning the Equitable Sharing in the Development, Conservation, and Use of Their Common Water Resources (adopted July 18, 1990). Available online at www.fao.org/docrep/W7414B/w7414b10.htm (accessed March 21, 2013).

13 See UN Economic Commission for Latin America and the Caribbean, 'Eighth Summit of the Niger Basin Authority (NBA) Heads of State and Government', *International Rivers and Lakes*, 2009; 47: 7–8. Available online at www.eclac.org/drni/noticias/circulares/1/35841/irl_47.pdf (accessed March 21, 2012).

14 UN Conference on Environment and Development (Rio de Janeiro, June 3–14, 1992), 'Rio Declaration on Environment and Development' (January 1, 1993), UN Doc A/CONF.151/26/Rev.1 (Vol. I), at 3. Available online at http://www.un.org/documents/ga/conf151/aconf15126-1annex1.htm (accessed March 21, 2013).

15 UN Conference on Environment and Development (note 14), at 9.

16 The importance of this section cannot be overemphasized. It should also be noted that the decision to use the word 'harmonization' is, *inter alia*, born from the second paragraph of Article 3 of the UNWC, which proposes the harmonization of agreements between watercourse states where necessary with its own basic principles.

The Senegal River Basin

The institutional framework for cooperation in the Senegal River Basin is the Senegal River Development Organization (OMVS). Before the organization was set up, Guinea, Mali, Mauritania and Senegal all belonged to the *Organisation des États riverains du fleuve Sénégal* (Senegal River Riparian States Organization). The Senegal River Riparian States Organization dissolved because of Guinea's withdrawal, and OMVS was subsequently created by the other three countries.[17] Guinea, which has the headwaters located within its borders, joined the OMVS in 2005.[18]

The OMVS is governed by a number of agreements,[19] including the Convention pertaining to the Creation of the Organization for the Management of the Senegal River (OMVS Convention),[20] the Convention relating to the Statute of the Senegal River (Senegal River Convention)[21] and the Charter of the Waters of the Senegal River (Senegal Water Charter).[22] The OMVS Convention is a multilateral treaty that was reached in 1972, focusing on institutional aspects, among other things. While Article 20, creating a standing committee on water, constituted an important step in the definition of principles and modalities for the allocation of water rights among states and different sectors, the OMVS Convention neither factored in the principle of equitable and reasonable utilization nor the rules that govern access to the rights of water. As a result, considering the fact that this is well laid out in the UNWC, the committee could benefit from the incorporation of such principles and procedures.

17   Uhlir, P. F. *Scientific Data for Decision Making toward Sustainable Development: Senegal River Basin Case Study* (National Academies Press, 2003).

18   See Haut Commissariat, 'États Membres de l'OMVS – Guinée, Mali, Mauritania, Sénégal' (OMVS). [French] Available online at www.omvs.org/fr/omvs/membres.php (accessed March 21, 2012).

19   See, e.g. Convention Relative a l'Aménagement General Du Bassin Du Fleuve Sénégal [Convention Relating to the General Development of the Senegal River Basin] (Guinea-Mali-Mauritania-Senegal) (signed July 26, 1963) (Convention of Bamako); Convention of Dakar (Guinea-Mali-Mauritania-Senegal) (signed January, 1970); Convention concluded between Mali, Mauritania, and Senegal Relative to the Legal Statute of Common Works (signed December 21, 1978). For a list of agreements on the Senegal River, see 'International Freshwater Treaties Database', in Transboundary Freshwater Dispute Database (Oregon State University). Available online at http://ocid.nacse.org/tfdd/treaties.php (accessed March 21, 2013).

20   Convention Portant Création de l'Organisation pour la Mise en Valeur du Fleuve Sénégal [Convention Pertaining to the Creation of the Organization for the Management of the Senegal River] (Guinea-Mali-Mauritania-Senegal) (adopted March 11, 1972). Available online at http://ocid.nacse.org/tfdd/tfdddocs/344FRE.pdf (accessed March 21, 2013).

21   Convention Relative au Statut du Fleuve Sénégal [Convention Relating to the Status of the Senegal River] (Guinea-Mali-Mauritania-Senegal) (signed March 11, 1972). Available online at http://faolex.fao.org/docs/texts/mul16004.doc (accessed March 21, 2013).

22   Charte des Eaux du Fleuve Sénégal [Senegal River Waters Charter] (Mali-Mauritania-Senegal), Organisation pour la Mise en Valeur Du Fleuve Sénégal (OMVS) Resolution 005, adopted by the Conference of Heads of State and Government (May 18, 2002) [French]. Available online at http://lafrique.free.fr/traites/omvs_200205.pdf (accessed March 21, 2013).

Regarding the settlement of disputes, when differences between states owing to interpretation or application of the OMVS Convention arise, according to Article 24, members shall first resolve them via conciliation and mediation, then to the Organization of African Unity[23] for arbitration and finally to the International Court of Justice (ICJ) as a last resort. This is in line with certain provisions under Article 33 of the UNWC, but is not as detailed. For instance, where there is deadlock, there is no suggestion on the way forward, as opposed to the submission of the dispute to an impartial fact-finding commission which is proposed under the UNWC.

The Senegal River Convention is quite old, and though it is lacking in some key areas as will be discussed, the substantive rule affirming the willingness of contracting states to develop close cooperation in order to rationally exploit resources of the Senegal River, laid out in Article 2, is a notable strength among its provisions. Be that as it may, this does not adequately incorporate the principle of equitable and reasonable utilization. In addition, the Senegal River Convention lacks any detailed provisions on environmental protection; it does not establish a general duty on data sharing and harmful conditions, and emergencies are not addressed.[24] However, some environmental protection does exist under Article 4, where projects may not change to an appreciable extent the characteristics of the river's fauna and flora, except when agreed to jointly by the parties. There is no equivalent condition requiring consensus among all basins states before the implementation of such measures under the UNWC.[25] Ultimately, the Senegal River Convention could learn from the umbrella UNWC by strengthening its provisions on environmental protection; elaborating the principle of equitable and reasonable utilization; providing more specific rules on the regular exchange of data by member states; establishing duties on emergencies and harmful conditions; and detailing rules pertaining to notification, consultation and negotiation in the event of planned measures.

The Senegal Water Charter is more detailed and structured than the OMVS and Senegal River Conventions. It establishes, in Article 2, principles and methods for water distribution and sharing of benefits, guidance for the evaluation and approval of planned measures, provision of rules for the protection of the environment and conditions for public participation in the decisions of stock management out of the water of the Senegal River. The Charter agrees with a number of substantive and procedural rules of the UNWC as in fact, it acknowledges in its Preamble section the need to respect general principles of water rights stemming from international law with particular mention of the UNWC as a backbone framework on rights relating to the use of transboundary waters. Evidence that a number of major provisions under the Charter were informed by the UNWC is visible: e.g. under Article 4, distribution of water is to be subjected to the condition of equitable and reasonable use, there is an obligation to preserve the environment and also to

---

23   Now known as the African Union.
24   Garane (note 11), at 14.
25   Ibid., at 15.

negotiate where conflict may arise.[26] Article 24 of the Senegal Water Charter also establishes the obligation to notify on planned measures. This Article includes a deadline of three months for states to answer notifications, with specific mention on type of data required. On the other hand, the Senegal Water Charter can still be complemented by incorporating a lot of essential provisions contained in the UNWC. Some of these include the obligation not to cause significant harm on the international watercourse, and exploration of other settlement measures, like the employment of a fact-finding commission. Even the provisions that are contained in the Charter can still be elaborated with those of the detailed UNWC.

The Gambia River Basin

Three countries share the Gambia River Basin: Gambia, Guinea and Senegal. The two main agreements governing them are the Convention relating to the Status of the River Gambia (Gambia River Convention)[27] and the Convention relating to the Creation of the Gambia River Basin Development Organization (OMVG Convention).[28] The Gambia River Convention primarily concerns navigation, as opposed to other types of usage.[29] It does, however, have some notable provisions, such as not implementing any project that is likely to bring about any serious trans-formation to the basic characteristics of the river without the approval of member states (Article 4), the setting up of a joint organization for cooperation and for proper coordination of developmental projects (Article 11), and conciliation and mediation through the African Union and the ICJ as a last resort when parties fail to reach an agreement (Article 18). However, generally, the Gambia River Convention does not incorporate obligations to prevent significant harm and equitable use of the river, there is no provision detailing the obligation to share data or to protect the environment, and, when planned measures as regards to settling disputes collapse for any reason, there is no arrangement as to how to make head-way. Therefore, the Gambia River Convention could benefit from incorporating the aspects dealt with under the substantive (e.g. environmental protection) and procedural rules (e.g. fact finding) of the UNWC.

As noted by Garane, '[t]he OMVG Convention mainly contains rules dealing with the institutional structure of the basin body it creates'.[30] This is also apparent

---

26 As regards settling of disputes, Article 30 of the Charter goes further on seizing the Commission of Conciliation and Arbitration of the African Union and the ICJ as a last resort.

27 Convention Relating to the Status of the River Gambia (Gambia-Guinea-Senegal) (signed June 30, 1978). Available online at www.fao.org/docrep/w7414b/w7414b0b.htm (accessed March 21, 2013) (Gambia River Convention).

28 Convention Relating to the Creation of the Gambia River Basin Development Organization (Gambia-Guinea-Senegal) (signed June 30, 1978). Available online at http://ocid.nacse.org/tfdd/ tfdddocs/400ENG.pdf (accessed March 21, 2013) (OMVG Convention). Guinea-Bissau became a party to this Convention on June 6, 1981, pursuant to Article 21.

29 Part III of the Gambia River Convention, which consists of Articles 6–10, is dedicated to Navigation and Transport.

30 Garane (note 11), at 16.

from the titles of the different parts of the Convention, and the fact that it only contains one substantive and procedural rule. Article 19 of the OMVG Convention charges the Permanent Water Commission (one of the permanent bodies of the organization for the development of the basin) with the responsibility of determining how water is to be utilized between member states and between the industrial, agricultural and transportation sectors. This Convention is in addition and similar to the Gambia River Convention.

The Lake Chad Basin

The main legal instrument guiding the utilization and management of the Chad Basin is the Convention and Statutes relating to the Development of the Chad Basin (Lake Chad Convention and Statutes).[31] The Convention establishes the Lake Chad Basin Commission (LCBC),[32] which is responsible for implementing the provisions contained in the Conventions itself. The Lake Chad Convention and Statutes were initially adopted by Cameroon, Niger, Nigeria and Chad, in Fort Lamy (now N'Djamena), and was joined by the Central African Republic in 1994.[33] The Convention does not contain the substantive rules provided for in the UNWC; and the procedural rules therein, such as notification and prior consultation for planned measures, are presented narrowly as compared to the UNWC. Disputes that may arise over the application and interpretation of the Convention are to be referred to the LCBC. If the LCBC is unable to resolve the disputes, they are to be submitted to the African Union.[34]

In order to supplement the Charter and Statutes, the LCBC concluded a Water Charter for the Lake Chad Basin in 2011.[35] In the Preamble, the Charter acknowledges the contribution of international agreements, including the UNWC. Accordingly, the Charter incorporates, adjusts, develops and adds specificity to many of the provisions of the UNWC.

For example, Articles 52–60, on planned measures, require the notifying state to follow essentially the same procedure as under Part III of the UNWC. However, instead of exchanging information with potentially affected states and making a decision on the planned measure, the notifying state provides all relevant data to the LCBC. The LCBC is then responsible for ensuring that affected states receive the information and for making a final decision based on the available information, in consultation with those states. The LCBC is also required, under Article 52, to maintain an updated list of actions that require notification of the Commission.

---

31 Convention and Statutes Relating to the Development of the Lake Chad Basin (signed May 22, 1964). Available online at www.fao.org/docrep/W7414B/w7414b05.htm (accessed March 21, 2013) (Lake Chad Convention and Statutes).
32 Ibid., at Chapter IV (Arts 8–17).
33 Jauro, A. B. 'Lake Chad Basin Commission (LCBC) Perspectives' (Conférence Internationale: Eau et Développement Durable, Paris, March 19–21, 1998).
34 Lake Chad Convention and Statutes (note 31), at Article 7.
35 Water Charter for the Lake Chad Basin (adopted June 2011) (on file).

The Appendices to the Charter add another layer of detail. Appendix 2 sets overall limits for abstractions from the entire basin; Appendix 3 establishes minimum flows at multiple points in the basin that must be maintained; and Appendices 4 and 5 deal with methods for obtaining and recording information, as well as the type of information to be exchanged.

The Charter also includes specific provisions on groundwater (Articles 19–20); biodiversity (Articles 28–29); soil degradation and vegetation cover (Articles 31–32); traditional rights, including over fisheries and herding (Articles 33–36); taxing schemes to help finance the basin's sustainable development (Articles 18); rights of basin populations (Articles 72–77); and 'promotional actions', such as creating civil society and grassroots organizations, building capacity and encouraging scientific research (Articles 78–81).

Therefore, by providing a framework for the elaboration of the Charter, the UNWC has already played an important role in the Lake Chad Basin. The Charter goes beyond the requirements found in the UNWC, and, as such, is a perfect example of the role and value of the UNWC for informing the adoption of watercourse agreements that apply and adjust its basic provisions.

The Volta River Basin

The Volta River Basin is one of the most important basins in West Africa. As presented in Table 9.2, six member countries share the basin.

Table 9.2 shows that not all the countries are comparable in size, and the surface of the basin area shared is not distributed equally. The implication of this diversity means that there has to be a well-coordinated framework in place to avoid conflicts and oversee the management of the basin. Barely more than four years ago, the member states of the Volta Basin adopted the Convention on the Statute of the Volta River and Setting up the Volta Basin Authority (Volta River Convention).[36] The Convention did not enter into force immediately after it was signed by the heads of states and governments because it required ratification by all states. Owing to the different internal procedures for the ratification process, countries did not ratify the Convention at the same time, but a lot has changed between that time and now. Only Burkina Faso ratified it towards the end of 2007; Mali followed in February 2008, Ghana in November 2008 and the last two states – Togo and Benin – ratified it in April and June 2009, respectively.[37]

The Volta River Convention, inter alia, established the Volta Basin Authority (VBA), and required that the Volta River, as a transboundary watercourse, be managed collectively. The VBA was actually set up through an evolution of events over the years such as the creation of a technical committee (in 2004), the signing of a Memorandum of Understanding by ministers of water resources of the

---

36 Convention on the Statute of the Volta River and Setting up the Volta Basin Authority (signed January 19, 2007) (Volta River Convention).
37 'About VBA: Review' (Volta Basin Authority, 30 March 2011). Available online at www.abv-volta.org/abv2/about/historique (accessed March 21, 2013).

Table 9.2 Share of countries in the Volta River Basin

Country	Area of basin (km²)	Basin area (%)	Country area (%)
Benin	13,590	3.41	12.1
Burkina Faso	171,105	42.95	62.4
Cote d'Ivoire	9,890	2.48	3.07
Ghana	165,830	41.63	70.1
Mali	12,430	3.12	1.0
Togo	25,545	6.41	45.0
Total	398,390	100	

Source: Reproduced with permission from Volta Basin Authority (www.abv-volta.org:10000/abv2/pays)

riparian countries (in 2005), and the approval of a Convention and Statutes for the VBA with its headquarters in Ouagadougou (in 2006).[38] Instruments on the principles of equitable and reasonable use, obligation to exchange information and not to cause harm, dispute settlement, and protection of the ecosystem were established in the Volta River Convention. The structuring of the provisions under the Volta River Convention appears to be well laid out and largely corroborates those that are also contained in the UNWC. However, the Volta River Convention is still not as detailed as the UNWC in a number of respects.

For instance, the Volta River Convention:

> fails to require reasonable and equitable participation in the protection of the Volta Basin, provides no guidance for the participation in the protection of the Volta Basin, provides no guidance for the application of its substantive rules, and does not explicitly require states to consult in good faith when necessary for the sound application of the rules.[39]

Additionally, the Volta River Convention does not mention rules pertaining to prevention of pollution and processes that are artificially caused and can cause harm to the basin. The VBA, per Article 6(2)–(3), is to promote the implementation of integrated water resources management and the sharing of benefits from the various uses in an equitable manner, and is also to authorize planned projects by member states that are likely to have a significant impact on the water resources. The Volta River Convention can learn from the UNWC in the development of more detailed procedures and timelines so that the member countries can be notified prior to the potential duration and scope of the approval process.

On the settling of disputes, the Volta River Basin states are to seek amicable settlement within the VBA at the first instant, and, similar to the Gambia River Convention, to resort to conciliation and mediation. This is to be taken further to the Economic Community of West African States and the African Union when

38 Ibid.
39 Garane (note 11), at 18.

agreement is not reached, and the ICJ as a final resort. The Volta Basin can therefore supplement these dispute settlement mechanisms with the UNWC by incorporating deadlines for states to reach an agreement and with procedures such as fact finding when settlements cannot be brokered.

Finally, the Code of Conduct for Sustainable and Equitable Management of the Volta Basin Water Resources (Volta Basin Code of Conduct),[40] containing up to 59 articles, was adopted by Burkina Faso and Ghana in 2006. Although the Code establishes the rules and principles provided by the UNWC, including the principle of equitable use (Article 11), prevention of pollution (Article 12), the principle of benefit sharing (Article 14), etc., it is applicable to only two of the six countries in the basin. This makes the Volta Basin Code of Conduct less important than the Volta River Convention.

The Corubal River Basin

The Corubal River flows through the southern half of Gabú – a region located in the northeastern area of Guinea-Bissau.[41] The river is quite important, in that the southern part constitutes an international border with neighboring Guinea, where the Boé hills are situated. The only mechanism that is applicable to the river basin is the Protocol of the Agreement on the Management of the Koliba-Korubal River (Corubal River Agreement).[42] The Agreement is quite brief, with only eight articles in total. Nonetheless, in Article 1, it sets as its objective the development of the river without prejudice to the interests of any of the states. Article 3 establishes a Permanent Technical Committee, which is to supervise proposed tasks and coordinate contributions of governments.

The Corubal River Agreement is devoid of any detailed substantive and procedural rules for the collaborative governance between Guinea and Guinea-Bissau. Even the proposed tasks outlined in Article 2, which mentions the development of a master plan in different areas (including protection of nature), do not go into any depth. Therefore, the Koliba-Korubal River Basin stands to gain extensively from the UNWC, which will go a long way in promoting better management of the river and cooperation between both countries. In addition, the specific provisions of the UNWC can also provide the Permanent Technical Committee with legal guidance for implementing its activities in a more efficient manner.[43] The Corubal

---

40 Code of Conduct for the Sustainable and Equitable Management of Shared Water Resources of the Volta Basin (Burkina Faso-Ghana) (signed July 2006). Available online at www.dialoguebarrages.org/dialoguebarrages/images/stories/downloads/final_code_of_conduct_j uly_2006_eng.pdf (accessed March 21, 2013) (Volta Basin Code of Conduct).

41 'Gabú', Encyclopædia Britannica Online (2013). Available online at www.britannica.com/ EBchecked/topic/1481470/Gabu (accessed 23 March 21, 2013).

42 Protocol of the Agreement between the Republic of Guinea and the Republic of Guinea Bissau on the Management of the Koliba-Korubal River (signed October 21, 1978) [French]. Available online at www.fao.org/docrep/w7414b/w7414b0t.htm (accessed March 21, 2013) (Corubal River Agreement).

43 Garane (note 11).

River Agreement incorporates into Article 8 the resolving of disputes with amicable settlement by both parties. Thus, the Agreement can supplement this provision with almost all of the dispute settlement mechanisms that are contained in the UNWC.

Conclusion

At the UNWC's adoption in 1997, Burkina Faso, Cameroon, Cote d'Ivoire and Nigeria voted in its favor. Ghana and Mali abstained, while Benin, Cape Verde, Mauritania, Niger and Senegal were absent.[44] Reference is not made to Guinea-Bissau, The Gambia and Togo in the voting records. No West African state voted against the Convention.[45]

Different explanations have been made as to why some states abstained or were absent from the UNWC's vote,[46] including low levels of awareness and understanding of the Convention.[47] But recent developments in the West African region show that awareness is increasing. One of these developments includes the 2007 Dakar Call for Action on the Ratification of the UNWC by West African States,[48] signed by Senegal, Togo, Niger, Nigeria, Mali, Guinea, Cote d'Ivoire, Ghana, Burkina Faso and Benin in September 2007.[49] In addition, Guinea-Bissau and Nigeria ratified the Convention in May and September 2010, respectively, followed by Burkina Faso in March 2011, and Benin in July 2012. Chad, which shares the Chad and Niger Basins with West African countries, acceded to the Convention in September 2012. Table 9.3 presents an excerpt of the signature, ratification and accession of West African states to the UNWC.

The watercourses in West Africa are transboundary in nature, making cooperation over them very important to the respective riparian states that share them. Because of the propensity for disagreements to occur because of conflicting interests as to how the international watercourses should be used, member states have not only adopted agreements to appropriately govern their basins, but they have also tried to modify them (in some cases) so as to adequately reflect any changing circumstances or include vital provisions that had not been earlier considered. Nonetheless, many West African agreements are still lacking relevant principles and

---

44 Loures, F. R., Rieu-Clarke, A. and Vercambre, M. L. *Everything You Need To Know About The UN Watercourses Convention* (WWF International, 2008), at 24.

45 UNGA, '99th Plenary Meeting' (May 21, 1997) UN Doc A/51/PV.99, at 2, 7–8.

46 Garane (note 11).

47 Loures, F. R. and Rieu-Clarke, A. 'Still Not in Force: Should States Support the 1997 UN Watercourses Convention?' *Review of European Community and International Environmental Law*, 2009; 18: 185–97.

48 West Africa Regional Workshop on the UN Watercourses Convention (Dakar, September 20–21, 2007), 'Dakar Call for Action on the Ratification of the 1997 UN Convention on the Law of the Non-Navigational Uses of International Watercourses' (on file).

49 The workshop encouraged West African governments to ratify the UN Watercourses Convention, in part through the promotion of its principles by river basin organizations and other subregional integration organizations.

Table 9.3 Status of states on the UNWC

Participant	Signature	Ratification (r), Accession (a)
Benin		July 5, 2012 (a)
Burkina Faso		March 22, 2011 (a)
Cote d'Ivoire	September 25, 1998	
Guinea-Bissau		May 19, 2010 (a)
Niger		February 20, 2013 (a)
Nigeria		September 27, 2010 (r)

Source: Adapted from UN Treaty Collection, 'Status: Convention on the Law of the Non-
Navigational Uses of International Watercourses' (UN, July 10, 2012). Available online at
http://treaties.un.org/pages/ViewDetails.aspx?src=TREATY&mtdsg_no=XXVII-
12&chapter=27&lang=en (accessed March 21, 2013).

mechanisms to sufficiently manage and protect the watercourses. In a number of
cases, it is either because the agreements have focused mainly on building institutional mechanisms or because the framework laid out has been narrowly
articulated.

Also, considering that many of these agreements were already in existence when
the UNWC was adopted in 1997, they have yet to benefit from the more comprehensive rules, principles and procedures that the Convention proffers as a standard
framework. While more needs to be done, from the trend with which West African
countries have subsequently pushed for its ratification, there is an indication that
there is more awareness and increasing interest in the UNWC in the region. If the
UNWC enters into force and eventually gets ratified across the region, it will
improve cooperation between member countries and assist in the reinforcement of
existing substantive and procedural rules, and help harmonize the existing basin
treaties.

10 Southern Africa

Daniel Malzbender and Anton Earle

When the UN Watercourses Convention (UNWC) was passed, all votes among Southern African Development Community (SADC) member states were in favor, except for the abstention from Tanzania, and the absence from the Democratic Republic of the Congo, Swaziland and Zimbabwe.[1] No SADC country voted against the Convention. Of the 15 SADC member states, to date, only Namibia and South Africa are parties to the Convention.

This chapter is a comparative analysis between the UNWC and the SADC Revised Protocol on Shared International Watercourses (Revised SADC Protocol),[2] a regional framework agreement for the management of shared watercourses concluded by SADC member states. It first highlights similarities and differences between the two instruments. The chapter then assesses the potential benefits for SADC states of adopting the UNWC with respect to basins shared between SADC states only and basins shared with neighboring non-SADC states.

The main conclusion is that, with respect to basins shared between SADC states only, the value of adopting the UNWC for SADC states would merely lie in interpretational guidance for some Revised SADC Protocol provisions, rather than creating a new or more comprehensive legal framework. It would, however, provide SADC states with a number of tangible benefits in relation to neighboring non-SADC states with which they share basins. It seems to be in the interest of SADC states, as well as their neighbors, to adopt the UNWC and to extend the harmonized legal framework that SADC states have created among themselves to basins that are shared with non-SADC neighbors.

The countries of Southern Africa, in this report taken to refer to the 15 SADC countries, depend to a large extent on shared rivers to meet their water needs. Water resources have been and will continue to be developed and managed in the region to promote agriculture, industry, mining, and power generation, thus contributing to regional socioeconomic development. Increasingly, it is recognized

---

1 UN Environment Programme, *Hydropolitical Vulnerability and Resilience along International Waters* (UNEP, 2005).
2 SADC Revised Protocol on Shared International Watercourses (adopted August 7, 2000, entered into force September 22, 2003) (2001) 40 ILM 321, Article 5(3)(c) (Revised SADC Protocol).

that water needs to be secured to sustain biodiversity and natural ecosystems, including wetlands, which are the basis for rural livelihoods and for tourism.[3]

In order to achieve a balance between the different water uses within as well as between countries and to ensure the sustainable development of the region's water resources, a comprehensive legal and institutional management framework is required. International water law, whether as customary international law or as treaty law, provides countries with a framework of rights and obligations with respect to the development and management of their shared water resources.

Although only two SADC states are party to the UNWC, most SADC states have signed and ratified the Revised SADC Protocol, which is in force and constitutes the primary legal instrument for the management of shared watercourses in the SADC region. The Revised SADC Protocol is drafted largely along the lines of the UNWC – yet some differences exist. Against this background, this study assesses whether there would be benefits to SADC states resulting from the entry into force of the UNWC.

In total, there are 16 shared watercourses in the SADC region. Most of these are shared between SADC member states only, but some – like the Pangani, the Congo and the Nile – are shared with neighboring, non-SADC member states (Table 10.1).

Compatibility between the UNWC and existing regional and national legal and policy tools

The UNWC and the Revised SADC Protocol

The most important regional legal instrument relating to shared watercourses is the Revised SADC Protocol. When comparing the UNWC with the Revised SADC Protocol, it is important to keep the history of the latter's development in mind.

The original SADC Protocol on Shared Watercourse Systems (as it was then called) was adopted in 1995, and was the first sectoral protocol following the signing of the SADC Treaty in 1992.[4] The 1995 SADC Protocol was greatly influenced by international water law instruments, such as the Helsinki Rules and the Dublin Principles, as well as Agenda 21.[5] The subsequent revision of the 1995 Protocol was influenced by two main factors. The first was that some member states had reservations about the contents of the Protocol, and the summit approved that their concerns should be addressed.[6] The second was the adoption of the UNWC in 1997.

During the revision period, the SADC member states used the opportunity to bring the Revised SADC Protocol into line with international water law, as

---

3   SADC, *Regional Water Policy* (SADC, 2005).
4   Ramoeli, P. 'The SADC Protocol on Shared Watercourses: Its Origins and Current Status' in Turton, A. and Henwood, R. (eds), *Hydropolitics in the Developing World – A Southern African Perspective* (African Water Issues Research Unit, University of Pretoria, 2002), at 105–12.
5   Ibid.
6   Ibid.

Table 10.1 International river basins of the Southern African Development Community (SADC)

Basin	SADC basin states	Other basin states	Basin-wide organization formed?
Buzi	Mozambique, Zimbabwe	n/a	Yes
Congo	Angola, DRC, Tanzania, Zambia	Burundi, Cameroon, Central African Republic, Congo (Brazzaville), Rwanda	Yes
Cunene	Angola, Namibia	n/a	Yes
Cuvelai	Angola, Namibia	n/a	Yes
Incomati	Mozambique, South Africa, Swaziland	n/a	Yes
Limpopo	Botswana, Mozambique, South Africa, Zimbabwe	n/a	Yes
Maputo	Mozambique, South Africa, Swaziland	n/a	Yes
Nile	DRC, Tanzania	Burundi, Egypt, Ethiopia, Eritrea, Kenya, Rwanda, Sudan, South Sudan, Uganda	Yes
Okavango	Angola, Botswana, Namibia	n/a	Yes
Orange-Senqu	Botswana, Lesotho, Namibia, South Africa	n/a	Yes
Pangani	Tanzania	Kenya	No
Pungwe	Mozambique, Zimbabwe	n/a	Yes
Ruvuma	Mozambique, Tanzania	n/a	Yes
Save	Mozambique, Zimbabwe	n/a	Yes
Umbeluzi	Mozambique, South Africa, Swaziland	n/a	No
Zambezi	Angola, Botswana, DRC, Malawi, Mozambique, Namibia, Tanzania, Zambia, Zimbabwe	n/a	Yes

Source: Adapted and updated from Earle, A. and Malzbender, D. *Water and the Peaceful, Sustainable Development of the SADC Region* (Safer Africa: Towards a Continental Common Position on the Governance of Natural Resources in Africa, 2007).

reflected in the UNWC. As a result, much of the Revised SADC Protocol, which was signed in August 2000 and entered into force in September 2003, is a verbatim reflection of the UNWC. Although only Namibia and South Africa have ratified the UNWC, SADC member states have expressed their consent with its key principles, albeit indirectly, through ratifying the Revised SADC Protocol which contains the identical key principles. These include, among others, the three principles that are considered to be accepted as customary international law: equitable and reasonable utilization; obligation to prevent significant harm; and obligation to notify of planned measures.

Despite the great similarity between the two agreements, there remain a few differences, some of which could result in rights and obligations that are substantially different, depending on whether the Revised SADC Protocol or the UNWC would apply. These differences include the factors considered when determining equitable and reasonable utilization, the extent of marine protection, the relationship between the principle of equitable and reasonable utilization and the obligation not to cause harm and dispute settlement mechanisms.

Factors for equitable and reasonable utilization

Article 6 of the UNWC and Article 3(8)(a) of the Revised SADC Protocol list factors to be taken into account for determining what is equitable and reasonable utilization. Whereas the list of factors in both agreements is otherwise identical, the Revised SADC Protocol lists a factor not to be found in the equivalent UNWC provision.

Both the UNWC (Article 6(1)(f)) and the Revised SADC Protocol (Article 3(8)(a)(vi)) list as one factor the 'conservation, protection, development and economy of use of the water resources of the shared watercourse and the costs of measures taken to that effect'. These provisions refer to the water resources of the shared watercourse itself; the social and economic needs of the watercourse state, beyond the specific watercourse in question, are listed as a separate factor in both agreements. With regard to the latter, Article 6(1)(b) of the UNWC specifies that the social and economic needs of watercourse states need to be considered. Article 3(8)(a)(ii) of the Revised SADC Protocol, on the other hand, adds to this the environmental needs of watercourse states.

It is debatable whether this indeed creates a difference in practice. The list of factors in both agreements is merely exemplary, and not exclusive, meaning that factors not specifically listed also need to be considered, where relevant. Arguably, the environmental needs of watercourse states thus could not be ignored under the UNWC as well – the Revised SADC Protocol may just be more specific.

Protection of aquatic environment vs. marine environment

Both the UNWC and the Revised SADC Protocol contain provisions dealing specifically with the protection and preservation of the environment. Whereas the rest of the provisions on this matter are identical in the two instruments, there is a difference between the UNWC's Article 23 and the equivalent Article 4(2)(d) of the Revised SADC Protocol.

Article 23 of the UNWC obliges states to take all necessary measures to protect and preserve the 'marine environment', including estuaries. Article 4(2)(d) of the Revised SADC Protocol, on the other hand, uses the term 'aquatic environment'. If the obligation set forth in this provision was meant to have the same scope as the one of Article 23 of the UNWC, there was no need to change the terminology, particularly since the Articles otherwise have identical wording. The replacement of the term 'marine' by the drafters of the Protocol therefore suggests that they preferred a more limited protection obligation compared with the one in the UNWC.

While there is no universally accepted definition for aquatic ecosystems, these are considered to include riverine systems, estuarine systems, coastal marine systems, wetland systems, floodplains, lakes and groundwater systems.[7] Following this definition, the Revised SADC Protocol obligation would extend only to coastal marine systems but not include impacts that occur in the open sea, i.e. beyond coastal areas.

The difference between the two provisions might, in practice, be reduced by the fact that many impacts that affect the marine environment in the open sea would also affect the estuary and coastal areas, in which case the protection obligation applies in any case. Where this is not the case, the two agreements seem to create different obligations for states, with the UNWC being the more demanding instrument as far as marine protection obligations are concerned.

Relationship of the 'equitable utilization' principle and the 'no harm' obligation

Article 5(1) of the UNWC and the equivalent Article 3(7)(a) of the Revised SADC Protocol incorporate the principle of reasonable and equitable use, stipulating that watercourse states shall, in their respective territories, utilize a shared watercourse in an 'equitable and reasonable manner'. At the same time, Article 7(1) of the UNWC and the equivalent Article 3(10)(a) of the Revised SADC Protocol codify the no-harm rule, obliging parties 'to take all appropriate measures to prevent the causing of significant harm to other watercourse states' when utilizing a shared watercourse. The no-harm rule is not 'an absolute obligation, but rather one of due diligence, or best efforts under the circumstances'.[8] The legal relationship between the two principles is dealt with in Article 7(2) of the UNWC. The equivalent provision in the Revised SADC Protocol is Article 3(10)(b); however, it is not clear how the two principles relate to each other.

The legal relationship between these two key principles is of great practical relevance. If the no-harm obligation enjoys precedence, developments cannot take place without the consent of the other riparian states if such developments are significantly harmful to other states. In other words, such developments would be automatically

7   Masundire, H. and Mackay, H. 'The Role and Importance of Aquatic Ecosystems in Water Resources Management' in Hirji, R., SADC (eds), *Defining and Mainstreaming Environmental Sustainability in Water Resources Management in Southern Africa, Volume 1* (World Bank, SADC, 2002), at Ch. 3.

8   McCaffrey, S. C. 'The Contribution of the UN Convention on the Law of the Non-Navigational Uses of International Watercourses', *International Journal Global Environmental Issues*, 2001; 1 (3/4): 250–63.

considered in violation of the principle of reasonable and equitable use. If, on the other hand, the equitable and reasonable utilization principle enjoys precedence over the no-harm obligation, there is no outright prohibition on planned developments capable of causing significant transboundary harm. Under certain circumstances, such developments may be considered reasonable and equitable.

It is today widely accepted among leading legal experts that the principle of equitable utilization takes priority over the obligation to prevent significant harm.[9] Arguably, the strongest argument in support of the above view is the wording of Article 7(2) of the UNWC, obliging a state that causes significant harm to another watercourse state to take all appropriate measures, *having due regard for the provisions of articles 5 and 6,*[10] to eliminate or mitigate such harm. The explicit reference that appropriate measures need to be taken with due regard to the equitable utilization principle confirms this principle's precedence over the no-harm obligation. Consequently, under the UNWC, developments that are equitable and reasonable and carried out in conformity with the due diligence duty of prevention established by the no-harm rule, but which nevertheless cause significant harm, could, in principle, be in compliance with the Convention.

The Revised SADC Protocol's equivalent to Article 7(2) of the UNWC is Article 3(10)(b), and it is drafted in a different manner. It specifically requires states to have due regard for the provisions of Article 3(10)(a) when taking appropriate measures, which is the Protocol's no-harm obligation itself. Thus, like the UNWC's Article 7(2), the Protocol does acknowledge that harm may be caused under certain circumstances. However, by obliging states to give due regard to the prevention of harm obligation in Article 3(10)(a) rather than the equitable utilization principle, it may be argued that the Revised SADC Protocol did not follow the UNWC's approach of giving precedence to the equitable utilization principle. Instead, the no-harm obligation seems to be strengthened and given precedence over the equitable utilization principle.

In addition to the text of a legal instrument, its Preamble also needs to be considered in an interpretation exercise, as determined by Article 31(2) of the Vienna Convention on the Law of Treaties.[11] In this sense, a different interpretation also seems possible on the basis of the Protocol's Preamble, which expressly refers to the UNWC as a legal source states were bearing in mind for drafting the Protocol. This provides a strong argument for concluding that the Protocol intended to reflect the relation between the substantive provisions at hand as codified by the UNWC. This view is supported by the fact that the minutes of the Protocol negotiations reveal that earlier drafts of the Protocol had used the same wording as used in the UNWC. This was later changed to the current wording, but there are no recorded discussions on the reasons for the change.

9   McIntyre, O. 'International Water Law: Factors Relating to the Equitable Utilisation of Shared Freshwater Resources' *SIDA International Training Programme in Transboundary Water Resources Management Paper* 2007; 1.2, at 255.
10 Emphasis added.
11 Vienna Convention on the Law of Treaties (adopted May 23, 1969, entered into force January 27, 1980) 1155 UNTS 331.

On the other hand, the circumstances of the Protocol's conclusion strengthen the view that the two instruments adopt the principle of reasonable and equitable use as their overriding substantive rule. The original 1995 SADC Protocol on Shared Watercourse Systems was revised exactly to harmonize it with the UNWC. Arguably, in this context, it is not reasonable to assume an intentional deviation from the Convention's primary substantive rules.

Dispute settlement mechanisms

The dispute settlement procedures provided for in the UNWC and the Revised SADC Protocol show some differences. The Revised SADC Protocol simply obliges member states to resolve disputes amicably. Where an amicable settlement is not possible, disputes shall be referred to the SADC Tribunal.

The UNWC, on the other hand, provides disputing parties with more dispute settlement options. In terms of Article 33(2), as a first step parties are obliged to enter into negotiations, at the request of (at least) *one* of the disputing parties. If agreement cannot be reached through negotiations, the disputing parties may *jointly* seek the good offices of, or request mediation or conciliation, by a third party or make use of any joint watercourse institution established between them. Alternatively, the disputing parties can agree to submit the dispute to arbitration or to adjudication by the International Court of Justice.

Arguably, the most significant difference between the UNWC and the Revised SADC Protocol is the procedure for resolving disputes. Article 33(3) of the UNWC provides for mandatory fact-finding if, after six months from the time of the request for negotiations, the parties have not been able to settle their dispute through negotiation or any of the other means referred to in Article 33(2). The inclusion in the UNWC of mandatory fact finding is due to the importance of facts in relation to the core obligations of the Convention.[12] Without establishing the facts, it will not be possible to determine whether harm occurring in one state was indeed caused by the other state or whether a specific use is equitable and reasonable.

Yet, the drafters of the Revised SADC Protocol have chosen not to make express reference to compulsory, impartial fact finding should the amicable settlement of disputes fail. One would argue that disputing parties can jointly agree on impartial fact finding as part of the efforts to achieve an amicable resolution of the dispute. However, could impartial fact finding, as an 'amicable' dispute settlement mechanism, take place under Article 7(1) of the Revised SADC Protocol at the request of only one party, as is the case under the UNWC? McIntyre[13] argues that the provisions regarding independent fact finding provide a *de minimis* standard for dispute settlement in the absence of specific binding provisions in regional agreements. The Revised SADC Protocol's applicable provisions simply require amicable settlement, and, if that is unsuccessful, adjudication by the SADC

12   McCaffrey (note 8).
13   Email communication with O. McIntyre (June 1, 2007).

Tribunal. But the Protocol does not contain detailed provisions on which procedure to follow to achieve amicable settlement. It could thus be argued that the *de
minimis* standard arguably set by the UNWC would have to be applied once the
latter is in force.

Regional and national benefits arising from the entry into force of the UNWC

Provisions of the UNWC that could support the interpretation of the Revised SADC Protocol

The Revised SADC Protocol covers most issues regulated in the UNWC and thus
creates a comprehensive legal framework for the management of shared watercourses in the SADC region; nonetheless, the UNWC, if in force for SADC states,
could support the interpretation of some provisions of the Revised SADC
Protocol.

Article 3(6) of the Revised SADC Protocol obliges states to exchange available
information and data regarding the hydrological, hydrogeological, water quality,
meteorological and environmental condition of shared watercourses. The UNWC
provides more detailed rules for instances where information or data is not readily
available. Article 9(2) of the UNWC provides that:

> if a watercourse State is requested by another watercourse State to provide data
> or information that is not readily available, it shall employ its best efforts to
> comply with the request but may condition its compliance upon payment by
> the requesting State of the reasonable costs of collecting and, where appropri-
> ate, processing such data or information.

Article 9(3) stipulates that 'Watercourse States shall employ their best efforts to
collect and, where appropriate, to process data and information in a manner which
facilitates its utilization by the other watercourse States to which it is communi-
cated'. If, in the context of the Revised SADC Protocol, there is uncertainty as to
the format in which data are presented and the responsibility of costs for the
collection and processing of data, these provisions of the UNWC would provide
valuable guidance for the interpretation of Article 3(6) of the Revised SADC
Protocol.

Likewise, the interpretation of Article 3(8) of the Revised SADC Protocol,
which lists factors for the determination of 'equitable and reasonable utilization',
could be supported by Article 10(2) of the UNWC. Article 10(2) makes specific
reference to the concept of vital human needs in the determination of 'equitable
and reasonable utilization'. Whereas the concept of vital human needs is increasingly being recognized in international water law as a key factor to consider in the
relationship between different uses, it is not explicitly mentioned in the Revised
SADC Protocol. If the UNWC became binding on SADC states, and could thus
be relied on more strongly in the interpretation of the Revised SADC Protocol,

the vital human needs factor would be strengthened in the application of the latter.

Article 28(4) of the UNWC could support Article 4(5) of the Revised SADC Protocol on emergency measures. Whereas Article 4(5) of the Revised SADC Protocol and Article 28(1)–(3) of the UNWC establish the same notification and mitigation obligations for states in cases of emergency, Article 28(4) goes further and requires, where necessary, the joint, cooperative development of contingency plans for responding to emergencies. The Revised SADC Protocol currently does not establish such an obligation, but, arguably, the joint development of contingency plans would be in the interest of the region. With the effects of climate change likely to become more relevant for the region over time, Article 28(4) of the UNWC could also provide valuable guidance in the development of adaptation strategies to respond to climate change impacts.

Revision of watercourse agreements

Both the UNWC (Article 3(1)) and the Revised SADC Protocol (Article 6 (1)) do not affect the rights and obligations of states resulting from existing agreements. At the same time, the Convention and the Protocol both encourage states to harmonize such agreements with their respective principles and substantive rules. Particularly in regard to pollution, Article 21(2) of the UNWC requires states to take steps to harmonize their policies with respect to the prevention, reduction and control of pollution in international watercourses, thus providing guidance for states to cooperate in those matters. The Revised SADC Protocol contains a similar but more far-reaching provision in this regard. Article 4(2)(b)(ii) determines that 'watercourse states shall take steps to harmonize their policies and legislation in this connection'. The UNWC's policy harmonization provisions would provide guidance in relation to non-SADC member states where the Revised SADC Protocol is not applicable. For SADC states sharing basins with non-SADC states, it is in their interest to harmonize their policies and legislation, not only with fellow SADC members, but also with non-SADC riparians, to create a harmonized approach to the management of their shared watercourses. Yet, in relation to non-SADC states, there is currently no agreement requiring such harmonization efforts. The entry into force of the UNWC, with its encouragement to harmonize policy, would provide SADC countries with additional arguments (and a guiding framework in the Convention itself) in relation to their non-SADC neighbors. Although not directly provided for in the UNWC, this might, in the long run, also aid the harmonization of legislation with non-SADC members states, based on the harmonized policies.

Dispute settlement and prevention

The adoption of the UNWC could possibly add benefits as far as the prevention or resolution of disputes between SADC member states is concerned. If one followed the interpretation that the fact-finding requirement set forth by the UNWC sets minimum standards for dispute resolution procedures, and could thus also be relied on in the SADC context, this would provide SADC member states with additional options for dispute resolution not provided for in the Revised SADC Protocol itself.

The adoption of the UNWC would, even more so, create benefits for dispute prevention and settlement with neighboring non-SADC states. There are no binding regional or basin-wide watercourse agreements in place between SADC states and their non-SADC neighbors. Consequently, there are no universally agreed procedures for the resolution of disputes over shared watercourses. The UNWC would provide such procedures and thus put the settlement of disputes related to watercourses shared between SADC and non-SADC member states on a solid legal footing.

A common legal framework among SADC and non-SADC member states

The ratification of the UNWC would be of great relevance for basins shared with non-SADC member states also in regards to substantive principles and rules. The Revised SADC Protocol is not applicable to non-SADC member states and thus cannot fulfill its guidance function beyond the SADC region in basins like the Nile and the Congo. Instead, this role could be played by the UNWC once it comes into force. Article 3(3)-(4) of the UNWC is equivalent to the provisions in Article 6(3)-(4) of the Protocol, and encourages watercourse states that want to conclude watercourse specific agreements to do so in line with the provisions of the UNWC.

It could be argued that the key principles of the UNWC are accepted as customary international law anyway and thus need to be adhered to even without the entry into force of the UNWC. Yet, with its entry into force (and the respective SADC and non-SADC states being party to it), these principles would become applicable treaty law between these countries, thus arguably giving them further weight and clarifying their scope and extent. Furthermore, only the three key principles of the UNWC are clearly accepted as customary international law. The coming into force of the UNWC would make the full set of substantive provisions, as well as its dispute settlement machinery, applicable law to state parties.

With the Revised SADC Protocol and the UNWC to a large extent setting forth the same principles, SADC member states would benefit from the same legal clarity and harmonized basin management framework that they enjoy with fellow SADC members also in relation to non-SADC member states. Given that SADC member states share not only Africa's largest river (Congo), but with the Nile river also one of the most politically and institutionally complex basins in Africa, the clarity and guidance provided by the UNWC would appear to be beneficial to all parties involved. In the Pangani, Congo and Nile Basins, the UNWC, once in force, would be the first comprehensive legal framework applicable on a basin scale. As the Revised SADC Protocol does for SADC states, the UNWC could provide direction for further development of a legal framework for these basins through its encouragement to harmonize existing agreements with the principles of the UNWC (Article 3(2)) and to enter into future watercourse agreements applying the provisions of the Convention to the characteristics of the watercourse in question (Article 3(3)).

Conclusion

The member states of the SADC are committed to managing their shared water resources in a cooperative manner within the framework of international law. To this end, they have concluded between them the Revised SADC Protocol on Shared Watercourses, which sets out key principles guiding the management of shared watercourses in the region. Whereas some differences remain, the provisions of the Revised SADC Protocol largely mirror the provisions of UNWC and fully endorse the principles of international water law enshrined in the latter. With the Revised SADC Protocol having precedence over the UNWC, the benefits of the entry into force of the UNWC for SADC states would merely lie in interpretational guidance (for some Revised SADC Protocol provisions, e.g. on the duty to address emergency situations), rather than creating a new or more comprehensive legal framework. In addition, the UNWC would offer the fact-finding procedures as an additional dispute settlement mechanism on which SADC member states could potentially rely.

The adoption of the UNWC would, on the other hand, provide SADC states with a number of tangible benefits in relation to neighboring non-SADC states with which they share basins. It would therefore seem to be in the interest of SADC states, as well as their neighbors, to adopt the UNWC and extend the harmonized legal framework that SADC states have created among themselves to basins that are shared with non-SADC neighbors.

11 Central America

Alexander López and Ricardo Sancho

Central America can be defined as a region of international river basins: nearly 3941 km of borders separate Guatemala, Belize, El Salvador, Honduras, Nicaragua, Costa Rica and Panama. In a small territory of 524,000 km², or 0.1 percent of the Earth's surface, there are 23 transboundary watersheds, representing 137,216.1 km², close to 40 percent of the region's land surface.[1] This is an area larger than any single country in the region, which, as we discuss below, urgently requires more robust transboundary water governance regimes.

Within this context, this chapter investigates if and to what extent the UN Watercourses Convention (UNWC) could be an important mechanism for strengthening the governance of Central America's international watercourses. We first look at the region's geography, focusing on the large number of international basins. We then examine the current governance structure for these internationally shared basins, as well as impediments to more integrated water management across international borders. The role of domestic water legislation is also considered. Overall, the chapter concludes that the UNWC can fill in many of the current gaps in the governance of international watercourses in Central America.

Central America: A region of international river basins

The existence of 23 international river basins implies a high hydrological interdependence that should promote cooperation. However, a river basin does not necessarily provide a good foundation for transboundary water cooperation simply by virtue of crossing international borders. Aspects such as institution building and availability of funds – which, in turn, promote the generation of positive externalities and allow for the distribution of cost and benefits – and the national political contexts in each riparian country can be more relevant in fostering interstate collaboration on water issues.

In Central America, the process of institution building has been weak and cooperation has taken place mainly informally in the context of isolated projects or as an afterthought within the framework of broader cooperation mechanisms dealing

---

1 Granados, C., Delgado, H., Hernández, A. and Herrera, E. R. *Cuencas Internacionales: Conflictos y Cooperación en Centroamérica* (FUNPADEM, 2000).

with boundary issues. In general, Central America's transboundary basins lack or have limited institutions and legal frameworks in place. In addition, no country in the region has joined the UNWC.

To understand this situation, there are three key aspects to keep in mind. First, the issue of benefit sharing is highly contentious, in part because of the belief that benefit sharing would require some sort of redistribution or compensation and direct payments. Some countries (mainly upstream), such as Guatemala, are very skeptical about this. Secondly, building watercourse agreements and transboundary institutions presume the reconceptualization of classical notions of national security, sovereignty, and territoriality. Central American states are reluctant to adopt schemes they see as capable of undermining state sovereignty. 'In Latin America, as in most parts of the world, environmental and resource management has tradition-ally been the preserve of national governments. Crucial environmental issues such as the use of rivers and watersheds remain largely within the regulatory ambit of states'.[2] Finally, there are structural considerations, which can either enable or constrain regime formation. In Central America, public policies and institutions relating to water have historically been weak and fragmented, as this chapter will outline. The above factors have implications for the potential adoption of water treaties among the region's countries, including the UNWC.

At the same time, there are other points to be considered when analyzing international river basins in Central America:

> [T]he internationalization of environmental problems and its impacts on national structures is having a profound effect on how such resources are managed, and therefore on how cooperation is understood and operational-ized. In [Central] America this internationalization and the new framework of environmental cooperation in most cases [are] the product of four factors: first, there is a new understanding of the international effect of the process of envi-ronmental change; second, environmental problems have become more international because the internationalization of the…economy has intensified pressures on national ecological systems; third, the existence of natural ecosys-tems shared by two or more states, such as a river basin, requires new frameworks for regional cooperation; and finally, the transborder externalities produced by the exploitation of such resources have contributed to the inter-nationalization of problems, and consequently these externalities have challenged the traditional means of cooperation.[3]

As Table 11.1 shows, most countries in Central America have a significant portion of their respective territories within international river basins. More than half of El Salvador's land area is located within three such basins – the Lempa, Paz and

---

2   López Ramírez, A. 'Environmental Transborder Cooperation in Latin America: Challenges to the Westphalia Order' in Matthew, R. A. and others (eds), *Global Environmental Change and Human Security* (Massachusetts Institute of Technology Press, 2009), at 291.

3   Ibid.

Table 11.1 Central America: Percentage of national territory located in international
basins by country

Country	National territory within international basins (%)
Belize	65.1
Guatemala	64.6
El Salvador	61.9
Honduras	18.5
Nicaragua	34.7
Costa Rica	34.3
Panamá	5.2

Source: adapted from Ulate, A. H., Ramírez, A. L. and Elizondo, A. J. *Gobernabilidad e Instituciones en las Cuencas Transfronterizas de América Central y México* (FLACSO, 2009), at 60.

Goascorán. El Salvador relies heavily on its international rivers, in particular on the Lempa River, which spans more than half of its territory. Yet, as the most downstream riparian on all three rivers, El Salvador depends on the goodwill of its neighbors to ensure that a sufficient quality and quantity of water flows through those watercourses.

Like El Salvador, much of Guatemala's territory falls within international basins. As Table 11.2 shows, Guatemala shares 13 river basins with its neighbors: three with Mexico, five with Belize, two with Honduras, one with El Salvador, one with both Mexico and Belize, and one with both Honduras and El Salvador. However, unlike El Salvador, Guatemala is the upper riparian on most of those international watercourses. This gives Guatemala an advantage, at least in terms of its reduced vulnerability to the overexploitation and pollution of its water resources by neighboring countries.

Those differences between the positions of Guatemala and El Salvador demonstrate that, although the region is one of international river basins, there is not an obvious regional motivation to ratify (or not to ratify) the UNWC.

Other important aspects of international river basin distribution in Central America include:

- Almost all river basins are divided between only two countries. The exceptions are the Usumacinta, Lempa and Hondo Rivers, which are shared by three countries [Table 11.2]. This fact is significant, since one might think that cooperation is easier when fewer states are involved.
- Some basins are divided nearly equally, such as the Goascorán River Basin (48.1 percent in El Salvador and 51.9 percent in Honduras) and the Paz River (47.4 percent in El Salvador and 52.6 percent in Guatemala). This is an interesting consideration, because one might expect that where the countries' participation is more equal, such as in the above river basins, the necessity of, and the possibility for, cooperation are greater.
- The opposite occurs with other basins – that is, when one single country possesses the majority of the basin. This is the case for the Chamelecon

River (98 percent in Honduras and two percent in Guatemala) and the Choluteca River (96.7 percent in Honduras and 3.3 percent in Nicaragua). Although these are international basins, (arguably), they basically function as domestic basins because of the overwhelming presence of one country in the basin.[4]

Table 11.2 International river basins of Central America, plus Mexico

River basin	Countries sharing the basin	Area (km²)
Usumacinta – Grijalva	Guatemala – Mexico – Belice	106,000
San Juan	Nicaragua – Costa Rica	38,569
Coco or Segovia	Nicaragua – Honduras	24,866.6
Lempa	El Salvador – Honduras – Guatemala	18,234.7
Motagua	Guatemala – Honduras	15,963.8
Belice	Belice – Guatemala	12,153.9
Choluteca	Honduras – Nicaragua	8132.6
Hondo	Guatemala – Belice – Mexico	7189
Chamelecón	Honduras – Guatemala	5154.9
Changuinola	Panama – Costa Rica	3387.8
Sixaola	Costa Rica – Panama	2839.6
Goascorán	Honduras – El Salvador	2745.3
Negro o Guasaule	Nicaragua – Honduras	2371.2
Paz	Guatemala – El Salvador	2647
Sarstún	Guatemala – Belice	2009.5
Suchiate	Guatemala – Mexico	1499.5
Coatán	México – Guatemala	1283.9
Colorado – Corredores	Costa Rica – Panama	1281.8
Moho	Belice – Guatemala	911.9
Temash	Belice – Guatemala	476.4
Jurado	Panama – Colombia	234.3
El Naranjo	Nicaragua – Costa Rica	50.7
Conventillos	Nicaragua – Costa Rica	17.5
Total	23 river basins	219,451.9

Source: Unidad de Investigación en Fronteras Centroamericanas, *Cuencas Internacionales: Conflicto y Cooperación en Centroamérica* (FUNPADEM and UCR, 2000); Cabrera, J. and Cuc, P. *Ambiente, Conflicto y Cooperación en la Cuenca del Río Usumacinta* (FUNPADEM, UNA and UCR, 2002); Comisión Nacional del Agua, *Consejo de la Cuenca de los Ríos Grijalva y Usumacinta*, at 5. Available online at ftp://ftp.consejosdecuenca.org.mx/pub/downloads/docs_basicos/ ejecutivos/24-RGU.pdf (accessed March 22, 2013); Organization of American States, *Gestion Integrada de los Recursos Hídricos y Desarrollo Sostenible de la Cuenca del Río San Juan y su Zona Costera* (Procuenca, 2004). Available online at www.oas.org/sanjuan/spanish/sobre/ descripcion.html (accessed March 22, 2013); World Bank, 'Plan Maestro y para el Desarrollo Integrado y Sostenible de la Cuenca Binacional del Río Paz' (World Bank). Available online at www.asb.cgiar.org/bnpp/docs/Rio_Paz1.doc (accessed March 22, 2013).

4   López Ramírez, A. 'Hydropolitical Vulnerability and Resilience in Central America and the West Indies' in UNEP, *Hydropolitical Vulnerability and Resilience Along International Waters: Latin America and the Caribbean* (UNEP, 2007), at 20.

Nevertheless,

> [t]o know the dimensions of a river basin and the countries with sovereignty over it does not provide enough information to draw conclusions. The more important factor is not simply how much one country possesses—but *how* a basin is divided and the level of dependency. For example, where one country possesses the upper basin and the lower part belongs to another, one might expect the latter country to have a greater role in the management of the basin, since it also has more at risk concerning deterioration. In the same way, if one of the countries depends heavily on the river basin in question (such as El Salvador on the Lempa), one can expect a higher level of involvement by that country in the management of such river basin.

Some countries with international basins show a low interdependency in comparison to their neighbours. This is the case for Panama: Panama shares the small basin of the Jurado River with Colombia. With Costa Rica, Panama shares two larger basins, the Changuinola and Sixaola, which belong in great manner to one of the two countries. Likewise, Guatemala and Honduras, with the Motagua and Chamelecon Basins, demonstrate very little bilateral participation. On the other hand, countries such as Guatemala and Belize, Honduras and El Salvador, and Guatemala and Mexico, are highly interdependent.[5]

International and domestic governance of international watercourses in Central America

Elhance has hypothesized that hydrological interdependence is the basis for regional cooperation.[6] However, as stated before, an international river basin does not necessarily provide a good or sufficient foundation for regional cooperation simply because it crosses a national border. For transboundary cooperation to be effective, solid institutions and legal frameworks should be developed, establishing rules of conduct, defining practices and assigning roles for grappling with collective problems. For the parties involved in the management of international river basins, this collective process implies sharing responsibility, both for making and implementing decisions, as well as providing a fair opportunity to either prevent or manage conflicts. The above assumption can be related to the premise that 'the likelihood and intensity of dispute rises as the rate of change within a basin exceeds the institutional capacity to absorb that change'.[7]

When it comes to governance structures in Central America, there has been a chain of efforts directed towards the better management of river basins within

---

5 Ibid., at 20–1.
6 Elhance, A. P. *Hydropolitics in the Third World: Conflict and Cooperation in International River Basins* (United States Institute of Peace, 1999).
7 Shira, Y., Fiske, G., Giordano, M., Giordano, M., Larson, K., Stahl, K. and Wolf, A. T. 'Geography of International Water Conflict and Cooperation: Data Sets and Applications', *Water Resources Research*, 2004; 40: 1–12, at 3.

countries through the development of new domestic legal frameworks for the water sector. However, the emergence of international institutional and legal frameworks is still limited. There have been several initiatives in some of these basins, such as the San Juan, Paz, Sixaola and Lempa. And yet such initiatives have been mostly part of concrete projects and have not led to the creation of international river basin agreements or organizations.

At the regional level, the Action Plan for the Joint Management of Water in the Central American Isthmus (PACADIRH) was adopted in 1999 to facilitate the governance process – in other words,

> [t]o promote and capture the added value offered by regional initiatives, concentrating on the resolution of the main water resource conflicts, through an integrated focus on conservation and sustainable management of this vital resource, articulating, in a complementary way, the actions being executed at the regional, national and local levels, considering the social, economic and environmental issues.[8]

PACADIRH was signed as part of a broader regional integration agenda, along with action plans on the environment and disaster preparedness. Despite the centrality of water management in all three plans, they were not adequately linked together and, as a result, their impact was mainly political.

The issues of water management, risk and the environment were taken up again in 2009, resulting in two new instruments: the Central American Strategy for the Integrated Management of Water Resources (ECAGIRH) and the Central American Action Plan for the Integrated Management of Water Resources (PACAGIRH). The former instrument outlines the strategic objectives and guidelines for the implementation of a regional policy over a period of ten years. The action plan sets out the actions necessary to implement the Strategy in its first three years.

As their names suggest, both instruments are focused on integrated water resources management. PACAGIRH defines this as 'a process that promotes the management and coordinated development of water, land and related resources, with the purpose of maximizing the resulting social and economic wellbeing in an equitable manner without compromising the sustainability of ecosystems'.[9] PACAGIRH aims to integrate water resources management into national and regional policies, but does little to address governance in shared basins beyond the concept of integrated water resources management. While it is an important element for managing shared basins, it does not, by itself, provide the necessary governance basis to assist states in sharing the watercourse. In this sense, arguably, the UNWC is complementary to PACAGIRH. For example, PACAGIRH Resultado 4.1

---

8 *Plan de Acción para el Manejo Integrado del Agua en el Istmo Centroamericano* (PACADIRH) (SG–SICA, 1999) [Spanish], at 51.

9 Sistema de la Integración Centroamericana (SICA), *Plan Centroamericano para la Gestión Integrada de Recursos Hídricos: PACAGIRH 2010–2012* (SICA, 2009), at 2 (PACAGIRH).

includes a goal of incorporating an integrated risk-management system into regional and national planning processes. Articles 27 and 28 of the UNWC require and outline how states are to work together to prevent harmful conditions related to an international watercourse, and to avoid, mitigate and eliminate harmful effects if an emergency occurs.

At the domestic level, each country's experience in Central America is unique in terms of national institutions and regulatory frameworks. For instance, in Honduras, water management frameworks have emerged as part of decentralization strategies. Among other provisions, Honduras's new water law mentions the creation of river-basin organizations for the country's international watercourses. In turn, Guatemala's remarkable experience in the establishment of domestic river basin institutions for the Amatitlán, Atitlán and Izabal lakes offers lessons and possibilities for the management of international watercourses.[10] However, this pioneering exercise has been restricted to the domestic level – Guatemala has yet to extend this experience to its shared river basins.[11]

Impediments to the good governance of international river basins in Central America

Most Central American boundaries were defined during the nineteenth and early twentieth centuries, within a context of high-pitched nationalistic rhetoric and strong traditional notions of sovereignty. This has contributed to permanent diplomatic confrontation with regard to border areas. At the same time, many of the region's remaining forests – including almost 40 percent of its protected areas – are located in borderlands, along with much of Central America's extraordinary biodiversity. Much of the region's fresh water flows through those political and biological hot-spots and often marks international borders, as outlined in Table 11.3.

Transboundary environmental activities in the region have thus been greatly influenced by border disputes involving natural resources:

> [i]n Central America, the constitution of borders among countries has been a long process derived from conflicts among states over differences in the demarcation of the boundaries. This is highly relevant for understanding the complexity involved in the establishment of [institutions and agreements for transborder environmental cooperation]. The Sarstún, Lempa and San Juan international river basins represent three cases where the creation of

---

10 See Diario de Centro América (DCA) Decreto No 64–96, Ley de Creación de la Autoridad para el Manejo Sustentable de la Cuenca y del Lago de Amatitlán, DCA, September 18, 1996 (Guatemala); DCA Decreto No 133–96, Ley de Creación de la Autoridad pare el Manejo Sustentable del Lago de Atitlaán y su Entorno, DCA, November 27, 1996 (Guatemala); DCA Decreto No 10–98, Ley de Creación de la Autoridad para el Manejo Sustentable de la Cuenca del Lago de Izabal, el Río Dulce y su Cuenca, DCA, February 11, 1998 (Guatemala).

11 Aragón, B., Rodas, O. and Hurtado, P. *Informe Nacional sobre la Situación del Manejo de las Cuencas de Guatemala* (FAO/REDLACH-PAFG, 2002), at 10.

Table 11.3 Borders marked by international river basins in Central America

River	Border	Border length (km)
Usumacinta	Guatemala-Mexico	286.00
Suchiate	Guatemala-Mexico	79.80
Hondo	Mexico-Belice	153.70
Sarstún	Belice-Guatemala	50.19
Motagua	Guatemala-Honduras	28.30
Paz	Guatemala-El Salvador	89.88
Lempa	Honduras-El Salvador	284.28
Goascorán	Honduras-El Salvador	82.11
Negro	Honduras-Nicaragua	46.13
Coco	Honduras-Nicaragua	593.30
San Juan	Costa Rica-Nicaragua	125.80
Sixaola	Costa Rica-Panama	74.15
Juradó	Panama-Colombia	29.95
Central America total		1893.64

Source: See López Ramírez, A. and Hernández, A. 'Fronteras y Ambiente en América Central: Los Desafíos para la Seguridad Regional' (La Agenda de Seguridad en Centroamérica seminar, San Salvador, July 14–15, 2005).

transboundary institutions [has] had to face the challenge of overcoming prevailing tensions. For instance, the Sarstún River is not officially recognized as frontier because of the territorial disagreement between Guatemala and Belize. In the Lempa River Basin, the war between Honduras and El Salvador in 1969 and the territorial tensions produced by the 'Bolsones' may eventually represent a challenge that must be overcome. Finally, in the case of the San Juan River, the rights and conditions for navigation are still a factor of dispute between Costa Rica and Nicaragua.[12]

The second problem to be solved is the type of institution required to cope with environmental change [as a] transnational threat. The starting point is that current structures in [Central America] do not correspond to the new reality, because most of them are based on national considerations. The transborder issues make necessary the adoption of new regulatory frameworks that in most cases reduce the internal territorial power of the sovereign state, but at the same time guarantee for the state an important role in the management of such shared ecosystem[s] at a regional level.[13]

Indeed, Central American states still feel a strong link with the notions of traditional sovereignty, territoriality and national interests. Yet, the reality is that 'transboundary environmental problems necessarily undermine state sovereignty. Hence, while states may claim sovereignty over natural resources, they have come

---

12 López Ramírez, (note 4), at 42.
13 López Ramírez, (note 2), at 293 (internal citations omitted).

under mounting pressure to manage their resources according to international norms'.[14] Nonetheless, the above issues remain serious challenges for the construction of new institutional and legal frameworks dealing with international river basins and interstate cooperation in the region and, therefore, for the widespread ratification of the UNWC.

The role of domestic governance structures

A key factor for understanding the challenge in developing institutional and legal frameworks for international river basins in Central America is the national context – the so-called structure problem. The national context can either enable or constrain transboundary water cooperation and, more specifically, the ratification and implementation of the UNWC.

Countries get involved in an international regime to better estimate costs and benefits and, therefore, avoid suboptimal outcomes.[15] However, in Central America, most legal and policy instruments governing freshwater resources and ecosystems fail to recognize the important benefits of having a trans-border environmental regime for dealing with international river basins.

Understanding the process of regional environmental governance requires an analysis of domestic structures. Such structures are likely to determine: a) the availability of channels into the political system for transnational actors; and b) the possibility of, and requirement for, winning coalitions to change policies. Domestic structures and international institutions are likely to interact in determining the ability for trans-border cooperation. Thus, the more an issue-area is regulated by international norms of cooperation, the more permeable state boundaries become for transnational activities. This seems to be one of the reasons why some Central American states may be reluctant to accede to the UNWC.

Although Central America's record of entering into international and regional environmental conventions is remarkable, the region's engagement is lacking in the water sector. For instance, even environmental policy makers are unfamiliar with the UNWC – as indicated by the lack of endorsement of the UNWC by any Central American country, and confirmed through meetings and workshops involving government officials from across the region.[16]

Domestic structures are also responsible for the lack or inadequacy of modern water legislation in Central America. Across the region, legal and institutional arrangements related to water resources are characterized by fragmentation and the disperse nature of norms and competencies. Laws, regulations and policies

14 Ibid.
15 Hasenclever, A., Mayer, P. and Rittberger, V. *Theories of International Regimes* (Cambridge University Press, 1997), at 45.
16 See, e.g. López, A. and Porta, M. A. 'Estudio Línea Base Convención de la ONU Sobre los Cursos de Aguas Internacionales: El Salvador' (2011) (on file with the authors); López, A. and Porta, M. A. 'Estudio Línea Base Convención de la ONU Sobre los Cursos de Aguas Internacionales: Honduras' (2011) (on file with authors); López, A. and Porta, M. A. 'Estudio Línea Base Convención de la ONU Sobre los Cursos de Aguas Internacionales: Costa Rica' (2011) (on file with the authors).

addressing water resources are not prominent in national agendas, despite constant political arguments in favor of protection of and access to water. Even in countries like Nicaragua and Honduras, which have approved updated regulatory frameworks, their effectiveness remains uncertain because of the complex challenges that the region faces in managing water and sanitation, in addition to other related areas, such as health, infrastructure, environment, natural disasters and poverty alleviation. In both cases, references to international river basins are marginal at best.

Water in Central American constitutions

Constitutional mandates related to water resources are limited in Central America. Although several environmental and social organizations have recently put forward actions aimed at fostering constitutional reforms that would guarantee access to water as a human right, the response has been slow. Still, constitutional pillars often emphasize state ownership of water and entail subsequent legal instruments for the establishment of water rights by ministries or decentralized institutions.

In the case of Costa Rica, the Constitution[17] does not specifically mention water as a public good. But the systematic interpretation of Articles 6, 50 and 121(14) of the Constitution, combined with Article 8 of the Mining Code,[18] Articles 1–3 and 17 of the Law of Waters,[19] and Article 50 of the Environmental Law[20] justifies the recognition of: a) fundamental rights to water; b) the public domain over such waters; and c) the right to a healthy and ecologically balanced environment.[21] Accordingly, in 1996, the Constitutional Court declared that water is a human right.

El Salvador's Constitution declares a social interest in the protection, restoration and use of natural resources, including national waters.[22] For example, expropriation may take place for the public good, such as supplying water, as per Article 106. In addition, Article 117 of the Constitution addresses environmental conservation and the rational use of water and other natural resources. Yet, the country's fragmented water-related legislation leads to an unclear definition of public competencies and thus to confusion.

The Honduran Constitution[23] does not contain explicit norms about water, either from environmental or human rights perspectives. The only references to water can be found in Articles 10 and 12; the former establishes the state's ownership over all territories situated within its internal waters; Article 12 declares

---

17   Constitución Política de la República de Costa Rica (Edición Actualizada, Publicaciones Jurídicas, 1994).

18   Asamblea Legislativa de Costa Rica Ley 6797, Código de Minería, October 4, 1982 (Costa Rica).

19   Asamblea Legislativa de Costa Rica Ley 276, Ley de Aguas, August 26, 1942 (Costa Rica).

20   Asamblea Legislativa de Costa Rica Ley 7554, Ley Orgánica del Ambiente, October 4, 1995 (Costa Rica). Article 50 states that water is in the public domain and that its conservation and use are of social interest. Article 51 sets criteria for the conservation and sustainable use of water.

21   Programa Hidrológico Internacional, *Marco Legal e Institucional en la Gestión de los Sistemas Acuíferos Transfronterizos en las Américas* (UNESCO/OEA ISARM, 2008), at 33–4.

22   Constitución Política de la República de El Salvador de 1983.

23   Constitución Política de la República de Honduras, Decreto No. 131, January 11, 1982.

Honduras' sovereignty over underground and territorial waters and the continental platform. Article 12 also reaffirms the hierarchy of international law with respect to the right of free navigation, and recognizes the validity of relevant international legal instruments.

Legislation related to water resources in Central America

In recent years, Nicaragua and Honduras have approved new legislation on water resources. The rest of the countries, especially El Salvador, Costa Rica and Guatemala, have been unable to push their respective bills through Congress, owing to political disputes and a perception that any change to water legislation would address privatization interests.

Honduras and Nicaragua: New acts under implementation[24]

Honduras' 2009 General Water Act[25] replaces the 1927 National Water Use Law, which, along with 20 other legal instruments, governed the nation's water resources for almost a century. Unlike previous reforms, the new water law introduced the integrated management of water resources. As a modern piece of legislation, the General Water Act promotes institutional coordination at national and local levels, and creates several bodies in charge of coordination, implementation and technical functions. However, because of the short time since its adoption and recent government changes in the country, it is difficult to assess the impact of the new law.

Honduras' General Water Law affirms the public domain of national waters, lakes, lagoons, underground waters and maritime platforms, as well as over natural canals, dams, continuing courses of water and aqueducts.[26] However, the issuance of regulations and the transfer of personnel and resources from previous public entities to new ones, such as the Water Authority, as required under Articles 102 and 104, have yet to take place.

The General Water Law is notable for four strategic innovations. First, it incorporates a number of guiding pillars and concepts of integrated water resources management that align with the Dublin Principles,[27] and reiterates the rules of ownership and water use that have been part of Honduras' legal tradition since 1927. Second, Title VI outlines the regulation, planning and sectoral strategies, and creates a registry of water resources. Third, the creation of an economic system, in Title VII, which includes payments for environmental services and a National Water Fund to support conservation projects, reflects the value of water.

---

24 This section focuses on Honduras' General Water Act, keeping in mind that Nicaragua's new legislation is quite similar.
25 La Gaceta Decreto No 181-2009, Ley General de Aguas, December 14, 2009 (Honduras).
26 Fiallos Rodas, M. I. [Member of Legislative Assembly] 'Report on the New Water Law' (National Parliament, August 2009) (on file).
27 International Conference on Water and the Environment (Dublin, January 26–31, 1992), *Dublin Statement on Water and Sustainable Development* (January 31, 1992). Available online at www.wmo.int/pages/prog/hwrp/documents/english/icwedece.html (accessed March 22, 2013).

Finally, the Act creates several new institutions. The National Water Resource Council assures the participation in decision making of the highest levels of government, the private sector, municipalities and universities. The Water Authority, as a decentralized body of the Ministry of Natural Resources and Environment, is responsible for enforcing water sector policies. Regional agencies and the National Institute of Water Resources have executive functions. At the local level, the role assigned to basin organizations is remarkable, especially with regard to the watershed councils.[28] For the purpose of international river basin management, the final paragraph of Article 22 states: 'In the river basin councils for shared management of frontier and cross-border rivers, national government representation should include a representative of the Secretary of State for Foreign Affairs'. This implies recognition of the need for the cooperative management of transboundary basins, and is thus an effective step toward ensuring the sustainability of such basins and their ecosystems.

Although in its infancy, Honduras' new water law is an important step toward improved water management, both in Honduras and regionally. By consolidating water management into one agency, the new law streamlines the process for adopting new legal instruments (like the UNWC) and amending existing ones. The new law also demonstrates political will to tackle the issue of transboundary water governance, going as far as to mention international watercourses – a unique provision among water laws in this region. How this law, and a similar law in Nicaragua, will work in practice remains to be seen, but their mere existence inspires confidence.

Costa Rica and El Salvador: Dispersion and absence of political decision

Water resources management in Costa Rica and El Salvador is dispersed throughout numerous laws and regulations, many of them outdated. Costa Rica has approximately 115 laws and executive orders governing water resources.[29] The 1942 Water Act[30] was issued when Costa Rica had less than a third of its current population and its main economic activity was agriculture. Since then, the main productive use of water has shifted to power generation.

These various laws empower, to some extent, different actors in the management of water resources, resulting in complementary, but also overlapping, functions. Attempts to consolidate water resources management have been unsuccessful thus far. In the first ten years of the 21st century, several initiatives on a water resources act have been presented and discussed, with no progress in Parliament. An initiative and public policy package incorporating principles of integrated water resources management released in mid-2006 had poor results.

---

28 La Gaceta Decreto No 181-2009, Ley General de Aguas, Chapter 2, Title II, Articles 10, 16–17, 19–24, December 14, 2009 (Honduras).
29 Aguilar, A. *Manual de Regulación Jurídica para la Gestión del Recurso Hídrico en Costa Rica* (CEDARENA, 2001).
30 Asamblea Legislativa de Costa Rica Ley 276, Ley de Aguas, August 26, 1942 (Costa Rica).

Like Costa Rica, El Salvador's legal framework for water resources is scattered throughout various pieces of legislation, despite a 1981 Act on Integrated Water Resources Management[31] seeking to coordinate the mandates of different entities. Several bills were under review by the Technical Secretariat of the Presidency of the Republic during the Saca Administration, which ended in 2009. The new authorities have shown interest in promoting a public policy for water resources, but no actions have resulted to date. The results of having several institutions responsible for managing water resources are jurisdictional disputes, overlapping functions and rivalries, all leading to poor management of the resource.

Without clarity as to the authority in charge of water resources at the national level, becoming a party to an international agreement on watercourses, such as the UNWC, would be challenging. Rather than just seeking approval from one or two ministries, the President must seek input from all ministries that might be affected by the instrument's ratification. Even determining which ministries those might be can be a challenge. Furthermore, the costs may fall on one ministry, while other ministries reap most of the benefits. In other words, in the absence of a widespread culture around integrated water resources management, supported by the appropriate legal and institutional arrangements, it may be difficult to take full advantage of and effectively implement the UNWC, even if the countries at hand were to accede to it. Thus, when raising awareness and knowledge of the UNWC in Central America, the national context must always be taken into account.

Potential role and relevance of the UN Watercourses Convention in Central America

In Central America, as indicated above, most international basins lack adequate legal protection.[32] As a region dominated by 23 international watercourses, with at least three of these basins within each country, Central America could benefit from the widespread ratification and effective implementation of the UNWC, as a solid legal framework for transboundary water cooperation. The UNWC lays out basic standards and rules for cooperation on the use, management and protection of international watercourses, thereby providing legal certainty and stability to the relations between co-riparian states. This is especially true, given that creating individual basin-level agreements for each of those watercourses is unrealistic at the moment.

The issue of lack of adequate legal protection for the region's international watercourses can be exemplified in various ways: a) binding water agreements are

---

31 Diario Oficial Decreto No. 886, Ley sobre Gestión Integrada de los Recursos Hídricos, DO, December 2, 1981 (El Salvador).

32 An exception to this may be the Trifinio Treaty for the Lempa River Basin. Tratado entre las Republicas de El Salvador, Guatemala y Honduras para la Ejecución del Plan Trifinio (adopted October 31, 1997, entered into force May 28, 1998). Available online at http://www.gaiaelsalvador.org/files/Region%20Trifinio/Tratado%20Trifinio.pdf (accessed March 22, 2013). This is discussed in greater detail in Chapter 17, as it relates to the UNWC.

largely lacking;[33] b) informal cooperation, within the framework of projects, ad hoc bi-national commissions and/or non-binding statements, prevails across the region; c) where cooperation frameworks have been adopted, most are of limited scope, focusing on border demarcation or the establishment of border institutions, rather than laying out the rules of the game for the use, management and protection of shared freshwaters.

In this sense, for example, the UNWC could add legal value to those frameworks in the region that simply create international governance bodies: bi-national commissions exist between Costa Rica and Panama,[34] El Salvador and Guatemala,[35] Mexico and Guatemala,[36] and Mexico and Belize.[37] El Salvador, Guatemala and Honduras have a tri-national commission guiding policies and projects along their shared border.[38] Most of these border commissions may have an international watercourse as part of their geographic scope, but were not designed to foster the transboundary water management of the basin itself as their central goal. While, therefore, the establishment of those bodies is an important step towards enabling water cooperation, the principles and procedures contained in the UNWC are still needed to guide the work of such bodies and determine how member states are to cooperate under their umbrella.

The UNWC also provides the basis for future agreements specific to individual watercourses, where such agreements are absent. As stated above, transboundary water governance in Central America is limited. The UNWC offers an opportunity to develop that field.

Furthermore, while the Convention focuses specifically on international watercourses, its principles – particularly its emphasis on cooperation – could apply

33 Examples of basins not covered by any cooperative frameworks include the Belice, Moho, Temash and Sarstún rivers, shared between Guatemala and Belize; the Chamelecón River, shared by Guatemala and Honduras; the Goascorán River, shared by El Salvador and Honduras; the Choluteca, Coco and Negro rivers, shared between Nicaragua and Honduras; the Conventillos and Naranjo Rivers, shared by Costa Rica and Nicaragua; and the Colorado-Corredores and Changuinola, shared between Costa Rica and Panama. See, e.g. Ulate, A. H., Ramírez, A. L. and Elizondo, A. J. *Gobernabilidad e Instituciones en las Cuencas Transfronterizas de América Central y México* (FLACSO, 2009), at 96.

34 Convenio entre el Gobierno de la Republica de Costa Rica y el Gobierno de la Republica de Panamá sobre Cooperación para el Desarrollo Fronterizo y Su Anexo (adopted May 3, 1992, entered into force July 24, 1995) at Article 2.

35 Memorandum de Entendimiento para la Creación de la Comisión Binacional Salvadoreña-Guatemalteca [Memorandum of Understanding for the Creation of the Salvadoran-Guatemalan Binatinal Commission] (adopted August 19, 2000, entered into force April 14, 2010) at Article 1.

36 Acuerdo para la Creación de la Comisión Internacional de Límites y Aguas entre los Estados Unidos Mexicanos y la República de Guatemala (adopted November 2, and December 21, 1961). Available online at www.sre.gob.mx/cilasur/images/stories/canjenotasmexguat.pdf (accessed March 22, 2013); Tratado para Fortalecer a la Comisión Internacional de Límites y Aguas (adopted July 17, 1990). Available online at www.sre.gob.mx/cilasur/images/stories/tratadocila.pdf (accessed March 22, 2013).

37 Canje de Notas que Crea la Comisión Internacional de Límites y Aguas entre México y Belice (adopted July 6 and November 1993). Available online at www.sre.gob.mx/cilasur/images/stories/canjenotasmexbel.pdf (accessed March 22, 2013).

38 For more information on this commission, see Chapter 17.

more broadly to the management of other transboundary natural resources, as they affect or may be affected by those river systems.

The lack of international river basin organizations in the region also supports broad ratification of the UNWC. While several basin initiatives exist, they have mostly been part of specific projects and do not address the broader use, management and protection of freshwater. The UNWC can fill this gap as a blanket Convention, until separate agreements have been adopted for all the region's transboundary watersheds. Having one common instrument across the region would also make implementation and enforcement easier and help to develop international water law in a coherent manner.

The UNWC may also have a role to play with regard to fostering greater regional integration among Central American countries. The Convention promotes cooperation among riparian states, and encourages the creation of joint bodies to monitor and manage shared watercourses. In Central America, tensions between neighbors are long-standing, transboundary institutions are limited and state sovereignty remains a preeminent concern. For all these reasons, regional integration has proceeded slowly. Implementation of the UNWC could help to build trust among countries progressively, and generate positive externalities beyond the environmental sector. For example, the joint construction of hydroelectric projects, applying the principles of the UNWC to ensure watercourses, ecosystems and dependent communities are protected, can provide social and economic benefits to all riparian countries.

The UNWC's provisions on dispute resolution are yet another argument in favor of its ratification and implementation. Many international rivers in Central America delineate borders, some of which are still contentious. For example, Costa Rica and Nicaragua have yet to resolve their dispute over navigation rights on the San Juan River, despite several rulings by the International Court of Justice. The Sarstún River is not recognized as the official border between Guatemala and Belize because of outstanding territorial claims. By providing a detailed process for resolving disputes over international watercourses, the UNWC can potentially provide guidance and strengthen foundational capacity and mutual confidence as contributions toward tackling existing border disputes. The Convention's dispute prevention and settlement mechanisms can also help prevent future disagreements from escalating, which is an important consideration in a region where border disputes have often led to the use of force. The region's growing economies and populations, as well as its rapid urbanization, will likely contribute to future disputes over water resources beyond border issues. Thus, the UNWC can help mitigate existing and future disagreements in the region by providing a detailed procedure for preventing and resolving disputes.

Finally, the UNWC can serve as a model for national legislation. Many water laws in the region are based on outdated concepts of navigation and sovereignty, and do not take account of more recent issues, such as pollution and sustainable water use. Despite this, parliaments in the region have generally postponed decisions on water policy in favor of other more pressing domestic issues. Arguably, the UNWC could impel national legislatures to update national water laws, regulations and policies to better enable interstate cooperation, regional integration and benefit-sharing. The mobilization of congressional groups, academics and non-profit organizations in a joint effort to facilitate the ratification of the UNWC would create a positive climate for other reforms and a proactive awareness of water governance.

Conclusion

The UNWC has an important role to play in Central America. Despite the large number of international river basins in the region, few transboundary institutions and agreements exist to protect and manage those resources. Historical tensions have further challenged the establishment of such institutions, while national governments have not yet embraced transboundary cooperation. Domestically, water policy is a low priority and dispersed among many agencies. The UNWC, as a flexible and overarching legal framework, can fill in these gaps, providing protection for vulnerable freshwater resources, as well as encouraging greater regional cooperation.

12 Nile River Basin

Musa Mohammed Abseno

Africa is a continent with a number of transboundary rivers and lakes, some of which are shared by more than ten countries.[1] Of more than 276 international watercourses and aquifers, 63, covering 64 percent of the continent's land mass and accounting for 93 percent of its total surface water resources, are found in Africa.[2] Following the adoption of the 1997 UN Watercourses Convention (UNWC), a number of watercourse agreements in Africa have incorporated a number of basic principles of the Convention. This chapter highlights transboundary challenges facing the Nile River Basin and examines the evolution of treaty laws. The chapter specifically focuses on examining the Nile Basin Cooperative Framework Agreement (CFA), in light of the UNWC, to provide insights on its contribution to a permanent legal and institutional arrangement.[3]

The Nile River Basin

Hydrology

The River Nile travels more than 6700 km, covering three million km^2 and criss-crossing eleven riparian countries, eight of which are East African States.[4] The water resources of the Nile Equatorial region include one of the world's great complexes of lakes, wetlands and rivers.[5] Accordingly, the Luvironzia, Ruvuvu and Kagera Rivers and Lakes Tanganyika, Kivu, Edward, Albert, Kyoga, and Victoria, all form the

---

1 Hoque, A. F. *Transboundary River/Lake Basins Water Development in Africa: Prospects, Problems and Achievements* (UN Economic Commission for Africa, 2000).

2 Abseno, M. 'How Does the Work of the General Assembly and the ILC on the Law of Non-Navigational Uses of International Watercourses Contribute Towards Basin-Wide Legal Framework for the Nile Basin' (LLM thesis, University of Dundee, 2009), at 2.

3 'Agreement on the Nile River Basin Cooperative Framework' (opened for signature May 14, 2010). Available online at www.internationalwaterlaw.org/documents/regionaldocs/ Nile_River_Basin_Cooperative_Framework_2010.pdf (accessed March 22, 2013).

4 The Nile River is shared by Burundi, the Democratic Republic of Congo, Egypt, Eritrea, Ethiopia, Kenya, Rwanda, South Sudan, Sudan, Tanzania and Uganda.

5 Nile Basin Initiative, 'The Nile Equatorial Lakes Subsidiary Action Programme' (NBI, November 2, 2010). Available online at http://nilebasin.org/newnelsap (accessed March 22, 2013).

great Nile River system.[6] In addition, the Abbay, Baro-Akobo, and the Tekeze Rivers in Ethiopia form the Eastern Nile system, which contributes 86 percent of total Nile flow.[7] Table 12.1 gives key statistical facts about the Nile Basin.

Table 12.1 Key statistical facts on the Nile Basin

Basin area	3173 × 103 km²
Location	−4°S to 31°N and 24°E to 40°E
Riparian states	Burundi, Democratic Republic of Congo, Egypt, Eritrea, Ethiopia, Kenya, Rwanda, Tanzania, South Sudan, Sudan and Uganda
Main tributaries	Victoria Nile/Albert Nile, Bahr El Jabel, White Nile, Baro Pibor-Sobat, Blue Nile, Atbara, Bahr El Ghazal
River length	6850 km
Estimated navigable length	4149 km
Major lakes within the basin	Lake Victoria, Lake Tana, Lake Kyoga, Lake Albert
Population (total in all the Nile countries)	437 million
Population within the Nile Basin	238 million (54%)
Precipitation	Max annual: 2060 mm/year in Uganda
	Min annual: 0 mm/year in Egypt
Mean annual flow (discharge) (km³/year) at Aswan	84 × 109 m³
Discharge/unit area	28 × 103 m³/km²
Main consumptive water use	Agriculture

Source: Information adapted from Nile Basin Initiative, 'The Nile River' (www.nilebasin.org/newsite/index.php?option=com_content&view=article&id=138%3Athe-nile-river&catid=36%3Athe-nile-river&Itemid=75&lang=en).

Conflict and cooperation

Conflicts over the use, development and protection of transboundary resources in the Nile persist, while institutional and legal coordination for their resolution in an equitable and sustainable manner remains a gradual process. One of the unique aspects surrounding the Nile is the large number of riparian countries sharing the river and its resources. The Nile is shared by eleven countries, two of them – Eritrea (1993) and South Sudan (2011) – newly independent, which has made the scope of cooperation and challenges dynamic and more complex. Despite enormous challenges,

---

6 Collins, R. *The Waters of the Nile: Hydropolitics and the Jonglei Canal, 1900–1988* (Markus Wiener Publishers, 1990), at 5.

7 Eastern Nile Technical Regional Office, *Water Atlas of the Blue Nile Sub-Basin* (ENTRO, 2006) (draft copy on file).

however, significant development has been made in the basin, through the Nile Basin Initiative (NBI), and subsidiary action programmes: namely the Eastern Nile Subsidiary Action Programme (ENSAP), the Nile Equatorial Subsidiary Action Programme (NELSAP) and the new legal and institutional framework, the CFA.[8]

Sub-basin initiatives function outside the existing NBI mechanism, such as the Lake Victoria Basin Commission, established under the 2003 Protocol for Sustainable Development of Lake Victoria Basin. These initiatives offer additional strength to the new trend in legal and institutional building.[9] These initiatives are the result of the outcome of the previous efforts that were able to establish institutions, such as Organization for the Management and Development of the Kagera River in 1977,[10] the Technical Cooperation Committee for the Promotion of the Development and Environmental Protection of the Nile Basin (TECCONILE), in 1996, and others.

Table 12.2 The Nile Basin countries

Country	Total country area (km²)	Basin area (km²)	Total population (millions)
Burundi	27,830	13,250	8.6
Democratic Republic of Congo	2,344,860	20,191	67.8
Egypt	1,001,450	303,084	82.5
Eritrea	121,890	24,578	5.4
Ethiopia	1,100,000	364,925	84.7
Kenya	580,370	70,248	41.6
Rwanda	26,340	20,823	10.9
Sudan	2,505,810	1,993,082	44.6
Tanzania	945,090	120,768	46.2
Uganda	235,880	239,468	35.5

Source: Adapted from FAO, *Irrigation Potential in Africa: A Basin Approach* (FAO, 1997); Central
  Intelligence Agency, 'The World Fact Book: Population'
  (https://www.cia.gov/library/publications/the-world-factbook/index.html)

Key challenges

A number of challenges exist in the realization of cooperation towards equitable use, protection and conservation of the Nile River Basin. These challenges include:

8 Extraordinary Meeting of Nile Council of Ministers (Dar es Salaam, Tanzania, February 22, 1999), 'Agreed Minutes' (1999).
9 East African Community, *Protocol for Sustainable Development of Lake Victoria Basin* (adopted November 29, 2003). Available online at www.internationalwaterlaw.org/documents/regionaldocs/Lake_Victoria_Basin_2003.pdf (accessed March 22, 2013) (2003 Lake Victoria Basin Protocol).
10 Agreement for the Establishment of the Organization for the Management and Development of the Kagera River Basin (adopted August 24, 1977) 1089 UNTS 165 (1977 Kagera River Basin Agreement).

- *poverty*: most of the Nile Basin countries live under extreme poverty with one of the lowest GDP per capita. The development of water resources and alleviation of food security is frustrated, owing to high variability in rainfall in many parts of the basin, and the lack of infrastructure for the storage and regulation of water for productive uses.[11]
- *instability*: hydropolitical insecurity is responsible for mistrust, lack of confidence-building measures and constant adverse positions over the resolution of controversial legal disputes.[12] Past and present political tensions and civil and political conflicts in the region have continued to have detrimental influence on the attainment of equitable and sustainable water use.[13]
- *population growth*: rapid population growth is evident in the Nile Basin, where countries such as Ethiopia stand as the second largest population in Africa, with an estimated population exceeding 90 million.[14] At the basin level, the current population that is dependent on its waters is estimated at more than 300 million – a number expected to double to 600 million in the next 25 years.[15]
- *environmental degradation and climate change*: despite the basin's diverse flora, fauna and rich ecosystems spanning several bio geographical areas, there are major environmental threats to the basin's resources, including deforestation, soil erosion, overgrazing, desertification, loss of biodiversity, flood, drought and water pollution.[16] In the Blue Nile, severe deforestation can be observed from the loss of Western Ethiopian forest cover, which as a whole decreased from sixteen percent to two percent of the land area between 1950 and the late 1980s.[17] Lake Victoria is under considerable pressure from a variety of interlinked human activities, resulting in enormous environmental changes. Factors that contribute to rapidly evolving changes in Lake Victoria and other shared

---

11 World Bank, 'Nile Basin Initiative Institutional Strengthening Project: Project Information Document', Report No AB3787 (World Bank, 2008). Available online at www-wds.worldbank.org/external/default/WDSContentServer/WDSP/IB/2008/10/02/000076092_2 0081003122217/Rendered/PDF/08100020Projec1nt010Appraisal0Stage.pdf (accessed March 22, 2013).

12 Rogers, P. 'Water Governance, Water Security and Water Sustainability' in Rogers, P., Llamas, R. and Martínez-Cortina, C. (eds), *Water Crises: Myth or Reality: Marcelino Botin Water Forum 2004* (Taylor and Francis/Balkema, 2006), 3–35, at 22.

13 Jacobs, L. 'Sharing the Gifts of the Nile: Establishment of a Legal Regime for Nile Waters Management', *Temporary International and Comparative Law Journal*, 1993; 7: 95–122, at 118.

14 CIA, 'The World Fact Book: Ethiopia' (CIA). Available online at https://www.cia.gov/library/publications/the-world-factbook/index.html (accessed March 22, 2013).

15 See World Bank, 'Project Appraisal Document: Efficient Water Use for Agricultural Production Project', Nile Basin Initiative: Shared Vision Programme, Report No 34084-AFR (World Bank, 2005), at 77, Available online at www-wds.worldbank.org/external/default/WDSContentServer/WDSP/IB/2005/11/29/000090341_20051129093151/Rendered/PDF/340840AFR0Shar1Proje ct0PAD01PUBLIC1.pdf (accessed March 22, 2013).

16 World Bank, *Nile River Basin: Transboundary Environmental Analysis* (World Bank, 2001), at 18.

17 Ibid., at 19.

waters of the Nile include overfishing, siltation, erosion, deforestation, alien species, pollution, eutrophication and climate change.[18]

- *the lack of legal and institutional frameworks*: as elsewhere in Africa, the legal regime in the Nile is dominated by the legacy of colonial treaties. There are no exemplary cooperative legal practices for resolution of existing and future water disputes. The absence of a permanent legal and institutional framework is particularly evident, although the recent transitional arrangement of the NBI has helped the process towards a basin-wide framework (CFA).[19] Among the relevant sub-basin legal arrangements in the region is the 2003 Protocol for Sustainable Development of Lake Victoria Basin, which established the Lake Victoria Basin Commission and incorporated a number of basic principles from the text of UNWC.[20] Therefore, while a comparative analysis of the UNWC and the current CFA can highlight key contributions by the UNWC, a further highlight on the evolution of existing treaty regime in the basin is essential.

The evolution of treaty law and institutions in the Nile

The Nile is one of the international rivers with numerous treaties pertaining to the right of use of its water resources, 25 of which are found registered under the Transboundary Freshwater Dispute Database.[21] A number of these consumptive water-use treaties were made by Great Britain as a contracting party on behalf of Egypt and Sudan. The terms of most of the treaties virtually forfeit the rights of the upper riparian states. Table 12.3 lists the existing treaties relating to the Nile Basin, dating back to 1891. A number of these treaties, however, have never been executed. However, a number of these treaties have never been executed, mainly as a result of the end of colonial rule in the basin, while their binding force became increasingly untenable under the development of the principles of modern international water law. However, the agreements, notably, the 1929 Nile Waters Agreement, continue to dictate the current legal relationships concerning the distribution and utilization of the Nile waters between lower and upper riparian states today.[22] One of the most controversial provisions of the 1929 Nile Waters Agreement stipulates that:

---

18 Odada, E., Olago, D. O., Kulindwa, K., Ntiba, M. and Wandiga, S. 'Mitigation of Environmental Problems in Lake Victoria, East Africa: Causal Chain and Policy Options Analyses', *Ambio*, 2004; 33 (1/2): 13–23, at 14.

19 Agreement on the Nile River Basin Cooperative Framework (note 3).

20 See 2003 Lake Victoria Basin Protocol (note 9), at Article 33.

21 'International Freshwater Treaties Database' (Oregon State University). Available online at http://ocid.nacse.org/tfdd/treaties.php (accessed March 22, 2013).

22 Brunnée, J. and Toope, S. 'The Changing Nile Basin Regime: Does Law Matter?', *Harvard International Law Journal*, 2003; 43: 105–59, at 123.

Table 12.3 Nile Basin treaties

Date	Basin/Sub-Basin	Signatories	Treaty Name
April 15, 1891	Nile	Great Britain, Italy	Protocol between Great Britain and Italy for the demarcation of their respective spheres of influence in Eastern Africa
March 18, 1902	Nile	Ethiopia, Great Britain	Exchange of notes between Great Britain and Ethiopia
May 15, 1902	Nile, Sobat	Ethiopia, Great Britain	Treaties between Great Britain and Ethiopia, relative to the frontiers between Anglo-Egyptian Soudan, Ethiopia, and Erythroea (railway to connect Soudan with Uganda)
May 9, 1906	Nile	Independent State of Congo, Great Britain	Agreement between Great Britain and the Independent State of the Congo, modifying the agreement signed at Brussels 12 May 1894, relating to the spheres of influence of Great Britain and the Independent State of the Congo in East and Central Africa
December 13, 1906	Nile	France, Great Britain, Italy	Agreement between Great Britain, France, and Italy respecting Abyssinia
December 20, 1925	Lake Tsana	Great Britain, Italy	Exchange of notes between the United Kingdom and Italy respecting concessions for a barrage at Lake Tsana and a railway across Abyssinia from Eritrea to Italian Somaliland
May 7, 1929	Nile	Egypt, Great Britain	Exchange of notes between His Majesty's government in the United Kingdom and the Egyptian Government in regard to the use of the waters of the River Nile for irrigation purposes
November 22, 1934	Nile	Belgium, Great Britain	Agreement between the United Kingdom and Belgium regarding water rights on the boundary between Tanganyika and Ruanda-Urundi
December 7, 1946	Nile	Egypt, Great Britain	Exchange of notes constituting an agreement between the United Kingdom of Great Britain and Northern Ireland and Egypt regarding the utilization of profits from the 1940 British government cotton buying commission and the 1941 joint Anglo-Egyptian cotton buying commission to finance schemes for village water supplies
May 31, 1949	Nile	Egypt, Great Britain	Exchanges of notes constituting an agreement between the government of the United Kingdom of Great Britain and Northern Ireland and the government of Egypt regarding the construction of the Owen Falls Dam, Uganda

Table 12.3 continued

Date	Basin/Sub-Basin	Signatories	Treaty Name
January 31, 1950	Nile	Egypt, Great Britain on behalf of Uganda	Exchange of notes constituting an agreement between the government of the United Kingdom of Great Britain and Northern Ireland on behalf of the government of Uganda and the government of Egypt regarding cooperation in meteorological and hydrological surveys in certain areas of the Nile Basin
July 31, 1952	Nile	Egypt, Great Britain	Exchange of notes constituting an agreement between the government of the United Kingdom of Great Britain and Northern Ireland and the government of Egypt regarding the construction of the Owen Falls Dam in Uganda
November 8, 1959	Nile	Sudan, United Arab Republic	Agreement between the government of the United Arab Republic and the government of Sudan for full utilization of the Nile waters
August 24, 1977	Kagera	Burundi, Rwanda, Uganda, Tanzania	Agreement for the establishment of the organization for the management and development of the Kagera River Basin (with attached map), concluded at Rusumo, Rwanda
May 18, 1981	Kagera	Burundi, Rwanda, Uganda, Tanzania	Accession of Uganda to the agreement pertaining to the creation of the organization for the management and development of the Kagera River Basin
July 1, 1993	Nile	Arab Republic of Egypt, Ethiopia	Framework for general co-operation between the Arab Republic of Egypt and Ethiopia
August 5, 1994	Lake Victoria	Kenya, Tanzania, Uganda	Agreement to initiate programme to strengthen regional coordination in management of resources of Lake Victoria
February 22, 1999	Nile	Burundi, Kenya, Egypt, Ethiopia, Rwanda, Sudan, Tanzania, Uganda	Agreed Minutes, Extraordinary Meeting of Nile Council of Ministers, Dar es Salam, United Republic of Tanzania
November 29, 2003	Lake Victoria	Kenya, Uganda, Tanzania	Protocol on Sustainable Development of Lake Victoria Basin

Source: Adapted from 'Transboundary Freshwater Dispute Database' (Oregon State University; www.transboundarywaters.orst.edu/database); Extraordinary Meeting of Nile Council of Ministers (Dar es Salaam, Tanzania, 22 Feb 1999), 'Agreed Minutes'.

[s]ave with the previous agreement of the Egyptian Government, no irrigation or power works or measures are to be constructed or taken on the River Nile and its branches, or on the lakes from which it flows, so far as all these are in the Sudan or in countries under British administration, which would, in such a manner as to entail any prejudice to the interests of Egypt, either reduce quantity of water arriving in Egypt, or modify the date of its arrival, or lower its level.[23]

According to the above text, the agreement grants a veto power to Egypt over developments by upstream states.[24] In 1959, Egypt and Sudan concluded an agreement which divided the entire Nile water between the two countries: 55.5 billion cubic metres (m^3) to Egypt and 18.5 billion m^3 to Sudan.[25] The shares are recognized by the two parties as their respective 'historical rights', with no mention of correlative entitlements to the use of water by the upper riparian states; as a result of this, the treaty was rejected by the upstream states.[26]

Among the post-colonial treaties in the basin which contributed to the evolution of institutional arrangements is the 1977 Kagera River Basin Agreement,[27] which was considered as a modern river basin structure, but which, however, lacked appropriate legal principles governing the rights and obligations of the riparian parties to the treaty.[28] Another sub-basin treaty is the 2003 Lake Victoria Basin Protocol, which operates independently of the NBI structure, and under which the East African Community has adopted a number of basic substantive and procedural rules from the UNWC.[29]

Institutional growth on transboundary waters in the basin has been evolutionary both in its essence and form. It began with limited objectives, such as gathering data and information in the Lakes Basin (Hydromet-1967)[30] and later to a basin-specific

---

23 1929 Nile Waters Agreement, at Para 4(b).

24 McCaffrey, S. C. *The Law of International Watercourses* (2nd edn, Oxford University Press, 2003), at 265.

25 Agreement between the Republic of the Sudan and the United Arab Republic of Egypt for the Full Utilization of the Nile Waters (adopted November 8, 1959, entered into force December 12, 1959) 453 UNTS 66 (1959 Nile Agreement).

26 The 1929 and1959 Nile Agreements were denounced by independent East African states and Ethiopia, which led to the launch of the Blue Nile Study, the antecedent of the current dam-building regime over the Blue Nile. See Collins, R. 'History of the Nile and Lake Victoria Basins through Treaties' in Howell, P. and Allan, J. (eds) *The Nile: Sharing A Scarce Resource: An Historical and Technical Review of Water Management and of Economical and Legal Issues* (Cambridge University Press, 1994), at 122.

27 1977 Kagera River Basin Agreement (note 10).

28 Godana, A. *Africa's Shared Water Resources: Legal and Institutional Aspects of the Nile, Niger and Senegal River Systems* (Frances Pinter, 1985), at 261.

29 See 2003 Lake Victoria Basin Protocol (note 9). For example, Article 4(6) of the Protocol, on equitable and reasonable utilization, largely follows the UNWC, but adds an explicit requirement that partner states 'keep the status of their water utilization under review in light of substantial changes and relevant factors and circumstances'.

30 For a detailed description, see Okidi, C. 'Review of Treaties on Consumptive Utilization of Waters of Lake Victoria and Nile Drainage System', *Natural Resources Journal*, 1982; 22: 161–99, at 185.

River Basin Action Plan (TECCONILE-1993),[31] which led to the development of existing basin wide arrangements – the NBI (1999) and the CFA. The NBI is a transitional process established by 'Agreed Minutes' in 1999, pending the establishment of a permanent legal and institutional framework agreement.[32] No formal rules or principles governing the management of water resources exist in the document establishing the NBI. However, a case-by-case approach is adopted in addressing emerging issues arising during the cooperative process, in the form of discursive decision-making processes by the Nile Council of Ministers (Nile-COM) and the Nile-Technical Advisory Committee (Nile-TAC). Moreover, the NBI has served as a vehicle to engage the Nile Basin states in planning projects, resolving disputes and, most importantly, helping to negotiate the legal content of the CFA.

The question of whether existing treaty law in the Nile River Basin can continue to work alongside the CFA is a matter of great controversy between the downstream and upstream riparian states. The imperative to move from treaties sanctioning 'historical rights' to internationally accepted principles apportioning equitable rights and obligations to all states over the utilization, development, conservation, and protection of their shared water resources depends on the adoption of the CFA.

The Cooperative Framework Agreement

The CFA provides a multidisciplinary legal and institutional arrangement of water resources development for all Nile River Basin states.[33] A number of core legal principles are contained in the CFA, such as equitable and reasonable utilization, no-significant harm and others, all of which have been imported from the UNWC.[34] The Framework is to create a legal relationship between the Nile Basin states, and to promote 'integrated management, sustainable development, and harmonious utilization of the water resources of the Basin, as well as their conservation and protection for the benefit of the present and future generations'.[35]

The CFA is the result of prolonged negotiations involving all the Nile Basin countries, except Eritrea, which chose to remain out of the process. The CFA, which was opened for signature on May 14, 2010, has been signed by six basin states to date, and provides an archetype for the current and future development of water law in the Nile River Basin.[36] The text of the CFA, as discussed below,

31 Waterbury, J. *The Nile Basin: National Determinants of Collective Action* (Yale University Press, 2002), at 78–9. See also TECCONILE, *The Nile River Basin Action Plan* (TECCONILE, 1995) (on file).

32 Extraordinary Meeting of Nile Council of Ministers (note 8).

33 Nile River Basin Cooperative Framework Project (D), 'Final Report' (PoE, Report 1.7, 2000) (on file).

34 Agreement on the Nile River Basin Cooperative Framework (note 3).

35 Ibid., at Preamble, Recital 4.

36 NBI, 'Burundi Signs the Nile Cooperative Framework Agreement' (NBI, October 26, 2010). Available online at www.nilebasin.org/newsite/index.php?option=com_content&view=article&id=70%3Aburundi-signs-the-nile-cooperative-framework-agreement-pdf&catid=40%3Alatest-news&Itemid=84&lang=en (accessed March 22, 2013).

exhibits similarities with the UNWC on the issues of scope, substantive and procedural rules, institutions and dispute settlement.

Scope and use of terms under international law and the CFA

Per Article 1(1), the UNWC 'applies to uses of international watercourses and of their waters for purposes other than navigation and to measures of protection, preservation and management related to the uses of those watercourses and their waters'. Navigation is not within the scope of the Convention unless other uses affect navigation or vice versa.[37] The CFA 'applies to the use, development, protection, conservation and management of the Nile River Basin and its resources and establishes an institutional mechanism for cooperation among the Nile Basin States'.[38] The CFA remains silent on the issue of navigational uses, although it is safe to assume that other uses affecting navigation or uses affected by navigation would fall within the scope of the CFA.

The UNWC, per Article 2, defines the term 'watercourse' as 'a system of surface waters and ground waters constituting by virtue of their physical relationship a unitary whole and normally flowing into a common terminus'. Along similar lines, Article 2(b) of the CFA defines the 'Nile River system' as 'the Nile River and the surface waters and groundwaters which are related to the Nile River'. However, a slight departure from the approach of the UNWC can be observed in CFA in its definition of the 'Nile River Basin' as 'the geographical area determined by the watershed limits of the Nile River system of waters'.[39] This expansive approach complements the scope of the CFA.[40] On the face of it, the explicit reference to the 'river basin' in the CFA adopts a much wider geographic remit as compared to the UNWC. However, it could also be maintained that, while the UNWC does not explicitly refer to a basin, an interpretation of its substantive norms supports a basin approach to the use, protection, preservation and management of international watercourses.[41]

Substantive norms

In terms of substantive norms, the UNWC and CFA contain very similar provisions. There are no variations between the text of the UNWC and the CFA on the issue of equitable and reasonable utilization, and the obligation not to cause significant harm, while both instruments adopted similar obligations on ecosystem

---

37   UNWC, at Article 1(2).
38   CFA (note 3), at Article 1.
39   The term 'Nile River Basin' is to be used 'where there is reference to environmental protection, conservation or development'. CFA (note 3), at Article 2(a).
40   'The present Framework applies to the use, development, protection, conservation and management of the Nile River Basin.' CFA (note 3), at Article 1.
41   Rieu-Clarke, A. S., Wouters, P. and Loures, F. 'The Role and Relevance of the UN Convention on the Law of the Non-Navigational Uses of International Watercourses to the EU and Its Member States', *British Yearbook of International Law*, 2008; 78: 398.

protection. However, on the rule of equitable and reasonable utilization, Article 4(1) of the CFA further stipulates that, 'each basin state is entitled to an equitable and reasonable *share* in the beneficial uses of the water resources of the Nile River System'. It is interesting to note that during the work of the International Law Commission (ILC) in developing the text of the UNWC, the 'sharing' concept was rejected on the grounds that it conveyed the idea of common or undivided property.[42] Both the UNWC and CFA provide a similar list of non-exhaustive factors that must be taken into account when determining whether a certain utilization of an international watercourse is equitable and reasonable. The CFA, in Article 4(2)(h)–(i), includes two additional factors, namely 'water contribution of each basin State' and the 'extent and proportion of the drainage area of each basin state'. The inclusion of these factors makes the CFA more inclusive of other factors from the 1966 Helsinki Rules, and demonstrates the freedom of basin states to include additional factors relevant to the characteristics of their basins.[43]

The issue of the proposed provisions on 'Water Security' and the question of their consistency with the principle of equitable and reasonable use require some examination. Article 2(f) of the CFA defines 'Water Security' as 'the right of all Nile Basin States to reliable access to and use of the Nile River system for health, agriculture, livelihoods, production and environment'. The text under Article 14 of the framework reads:

> [h]aving due regard for the provisions of Articles 4 and 5, Nile Basin States recognize the vital importance of water security to each of them. The States also recognize that cooperative management and development of the waters of the Nile River System will facilitate achievement of water security and other benefits. Nile Basin states therefore, agree, in a spirit of cooperation:
> (a) to work together to ensure that all states achieve and sustain water security;
> (b) the unresolved Article 14(b) is annexed to be resolved by the Nile River Basin Commission within six months of its establishment.

The proposed text for Article 14(b) is annexed to the agreement. The version agreed to by all countries except Egypt and Sudan reads: 'not to significantly affect the water security of any other Nile Basin State'. The counterproposal from Egypt and Sudan reads: 'not to adversely affect the water security and current uses and rights of any other Nile Basin State'. The lack of consensus on the issue prompted the Extraordinary Meeting of the Nile Council of Ministers held in Kinshasa, the Democratic Republic of Congo, on May 22, 2009 to pass a resolution that the issue on the Article 14(b) be annexed and resolved by the Nile River Basin Commission within six months of its establishment.[44]

42 Ibid. See also ILC, 'Report of the ILC on the Work of its 32nd Session' (May 5–July 25, 1980) UN Doc A/35/10(Supp) (1980).

43 ILA, 'Helsinki Rules on the Uses of the Waters of International Rivers' in *Report of the 52nd Conference* (Helsinki, 1966) (1966 ILA Report), at Article V(a)–(b).

44 See CFA (note 3), at Article 14(b).

Despite the signature of the CFA by six countries, the question of 'water security' remains the most critical issue in attaining an all-inclusive framework agreement. Clarity on the meaning of 'water security' and consistency with the concept of equitable and reasonable utilization should be properly addressed in order to build consensus for resolving disputes over the issue. The concept of 'water security' requires pursuit of an agreed definition, which cannot be envisaged at present, leaving states to work within the parameters of the existing definitions, which remain vague and broad.[45]

The inclusion of the 'right to a reliable access and use' in the definition of 'water security' in Article 2(f) could be taken as reflection of 'basic human needs', and the direct reference to Articles 4 and 5 supports its consistency with the spirit of the principles of equitable and reasonable utilization and the obligation of no significant harm. A more in-depth examination of the concept of water security follows in Chapter 27.

Procedural norms: Planned measures

Both instruments provide for an obligation to cooperate on the regular exchange of data and information. The UNWC lays down a procedure for notification, whereas Article 8 of the CFA stipulates that the Nile Basin Commission will develop detailed rules and procedures for notification once the instrument enters into force. Conversely, Article 9(1) of the CFA goes further than the UNWC by requiring the Nile Basin states to conduct 'a comprehensive assessment' of the environmental impacts of any 'planned measures that may have significant adverse environmental impacts'.

Institutional arrangements

Despite increased calls for joint institutions to manage international watercourses, the content of the provisions under the UNWC dealing with joint institutions is deemed weak, as a result of the framework nature of the instrument.[46] However, it could be argued that, in the majority of transboundary basins, effective fulfillment of many of the substantive and procedural norms contained in the UNWC could only be achieved through some sort of joint institutional arrangement. Conversely, the CFA provides a detailed institutional structure for the management of the Nile River Basin, in the form of the Nile River Basin Commission, as established in Article 15. Articles 16–32 lay out its purpose and objectives; the various organs, legal status and functions; and subsidiary institutions, including sub-basin and national institutions.

---

45   Wouters, P.,Vinogradov, S. and Magsig, B. O. 'Water Security, Hydrosolidarity and International Law: A River Runs through It ...' *Yearbook of International Environmental Law*, 2009; 19: 97–134, at 104.

46   Boisson de Chazournes, L. 'The Role of Diplomatic Means of Solving Water Disputes: A Special Emphasis on Institutional Mechanisms' in International Bureau of the Permanent Court of Arbitration (eds), *Resolution of International Water Disputes* (Kluwer Law International, 2003), at 102–3.

Dispute settlement mechanisms

The UN Charter calls for the settlement of disputes among states through peaceful means.[47] The UNWC and the CFA adopt much the same dispute settlement mechanisms: the peaceful resolution mechanisms include negotiation, good offices, mediation, conciliation or other third-party arbitration or submission to the International Court of Justice. Both instruments require the setting up of a compulsory fact-finding commission if the parties fail to resolve their disputes by all other means.[48]

Conclusion

Overcoming the existing challenges in the Nile River Basin requires a coordinated approach involving full participation of all the basin states for the realization of a basin-wide legal and institutional structure. The role and relevance of the UNWC centers on the need to forge closer collaboration with both individual basin states and regional and subregional institutions. In this regard, awareness creation with respect to the role and relevance of the Convention within the context of emerging basin-wide agreements, and the critical nexus for harmonization of the national laws and implementation, have to be given special attention.

One of the results of the 1929 and 1959 Nile Agreements is that they have resulted in the implementation of huge unilateral projects in downstream states. Upstream countries such as Ethiopia have placed significant emphasis on the need for new cooperative irrigation and hydropower development projects. Threats, such as climate change, population increase and environmental degradation have heightened the sense of urgency for a cooperative solution based on the CFA to emerge.

The current impasse relating to existing treaties depends on resolving the core dispute over the concept of 'water security' under the CFA. The reservations by Egypt and Sudan based on demands for the recognition of 'current uses and rights' in the text of existing agreements remain at variance not only with the views of upstream positions, but the spirit of the CFA. Striking a balance between existing agreements and the new Framework requires persuading all the basin states – in particular, Egypt and Sudan – on the provisions of the CFA as legal instruments providing guarantees for balancing competing uses and reconciliation in an equitable and reasonable manner.

The UNWC has played an important role in informing negotiations in the Nile River Basin, ensuring the CFA addresses key controversial issues for sound transboundary water management.

---

47 Charter of the UN (adopted June 26, 1945, entered into force October 24, 1945), 1 UNTS XVI, Article 2(3).
48 CFA (note 3), at Annex on Fact Finding Mission. See also UNWC, at Article 33(4).

13 Aral Sea Basin

Dinara Ziganshina

Activities related to the transboundary waters of the Aral Sea Basin, shared by Afghanistan, Kazakhstan, Kyrgyzstan, Tajikistan, Turkmenistan and Uzbekistan, are subject to a diverse patchwork of bilateral and multilateral treaties, inherited from the Soviet past and adopted in recent decades. These legal arrangements provide incomplete solutions to a myriad of water-related problems in the region. This chapter explores whether the 1997 UN Watercourses Convention (UNWC), the only global water treaty, has a role to play in assisting countries in building and maintaining effective and peaceful management systems for their shared waters. It is based on findings from a regional assessment of the role and relevance of the UNWC for the Aral Sea Basin countries undertaken within a research programme between the IHP-HELP Centre for Water Law, Policy and Science (under the auspices of UNESCO), University of Dundee and WWF.

The chapter starts with an overview of the Aral Sea Basin and challenges and opportunities for building water cooperation in the region. Then, the provisions of the existing treaty law in the basin and the UNWC are analyzed under issues related to scope, substantive and procedural norms, joint bodies, compliance review and dispute settlement. The study concludes with a summary of potential roles that the UNWC can play in assisting the Aral Sea Basin countries to address contemporary water related challenges in a peaceful manner.

The Aral Sea Basin – challenges and opportunities

The Aral Sea Basin, covering an area of 1,550,000 km², is the largest basin in Central Asia.[1] It embodies the entire territory of Tajikistan and Uzbekistan, the major part of Turkmenistan (excluding the Krasnovodsk region), southern part of Kazakhstan (the Kzyl-Orda and the South Kazakh regions), three regions of Kyrgyzstan (Osh, Djalalabad and Naryn), part of northern Afghanistan and

---

1 'Aral Sea Basin' (CAWATERinfo). Available online at http://cawater-info.net/aral/geo_e.htm (accessed March 23, 2013).

north-eastern Iran.[2] The Amudarya and Syrdarya are the two major rivers belonging to the basin of the Aral Sea. The Amudarya is the biggest river in Central Asia in terms of annual water runoff (79.4 km³/year). The Amudarya River originates in Tajikistan (where 74 percent of flow is formed), Kyrgyzstan (two percent), Afghanistan and Iran (13.9 percent),[3] then forms the border between Afghanistan and Uzbekistan (8.5 percent), crosses the territory of Turkmenistan (1.7 percent) and returns to Uzbekistan where it discharges into the Aral Sea. It is 2540 km long from the headwaters of the Pyandzh, its main tributary, to the Aral Sea, and has a catchment area of 309,000 km².

The Syrdarya is the longest river in Central Asia. The Syrdarya River is formed by the confluence of the Naryn and Karadarya rivers. About 75.2 percent of the Syrdarya flow originates in Kyrgyzstan. The river then flows across Uzbekistan and Tajikistan and discharges into the Aral Sea in Kazakhstan. About 15.2 percent of the flow of the Syrdarya is formed in Uzbekistan, about 6.9 percent in Kazakhstan, and about 2.7 percent in Tajikistan. The Syrdarya is up to 3,019 km long from its main tributary, the Naryn headwaters, to the Aral Sea, and has a catchment area of 219,000 km².

The social and economic development of the Central Asian republics and Afghanistan depends on the waters of the Amudarya and Syrdarya, and there are many competing claims over the use of their waters for hydropower, irrigation and the environment. Water resources are an important source of electricity generation in the Aral Sea Basin countries, constituting 27.3 percent of their total electricity consumption.[4] Irrigated agriculture accounts for more than 90 percent of the total water withdrawals in the basin, and plays a vital role in supporting the economy and livelihood.[5] The difficulties arise, however, in accommodating the different seasonal requirements of these two dominant water uses, aggravating competition for resources between sectors and ultimately between upstream and downstream countries trying to reach energy and food self-sufficiency.[6]

2   Ibid; Severskiy, I., Chervanyov, I., Ponomarenko, Y., Novikova, N. M., Miagkov, S. V., Rautalahti, E. and Daler, D. *Global International Waters Assessment Aral Sea: GIWA Regional Assessment 24* (UN Environment Programme, 2005). Available online at www.unep.org/dewa/giwa/publications/r24.asp (accessed March 23, 2013). Given that the portion of Iran in the basin is insignificant, the present study does not address the involvement of Iran in regional water cooperation.

3   Estimates on Afghanistan's contribution to the flow of the Amudarya River vary. Variations in the estimates primarily depend on which parts of northern Afghanistan and its various sub-basins are considered as part of Amudarya Basin. See Ahmad, M. and Mahwash, W. *Water Resource Development in Northern Afghanistan and Its Implication for Amu Darya Basin,* World Bank Working Paper No 36 (World Bank, 2004); Rycroft, D. W. and Wegerich, K. 'The Three Blind Spots of Afghanistan: Water Flow, Irrigation Development, and the Impact of Climate Change', *China and Eurasia Forum Quarterly,* 2009; 7 (4): 115–33.

4   Ibatullin, S., Yasinsky, V. and Mironenkov, A. *Impact of Climate Change to Water Resources in Central Asia* (Eurasian Development Bank, 2009), at 7.

5   Granit, J., Jägerskog, A., Löfgren, R., Bullock, A., de Gooijer, G., Pettigrew, S. and Lindström, A. *Regional Water Intelligence Report Central Asia,* Paper 15 (Stockholm International Water Institute, 2010), at 15.

6   Dukhovny, V. and Sokolov, V. *Lessons on Cooperation Building to Manage Water Conflicts in the Aral Sea Basin* (UNESCO-IHP, 2003).

The changes in river flow following the construction of existing reservoirs and the excessive water withdrawals for irrigation have profoundly affected the quality of water and modified the deltas of the rivers and floodplain environment. One of the most telling examples of this is the devastating degradation of the Aral Sea, its ecosystem and surrounding areas.[7] The shrinkage of the Aral Sea also affected the climate of the region, with both summer and winter temperatures becoming more extreme. Global climate change exacerbates the situation further, intensifying the aridity of the region, which is caused by glacial retreat that causes flood events in the short term, and declines in long-term water availability.[8] Dam safety[9] and uranium tailings located close to rivers[10] are other issues occupying a special place in regional water cooperation.

A complex web of water, energy and environmental problems in the region can be addressed in a holistic, mutually beneficial and peaceful manner only through collaborative actions, whether that is cooperation to increase economic benefits or cooperation to mitigate negative effects. The Central Asian republics have recognized the need to address these problems in a coordinated way at the top political level, including by establishing the Aral Sea Basin Programme (ASBP) – a programme of concrete actions to improve the environmental and socioeconomic condition in the Aral Sea Basin and attract much-needed investment.[11] Additionally, in the 1990s and 2000s, the countries adopted a number of subregional agreements, established new regional institutions and joined regional and global environmental and water-related treaties. A brief overview of these treaties follows.

Legal developments and challenges – role for the UNWC

The legal architecture of transboundary water cooperation in the Aral Sea Basin is composed of numerous agreements at bilateral, subregional, regional and global levels (Table 13.1). Structured around the core elements of a legal regime for the management of transboundary waters, namely scope, substantive and procedural rules, compliance and dispute settlement mechanisms, this section explores similarities and differences between the existing legal provisions regulating transboundary waters in the Aral Sea Basin and the UNWC.

Scope

The substantive scope of existing agreements varies from specific arrangements for water and energy trade-off regulation to general environmental issues, with the

---

7   Severskiy *et al.* (note 2), at 24, 30.
8   Ibatullin *et al.* (note 4), at 16, 28 and 30.
9   UN Economic Commission for Europe, *Dam Safety in Central Asia: Capacity-Building and Regional Cooperation*, Water Series No. 5 (UNECE, 2007), at 1.
10   Granit *et al.* (note 5), at 7.
11   The Aral Sea Basin Programme-1 (ASBP) was adopted in 1994, the ASBP-2 in 2002; the ASBP-3 is currently under preparation. See 'ASB Programme' (EC IFAS). Available online at www.ec-ifas.org (accessed March 23, 2013).

Table 13.1 Existing treaty law in the Aral Sea Basin

Treaty	Countries signed					
	Afg	Kz	Kg	Tj	Tm	Uz
Central Asia						
Agreement on Cooperation in the Field of Joint Management of the Use and Conservation of Water Resources of Interstate Sources (1992, Almaty)	–	✓	✓	✓	✓	✓
Agreement on Joint Actions for Addressing the Problems of the Aral Sea and its Coastal Area, Improving the Environment, and Ensuring the Social and Economic Development of the Aral Sea Region (1993, Kzyl-Orda)	–	✓	✓	✓	✓	✓
Agreement on Cooperation over Water Management Issues (1996, Chardjev)	–	–	–	–	✓	✓
Agreement on the Use of Fuel and Water Resources, Construction and Operation of Gas Pipelines in Central Asian Region (1996, Tashkent)	–	✓	✓	–	–	✓
Agreement on the Use of Water and Energy Resources of the Syrdarya Basin (1998, Bishkek)	–	✓	✓	✓	–	✓
Agreement on Cooperation in the Area of Environment and Rational Nature Use (1998, Bishkek)	–	✓	✓	–	–	✓
Agreement on the Parallel Operation of the Energy Systems of Central Asian States (1999, Bishkek)	–	✓	✓	✓	–	✓
Agreement on Cooperation in the Sphere of Hydromet (1999, Bishkek)	–	✓	✓	✓	–	✓
Agreement about the status of IFAS and its organizations (1999, Ashgabad)	–	✓	✓	✓	✓	✓
Framework Convention on Environmental Protection for Sustainable Development in Central Asia (2006, not in force, Ashgabad)	–	–	s	s	s	–
Commonwealth of Independent States (CIS)						
Charter of the CIS (1993, Minsk)	–	✓	✓	✓	✓	✓
Agreement on Interaction in the Field of Ecology and Environmental Protection (1992, Moscow)	–	✓	✓	✓	✓	✓
Agreement on the Main Principles of Interaction in the Field of Rational Use and Protection of the Transboundary Water Bodies (1998, Moscow)	–	s	–	✓	–	–
Agreement on Informational Cooperation in the Field of Ecology and the Environmental Protection (1998, Moscow)	–	✓	✓	✓	–	–

Table 13.1 continued

Treaty	Countries signed					
	Afg	*Kz*	*Kg*	*Tj*	*Tm*	*Uz*
UN Economic Commission for Europe						
Convention on Environmental Impact Assessment in a Transboundary Context (1991, Espoo)	–	✓	✓	s	–	–
Convention on the Protection and Use of Transboundary Watercourses and International Lakes (1992, Helsinki)	–	✓	–	–	–	✓
Convention on the Transboundary Effects of Industrial Accidents (1992, Helsinki)	–	✓	–	–	–	–
Convention on Access to Information, Public Participation in Decision-Making and Access to Justice in Environmental Matters (1998, Aarhus)	–	✓	✓	✓	✓	–
Global conventions						
UN Convention on the Law of the Non-Navigational Uses of International Watercourses (1997, New York)	–	–	–	–	–	✓
Convention on Wetlands of International Importance especially as Waterfowl Habitat (1971, Ramsar)	–	✓	✓	✓	✓	✓
Convention on Biological Diversity (1992, Rio de Janeiro)	✓	✓	✓	✓	✓	✓
UN Framework Convention on Climate Change (1992, New York)	✓	✓	✓	✓	✓	✓
UN Convention to Combat Desertification in Those Countries Experiencing Serious Drought and/or Desertification, Particularly in Africa (1994, Paris)	✓	✓	✓	✓	✓	✓

Selected agreements between the Soviet Union and Afghanistan
Frontier Agreement (1946, Moscow)
Treaty Concerning the Regime of the Soviet-Afghan State Frontier (1958, Moscow)
Protocol on the Joint Execution of Works for the Integrated Utilization of the Water Resources in the Frontier Section of the Amudarya (1958, Kabul)

1992 Almaty Agreement[12] – the most relevant of them at the basin level – laying a foundation for the 'joint management of interstate water resources use and protection' in the region. In terms of the geographic or hydrologic extent of the waters

---

12 Agreement on Cooperation in the Field of Joint Management of the Use and Conservation of Water Resources of Interstate Sources (adopted February 18, 1992) (1992 Almaty Agreement). An English translation can be found online at www.icwc-aral.uz/statute1.htm (accessed March 23, 2013).

covered in treaties, Article 1 of the 1992 Almaty Agreement recognizes 'water resources of interstate sources' as 'common and integral' for the region, revalidating the catchment area subject to cooperation as specified in the Soviet schemes, but falling short of including Afghanistan and its catchment area into joint management framework. The existing instruments do not extend the scope of regulation to transboundary groundwater, and only partly include freshwater ecosystems, which will be discussed in more detail below. In this context, countries could benefit from the UNWC and the UN Economic Commission for Europe (UNECE) Water Convention, which reflect contemporary approaches to water use and protection, inter alia, by defining a hydrological scope based on the concepts of a 'watercourse' (or a river system) and an 'ecosystem'.[13]

Substantive obligations

There are two substantive norms considered to be the basic rules of international water law: the principle of equitable and reasonable use and the no-harm rule. Emerging as specific applications of those rules are procedures and obligations related to the sustainable and optimal management, use and protection of international watercourses and their ecosystems.

The aforementioned agreements applicable to the Aral Sea Basin contain no explicit provisions on equitable and reasonable use, which is broadly recognized as 'a general rule of law for the determination of the rights and obligations of States' with respect to international watercourses.[14] Some preambular recitals in such treaties vaguely refer to 'adherence to the principle of international water law',[15] 'established international practice',[16] 'equitable solution in utilization of water and

---

13 ILC, 'Draft Articles on the Law of the Non-navigational Uses of International Watercourses and Commentaries thereto and Resolution on Transboundary Confined Groundwater' (1994 Draft Articles) in ILC, *Report of the International Law Commission on the Work of its Forty-sixth Session, Official Records of the General Assembly, Forty-ninth Session, Supplement No. 10* (May 2–July 22, 1994) UN Doc A/49/10 (ILC, 1994), at 90–91. Available online at http://untreaty.un.org/ilc/documentation/english/A_49_10.pdf (accessed March 23, 2013) (1994 ILC Report); UNECE, 'Draft Guide to Implementing the Convention: Meeting of the Parties to the Convention on the Protection and Use of Transboundary Watercourses and International Lakes' (November 10–12, 2009) ECE/MP.WAT/2009/L.2, at 23.

14 1994 Draft Articles (note 13), at 98.

15 Agreement between the Republic of Kazakhstan, the Kyrgyz Republic, the Republic of Tajikistan, Turkmenistan, and the Republic of Uzbekistan on Joint Actions for Addressing the Problems of the Aral Sea and Its Coastal Area, Improving the Environment, and Ensuring the Social and Economic Development of the Aral Sea Region (signed March 26, 1993) (1993 Kzyl-Orda Agreement). An English translation can be found online at http://ocid.nacse.org/tfdd/tfdddocs/517ENG.pdf (accessed March 23, 2013).

16 Agreement between Government of the Republic of Belarus, the Government of the Russian Federation, the Government of the Republic of Kazakhstan and the Government of the Republic of Tajikistan on the Main Principles of Interaction in the Field of Rational Use and Protection of the Transboundary Water Bodies (adopted September 11, 1998, entered into force for Belarus, Russian Federation and Tajikistan June 6, 2002) (1998 CIS Agreement on Transboundary Water).

energy resources... in accordance with norms of international law'[17] and 'solving the issues of joint management of water resources on the basis of common regional principles and equitable regulation of their consumption'.[18] Those statements fall short of specifying either what the equitable and reasonable use rule implies in general or what it would require in the context of the Aral Sea Basin.

In contrast, Article 6(2) of the UNWC presents the rule of equitable and reasonable use as being responsive to the necessities of time and place, and as providing a flexible all-encompassing approach to reconciling a broad range of existing and new economic, social and environmental issues – which ultimately provides for a legal framework for such discussions and, if necessary, adjustments. The UNWC can also be useful in regulating the relations between the Central Asian republics and Afghanistan, since the provisions of the existing treaties on Afghanistan's water use are rather limited.[19]

The no-harm rule, another fundamental substantive norm applicable to international watercourses, derives its normative foundation from *sic utere tuo ut alienum non laedas*, or the good neighborliness principle.[20] The existing treaty law in the Aral Sea Basin incorporates the no-harm rule in a way imposing significant restrictions on the activities within the territories of states unless these actions are coordinated with the affected parties. For example, Article 3 of the 1992 Almaty Agreement requires parties 'to prevent actions on its territory which can infringe on the interests of the other Parties and cause damage to them, lead to deviation from agreed values of water discharges and pollution of water sources'. Read in line with the title and preambular recitations of the agreement, this provision clearly limits the unilateral actions of the states within their jurisdiction unless any possibly harmful activity is coordinated and concerted among the parties.[21]

A similar provision requiring countries to coordinate their actions is enshrined in the 1998 Environmental Cooperation Agreement[22] (Article 2) and aimed to be

---

17   Agreement between the Governments of the Republic of Kazakhstan, the Kyrgyz Republic and the Republic of Uzbekistan on the Use of Water and Energy Resources of the Syrdarya Basin (signed March 17, 1998, Republic of Tajikistan joined in 1999) (1998 Syrdarya Agreement). An unofficial English translation can be found online at www.ce.utexas.edu/prof/mckinney/papers/aral/agreements/syrdaryaagr-mar17-98.pdf (accessed March 23, 2013).

18   1992 Almaty Agreement (note 12), at Preamble.

19   See, e.g. Frontier Agreement between Afghanistan and the Union of Soviet Socialist Republics (Including Exchange of Notes) (signed June 13, 1946, entered into force January 17, 1947) 31 UNTS 158, at Note III; Treaty between the Government of the Union of Soviet Socialist Republics and the Royal Government of Afghanistan Concerning the Regime of the Soviet–Afghan State Frontier (with Annexes and Protocols) (signed January 18, 1958) 321 UNTS 166, at Article 16 (1958 Soviet–Afghan Frontier Treaty).

20   Tanzi, A. and Arcari, M. *The UN Convention on the Law of International Watercourses: A Framework for Sharing* (Kluwer Law International, 2001), at 75; McCaffrey, S. C. *The Law of International Watercourses* (2nd edn, Oxford University Press, 2007), at 415.

21   The Title and Preamble of the 1992 Almaty Agreement reflect its spirit of 'joint management' and 'consolidation and coordination of actions'. See Vienna Convention on the Law of Treaties (adopted May 23, 1969, entered into force January 27, 1980) 1155 UNTS 331, at Article 31(3)(c).

22   Agreement on Cooperation in the Area of Environment and Rational Nature Use (adopted March 17, 1998) (1998 Environmental Cooperation Agreement). An unofficial English translation is available online at http://ocid.nacse.org/tfdd/tfdddocs/593ENG.pdf (accessed March 23, 2013).

detailed through establishing rules and procedures concerning preventive measures in the 2006 Sustainable Development Convention[23] (Article 9(3)(e)). Treaties adopted under the auspices of the Commonwealth of Independent States (CIS) also contain harm prevention provisions, seeking to constrain any activities that may cause negative impact on the environment.[24] Finally, treaties with Afghanistan require the avoidance of harm arising from the use of frontier rivers, and states that the parties shall reach an agreement before introducing any changes that may influence the flow of water or cause other damages.[25]

The above provisions suggest that, by committing to the UNWC, the Central Asian republics will not be under any stricter requirements than arise for them from the existing treaties. Under the UNWC, states do not have an absolute obligation to guarantee that no significant transboundary harm occurs. Rather, per Article 7(1), states are required to 'take all appropriate measures to prevent the causing of significant harm to other watercourse States'.[26] At the same time, the UNWC can introduce more clarity in the legal relationship between the no-harm obligation and the equitable and reasonable use rule,[27] which is lacking in the existing applicable legal framework, and can help to decipher the requirement of 'consolidation and coordination of action', essentially through its procedural system discussed below.

The obligations to protect international watercourses and their ecosystems are reflected in a range of multilateral, regional and basin agreements establishing due diligence requirements for the Aral Sea Basin states. Article 1 of the 1993 Kzyl-Orda Agreement[28] explicitly recognizes the Aral Sea itself and its deltas as a legitimate water user. Further, Article 3 of the 1993 Kzyl-Orda Agreement, in rather hortatory terms, stipulates other 'common objectives' relating to the protection of the environment. Provisions related to the environmental protection also are present in the 1998 Environmental Cooperation Agreement (Article 2), the 2006 Framework Convention on Sustainable Development (Article 3), the 1958 Soviet–Afghan Frontier Treaty[29] (Articles 13 and 22(1)(a)), the 1992 CIS

23 Framework Convention on Environmental Protection for Sustainable Development in Central Asia (signed November 22, 2006 by the Kyrgyz Republic, Tajikistan and Turkmenistan, not yet in force) (2006 Sustainable Development Convention). An unofficial English translation is available online at www.ecolex.org/server2.php/libcat/docs/TRE/Multilateral/En/TRE143806.pdf (accessed March 23, 2013).

24 1998 CIS Agreement on Transboundary Waters (note 16), at Article 2; Agreement on Interaction in the Field of Ecology and Environmental Protection (signed February 8, 1992), at Preamble (1992 CIS Agreement on Environmental Interaction). All Central Asian republics are parties to this agreement.

25 1958 Soviet–Afghan Frontier Treaty (note 19), at Articles 9(1), 9(3), 19, and 24(2).

26 See also UNECE Convention on the Protection and Use of Transboundary Watercourses and International Lakes (adopted March 17, 1992, entered into force October 6, 1996) 1936 UNTS 269, Art 2(1) (UNECE Water Convention).

27 Article 7(2) of the UNWC has subordinated the principle of significant harm to the principle of equitable and reasonable utilization.

28 1993 Kyzl-Orda Agreement (note 15).

29 1958 Soviet Afghan Frontier Treaty (note 19).

Agreement on Environmental Interaction[30] (Preamble and Article 2) and the 1998 CIS Agreement on Transboundary Waters[31] (Articles 1 and 2). The language of these agreements is relatively precise in laying down guiding principles, but needs further specification of the practical measures to be taken, including developing national laws and policies to implement the stated environmental objectives.

Additionally, the 1992 UNECE Water Convention[32] and other multilateral environmental agreements introduce the wide spectrum of obligations on environmental protection and contribute substantially to the protection of the ecosystems of international watercourses in the Aral Sea Basin.

The value of the UNWC with respect to the environmental protection lies in incorporating the environmental factor into 'virtually all provisions of the [C]onvention'.[33] What is more, the UNWC sets forth a separate obligation to protect and preserve the ecosystems of international watercourses in Article 20, which ultimately can serve as a basis for the implementation of equitable and reasonable use.[34] Because this obligation has probably not yet become a customary norm of international law, by joining the UNWC, the Central Asian republics and Afghanistan may not only strengthen the legal framework for the protection of freshwater ecosystems within their shared basins, but also contribute to strengthening the law of international watercourses in general.

Procedural obligations and joint bodies

Procedural cooperation between states and through joint bodies is an important means of giving concrete content to the stated obligations and of ensuring that these are consistently implemented and complied with. Procedural rules and strategies of cooperation over transboundary waters may relate to the establishment of joint bodies, the exchange of information among riparian states, interstate consultations, notification about proposed activities, impact assessments and monitoring and responses to emergency situations.

It is widely recognized that transboundary water management is most effectively accomplished through joint bodies established by the states concerned.[35] Treaties

---

30 1992 CIS Agreement on Environmental Interaction (note 24).
31 Agreement between Government of the Republic of Belarus, the Government of the Russian Federation, the Government of the Republic of Kazakhstan and the Government of the Republic of Tajikistan on the Main Principles of Interaction in the Field of Rational Use and Protection of the Transboundary Water Bodies (adopted September 11, 1998, entered into force for Belarus, Russian Federation and Tajikistan June 6, 2002) (1998 CIS Agreement on Transboundary Water).
32 UNECE Water Convention (note 26), Article 2(1).
33 Tanzi and Arcari (note 20), at 225.
34 McCaffrey (note 20), at 451.
35 McCaffrey, S. C. (Special Rapporteur), '6th Report on the Law of the Non-navigational Uses of International Watercourses' (February 23 and June 7, 1990), (1990) II(1) Yearbook of the International Law Commission 41, at 44, UN Doc A/CN.4/427 and Corr 1 and Add 1 (McCaffrey 6th Report). See also *Case Concerning Pulp Mills on the River Uruguay (Argentina v Uruguay)* (Judgment) (2010). Available online at www.icj-cij.org/docket/files/135/15877.pdf (accessed March 23, 2013 (*Pulp Mills*).

specific to the Aral Sea Basin envisage the need for joint bodies in strong language. The 1992 Almaty Agreement established the Interstate Commission for Water Coordination (ICWC) in Central Asia as a body responsible for water management policy in the region (Article 7), and subordinated to it two basin-water organizations for the Amudarya and the Syrdarya (Article 9). Subsequent agreements established the International Fund for Saving the Aral Sea (IFAS), which organizational structure, in addition to the ICWC, includes the Interstate Commission for Sustainable Development (ICSD).[36] The Central Asian republics are currently working on strengthening this institutional framework to improve its performance and enable better coordination among its organizations.[37] The UNECE Water Convention, which spells out the basic tasks of joint bodies (Article 9(2)), can be useful in this endeavour. Although the UNWC, under Article 24(1), does not mandate that states set up joint bodies, the fact that they have already been established in the region is an advantage, as such bodies could be tasked with the Convention's implementation.

Regular information exchange and consultations have become key elements of the international legal body of procedural rules governing transboundary waters. They serve as tools for states to reach and maintain an equitable balance of uses and benefits, and prevent transboundary harm.[38]

The existing CIS and subregional agreements mostly require parties to promote rather than to ensure the exchange of information related to international watercourses.[39] In contrast, the UNWC (Article 9) and the UNECE Water Convention (Article 13) codify and clarify a clear duty of information exchange. If taken together, the provisions of the existing agreements in the Aral Sea Basin are useful in defining the content of the information to be exchanged.[40]

36 1993 Kzyl-Orda Agreement (note 28), at Article 2; Agreement between the Republic of Kazakhstan, the Kyrgyz Republic, the Republic of Tajikistan, Turkmenistan, and the Republic of Uzbekistan on the Status of IFAS and Its Organizations (signed April 9, 1999) (1999 IFAS Agreement). See the 1999 Statute of the IFAS. In 2008, the IFAS was granted observer status in the UN General Assembly. UNGA, Res 63/133 'Resolution on Observer Status for the International Fund for Saving the Aral Sea in the General Assembly' (December 11, 2008) UN Doc A/Res/63/133 at 1. See also Statute of the Interstate Commission for Sustainable Development (ICSD) (approved by the Decision of ICSD on October 18, 2000).

37 See UNECE, 'Project "Regional Dialogue and Cooperation on Water Resources' Management in Central Asia"' (UNECE). Available online at http://unece.org/env/water/cadialogue/cadwelcome.htm (accessed March 23, 2013). The project is implemented by the IFAS Executive Committee and UNECE, and financed by the Government of Germany through GIZ.

38 McCaffrey (note 20), at 478; 1994 Draft Articles (note 13), at 107–8.

39 1992 Almaty Agreement (note 12), Article 5; 1998 CIS Agreement on Transboundary Waters (note 16), Article 2.

40 1958 Soviet–Afghan Frontier Treaty (note 19), at Article 17; 1991 Statement by the Heads of the Central Asian republics' Water Authorities (signed October 12, 1991), para. 6; 1992 Almaty Agreement (note 12), at Article 5; 1998 CIS Agreement on Transboundary Waters (note 16), at Articles 2–3; Agreement on Informational Cooperation in the Field of Ecology and the Environmental Protection (September 11, 1998) at Articles 1 and 3 (1998 CIS Agreement on Informational Cooperation).

However, as *regional* and *basin-specific* treaties, the instruments at hand could have been even more specific in exemplifying the relevant data to be shared, according to the characteristics and needs of the basin. This would have better provided the riparian countries with the material necessary to comply with their substantive obligations. Such specification would have been of particular relevance in the further development of a regional information exchange system, which the Central Asian republics have identified as a priority step towards improving water management.[41] The Central Asia Regional Water Information Base Project was developed in response to such a provision, and aims to ensure transparency and public awareness, and support decision-making in the water sector.[42] The provisions of the UNWC, which detail what information should be exchanged, when and how, can further support these practical endeavors.

By joining to the UNWC, the countries would also have clearer guidelines on consultations with each other with respect to their shared waters. To date, the agreements concluded under the umbrella of the Commonwealth and within the Aral Sea Basin make no direct reference to regular consultations. The only exception is Article 3 of the 1998 CIS Agreement on Transboundary Waters, which requires the parties to enter into mutual consultations when they develop water protection measures. For the rest, the regional and subregional agreements encompass consultation mainly as a means of dispute settlement.[43] By contrast, the UNWC envisages a set of obligations on consultations between riparians[44] and the UNECE Water Convention further assigns an important role to river basin commissions by requiring that '[a]ny such consultations shall be conducted through a joint body... where one exists' (Article 10).

41 'Database and Management Information System for Water and Environment' project was listed as a priority project in the ASBP-1. The ASBP-2 included as one of the activities 'Strengthening Material/Technical and Legal Basis for Interstate Organizations, Development of Regional Information System Designed to Manage Water Resources of the Aral Sea Basin'. See 'The Aral Sea Basin Programme-1' (note 11).

42 The project is funded by the Swiss Agency for Development Cooperation and implemented by the Scientific Information Centre of the ICWC with the assistance of UNECE and the UNEP/GRID-Arendal Office in Geneva, in close cooperation with five national water management organizations. For more information, see 'About project' (CAWater-info)'. Available online at www.cawater-info.net/about_e.htm (accessed March 23, 2013).

43 See, e.g. 1998 CIS Agreement on Transboundary Waters (note 16), at Article 13; 1998 CIS Agreement on Informational Cooperation (note 40), at Article 7; 1999 IFAS Agreement (note 36), at Article 14; Agreement between the Government of the Republic of Kazakhstan, the Government of the Kyrgyz Republic, the Government of the Republic of Tajikistan and the Government the Republic of Uzbekistan on Cooperation in the Sphere of Hydromet (signed June 17, 1999), Article 7 (1999 Agreement on Hydromet).

44 UNWC, at Articles 3(5), 4, 7(2), 21(3), 24(1), 26(2) and 30.

The Central Asian republics and Afghanistan have not agreed on detailed procedures to be invoked in the case of planned measures on an international watercourse. The language of subregional agreements only go as far as to suggest that water management projects are subject to 'joint consideration' by the parties concerned or an agreement between them.[45] Existing joint bodies also lack a clear mandate that would stipulate their role in procedures concerning planned measures.[46]

Therefore, the UNWC, with its sound and detailed procedural framework to guide countries in the case of planned measures, is of exceptional relevance for the countries in the Aral Sea Basin. The accession to the UNWC might not impose additional burdens with respect to the consultation procedures on planned measures for Kazakhstan and Kyrgyzstan, as parties to the 1991 Espoo Convention, a regional stand-alone procedural mechanism on notification and consultation on all major projects under consideration that might have an adverse environmental impact across borders.[47]

As far as environmental impact assessment (EIA) is concerned, the UNWC, prescribing to include the results of EIAs in the package of notification documents, seems to impose a less strict obligation than those found in the regional and subregional agreements. The 1991 Espoo Convention and the UNECE Water Convention (Article 3(1)(h)) formulate EIA provisions in a more robust way and specifically require parties to undertake EIAs. EIA-related provisions in the 1992 CIS Agreement on Environmental Interaction, the 1998 CIS Agreement on Informational Cooperation, and the 2006 Framework Convention on Sustainable Development, also require parties to conduct assessments, harmonize national EIA procedures and exchange information about those assessments.[48]

---

45 1958 Soviet–Afghan Frontier Treaty (note 19), at Article 19(2)–(3); 1992 Almaty Agreement (note 12), at Preamble; Agreement between the Government of the Republic of Kazakhstan, the Government of the Kyrgyz Republic and the Government of the Republic of Uzbekistan on Cooperation in the Area of Environment and Rational Nature Use (signed March 17, 1998), Article 2(z). An unofficial English translation can be found online at http://iea.uoregon.edu/pages/view_treaty.php?t=1998-EnvironmentRationalNatureUse.EN.txt&par=view_treaty_html (accessed March 23, 2013); 2006 Sustainable Development Convention (note 23), Article 4.

46 Statute of the Interstate Commission for Water Coordination of Central Asia (signed December 5, 1992), at para. 2.5. Available online at www.icwc-aral.uz/statute12.htm (accessed March 23, 2013) (1992 ICWC Statute); Statute of the Interstate Commission for Water Coordination of Central Asia (signed September 18, 2008), at paras 2.18–2.19. Available online at www.icwc-aral.uz/statute4.htm (accessed March 23, 2013) (2008 ICWC Statute).

47 UNECE Convention on Environmental Impact Assessment in a Transboundary Context (adopted February 25, 1991, entered into force September 10, 1997) 1989 UNTS 309 (1991 Espoo Convention).

48 1992 CIS Agreement on the Environmental Interaction (note 24), at Articles 2–3; 1998 CIS Agreement on Informational Cooperation (note 40), at Articles 2(1) and 3; 2006 Sustainable Development Convention (note 23), at Articles 2(b), 4(6) and 7.

The need for both prior and ongoing assessments of the effect of activities on transboundary waters also finds support in regional treaty practice.[49] In turn, the practices of international financial institutions illustrate an inclusive approach to impact assessment. For instance, the World Bank is presently supporting the preparation of an Assessment Study for the proposed Rogun hydropower plant in Tajikistan to assess: (a) techno-economic aspects and dam safety; and (b) environmental/social impacts,[50] which includes the assessment of Tajikistan's energy policy from environmental and social perspectives (strategic impact assessment)[51] and in terms of riparian and cross-border impacts (regional impacts).[52]

The devastating effects of natural and human-induced disasters on people and the environment ignore political boundaries and call for active cooperation between countries to prevent, reduce and eliminate such effects. The UNWC treats these cases as a matter of 'emergency', understood as 'a situation that causes, or poses an imminent threat of causing, serious harm to... States... and that results suddenly from natural causes, such as floods, the breaking up of ice, land-slides or earthquakes, or from human conduct, such as industrial accidents' (Article 28).

Existing treaty law envisages preventive and responsive obligations dealing with

49 2006 Sustainable Development Convention (note 23), at Articles 2(b) (regular assessments) and 7 (monitoring); UNECE Water Convention (note 26), at Article 11 (joint monitoring and joint or coordinated assessments); 1998 CIS Agreement on Transboundary Waters (note 16), at Article 4 (common monitoring system); 1991 Espoo Convention (note 47), at Article 7 (post-project analy-sis); UNWC, at Article 24 (management of an international watercourse). See also *Case Concerning the Gabčíkovo-Nagymaros Project (Hungary v Slovakia)* (Judgment) [1997] ICJ Rep 7 at 88 (Separate Opinion of Judge Weeramantry); *Pulp Mills* at Para 281; *Case Concerning Land Reclamation by Singapore in and around the Straits of Johor (No 12) (Malaysia v Singapore)* (Provisional Measures, Order of October 8, 2003) ITLOS Reports 2003, Paras 22–23; Espoo Inquiry Commission, *Report on the Likely Significant Adverse Transboundary Impacts of the Danube-Black Sea Navigation Route at the Border of Romania and the Ukraine* (UNECE, July 2006). Available online at www.unece.org/fileadmin/DAM/env/eia/documents/inquiry/Final%20Report%2010%20July%2 02006.pdf (accessed March 23, 2013).
50 World Bank, 'Assessment Studies for Proposed Rogun Regional Water Reservoir and Hydropower Project in Tajikistan'. Available online at http://go.worldbank.org/ZQXIA8J0H0 (accessed March 23, 2013); World Bank, *Techno-Economic Assessment Study (TEAS) for the Rogun Hydroelectric Power Plant Construction Project: Terms of Reference* (World Bank, 2010), at 4–5 (ToR for the Rogun TEAS); World Bank, *Environmental and Social Impact Assessment Study (ESIA) for the Rogun Hydroelectric Power Plant Construction Project: Terms of Reference* (World Bank, 2010), at 5–21 (ToR for the Rogun ESIA).
51 ToR for the Rogun ESIA (note 50), at 5.
52 ToR for the Rogun TEAS (note 50), at 4. See also ToR for the Rogun ESIA (note 50), at 4, 7–8, 14, 17, 22–26.

emergency situations in forceful language.[53] The tasks of the ICWC contain emergency-related functions, including the development of joint contingency plans to prevent emergencies and natural disasters and eliminate their consequences.[54] The presence of emergency-related obligations in the regional and subregional agreements is laudable and does credit to the countries' intentions to cooperate in critical situations. It also makes it easier for the countries to commit to largely similar obligations under the UNWC (Articles 27–28). This is especially so for Kazakhstan, which is also a party to the 1992 UNECE Convention on the Transboundary Effects of Industrial Accidents – designed to protect human beings and the environment against industrial accidents[55] – and of the UNECE Water Convention, with emergency-related provisions in Articles 3(j) and 14.

In terms of benefits from joining the UNWC, since the relevant provisions dealing with emergencies are scattered in various regional and subregional agreements, and not always water treaties, the Convention can serve as a single reference point for their application to transboundary waters. In addition, in fleshing out the anticipatory and responsive actions in the case of emergencies, the UNWC establishes linkages between these and other obligations under the Convention.

Compliance review and dispute settlement

The subregional agreements in the Aral Sea Basin do not provide for a compliance review procedure. Article 2 of the 1992 Almaty Agreement prescribes that the parties shall ensure that the agreed regime is 'strictly observed', but it remains unclear how non-compliance shall be detected and monitored. The agreement further stipulates that, by 1992, parties should have elaborated economic and other measures to deal with the cases of non-compliance with the established regime and limits of water use. However, such mechanisms are still lacking. Some disjointed

---

53 1958 Soviet–Afghan Frontier Treaty (note 29), at Article 17; 1992 CIS Agreement on the Environmental Interaction (note 24), at Articles 2–4; Agreement between the Government of the Republic of Kazakhstan, the Government of the Kyrgyz Republic, the Government of the Republic of Tajikistan and the Government the Republic of Uzbekistan on Joint Activities on the Rehabilitation of the Tailings and Rock Dump that Have Transboundary Effect (signed April 5, 1996), Articles 1–5; Agreement between the Government of the Republic of Kazakhstan, the Government of the Kyrgyz Republic and the Government of the Republic of Uzbekistan on the Use of Fuel and Water Resources, Construction and Operation of Gas Pipelines in Central Asian Region (signed April 5, 1996), Article 1; 1998 CIS Agreement on Informational Cooperation (note 40), at Article 3; 1998 CIS Agreement on Transboundary Waters (note 31), at Articles 1 and 6; 1999 Agreement on Hydromet (note 43), at Preambular Recitals and Article 3 on the forms of cooperation; Agreement between the Government of the Republic of Kazakhstan, the Government of the Kyrgyz Republic, the Government of the Republic of Tajikistan and the Government of the Republic of Uzbekistan on the Parallel Operation of the Energy Systems of Central Asian States (signed June 17, 1999), Article 8.

54 1992 ICWC Statute (note 46), at Article 2(9); 2008 ICWC Statute (note 46), at para. 2.14.

55 UNECE Convention on the Transboundary Effects of Industrial Accidents (adopted March 17, 1992, entered into force April 19, 2000) 2105 UNTS 457.

attempts to monitor and facilitate compliance have been undertaken under the 1998 Syrdarya Agreement.[56] Article 5 stipulates that parties shall take appropriate measures to ensure compliance with the provisions of the agreement through various forms of guarantees, such as credit lines, security deposits and others.

The UNWC does not require compliance monitoring, but does establish various provisions to facilitate it. These include Articles 8 (general obligation to cooperate), 9 (regular exchange of data and information), 11–19 (planned measures) and 24 (management). Another issue is the lack of government mechanisms under the Convention, such as a meeting of the parties or secretariat, to facilitate and review compliance.[57] This does not preclude the parties to the Convention establishing such institutions in the future if they so decide, following, for instance, the model of the Ramsar Convention[58] or Conventions under the umbrella of UNECE that provide a sound institutional support to facilitate implementation and compliance with their requirements through meetings of the parties, secretariats, implementation and compliance committees and various working groups and boards.[59] Some of these agreements also set up compliance review and monitoring systems.[60]

Dispute settlement mechanisms relating to transboundary waters are largely undeveloped in the Aral Sea Basin. For example, the 1992 Almaty Agreement refers any dispute that could arise between the Parties to the Ministers of Water Resources for the five Central Asian republics. In other words, disputes shall be resolved internally within the ICWC, the body responsible for the implementation of this agreement. As such, the ICWC acts to prevent and resolve emerging controversies and provides a forum where representatives of the five basin states can meet, discuss, and make binding decisions on contentious issues. The 1992 Almaty

56 1998 Syrdarya Agreement (note 17).

57 See Chapter 22 of this collection.

58 Convention on Wetlands (Ramsar, Iran, 1971) 2nd Meeting of the Conference of the Contracting Parties, Groningen, The Netherlands (May 7–12, 1984), see 'Recommendations of the 2nd Meeting of the Conference of the Contracting Parties', at 2.3. Available online at www.ramsar.org/pdf/rec/key_rec_2.03e.pdf (accessed 3 Apr 2012).

59 See, e.g. UNECE Water Convention (note 26), at Articles 17 and 19; 1991 Espoo Convention (note 47), at Articles 11 and 13; UNECE, 'Decision III.2 Adopted by the 3rd Meeting of the Parties to the Espoo Convention' (2004) ECE/MP.EIA/6, Annex II. Available online at www.unece.org/fileadmin/DAM/env/documents/2004/eia/decision.III.2.e.pdf (accessed March 23, 2013) (revising UNECE, 'Decision II.4 Adopted by the 2nd Meeting of the Parties to the Espoo Convention' (2001) ECE/MP.EIA/4, Annex IV. Available online at www.unece.org/fileadmin/DAM/env/documents/2001/eia/decision.II.4.e.pdf (accessed March 23, 2013).

60 See, e.g. UNECE Convention on Access to Information, Public Participation in Decision-Making and Access to Justice in Environmental Matters (adopted June 25, 1998, entered into force October 30, 2001) 2161 UNTS 447, Article 10(2); 'Decision II.10 Adopted by the 2nd Meeting of the Parties to the Espoo Convention' (2001). Available online at www.unece.org/env/eia/decisions.html (accessed March 23, 2013).

Agreement further states that 'if necessary, an impartial third party can be involved' (Article 13), but fails to detail the procedure for such a dispute settlement and further measures if a dispute cannot be resolved in this manner.

Hence, as far as dispute settlement is concerned, the UNWC, under Article 33, has much to offer to supplement the insufficient provisions of the existing subregional instruments. Of special interest is the Convention's innovative mechanism of an impartial fact-finding commission to resolve a dispute, which can be triggered if the parties concerned have not been able to settle their dispute through negotiation or any other means within six months from the time of the request for negotiations (Article 33(3–10)).

Conclusion

The foregoing analysis has shown that the Aral Sea Basin does not suffer from the lack of regulatory endeavors; the existing legal framework is rather overwhelmed by numerous instruments seeking to govern the countries' relations over the use and protection of shared waters. The problem lies in the fact that the normative quality of these treaties, most of which have been adopted with no links to each other, fall short incorporating the contemporary principles of international water law and best water management practice and neglect the significance of establishing a sound procedural system of transboundary water cooperation. In this context, the need for the improvement of the existing legal framework is undeniable, although the caution of the Aral Sea Basin countries towards new regulatory instruments is also understandable.

The UNWC can serve for the Central Asian republics and Afghanistan as a *common platform for the negotiation of future agreements* in the Aral Sea Basin, since this global framework instrument does not preclude or dismiss the need for watercourse agreements.[61] Existing legal arrangements in the basin were not designed to accommodate changing circumstances, nor can they be easily amended. As a result, many treaties have become stagnant and lost their effectiveness. The negotiations of new agreements have not succeeded so far. Hence, the Central Asian republics and Afghanistan may want to join the UNWC to have an agreed common framework at the global level, and later on they can strengthen their commitments and/or adjust them to the characteristics of their watercourses. It might be also useful to have the UNWC as a background for interpretation of bilateral treaties and arrangements increasingly emerging in the region.[62]

---

61   UNWC, at Article 3(1).
62   Granit *et al.* (note 6), at 17–18.

14 Amazon Basin

Joshua Newton

No country in the Americas has ratified the UN Watercourses Convention (UNWC). This includes the countries that fall within the Amazon River Basin, one of the most important watersheds in the world. This does not mean, however, that countries in the region did not support the Convention at the time of its adoption: Paraguay and Venezuela have signed it, although they have not yet completed the ratification process; within the continent, there were 13 states that voted in favor of the Convention,[1] of which ten were also among its co-sponsors.[2]

Nonetheless, the question remains as to the Convention's relevance and applicability in the region. This chapter contributes to this discussion, with a focus on the Amazon Basin. The Amazon has several treaties and an international organization that govern its water and other natural resources, yet there are potential elements within the UNWC that could help strengthen cooperation between the basin states and mitigate future conflict.

The chapter starts with an overview of water resources in South America, before describing the general characteristics of the Amazon Basin. The chapter then gives an overview of the applicable water governance regime. An analysis of the gaps and weaknesses in such a regime follows, along with an assessment of how the UNWC could fill in the identified holes to provide for a more robust governance scheme.

Regional context

South America is one of the richest hydrological regions of the world, with over 30 percent of global water resources. This, combined with a low population, makes the per capita water availability in the region over 47,000 km³ per person, more than twice any other continent.[3]

South America's transboundary waters play an important role in the hydrological regime of the continent. There are 38 international basins, which cover almost

---

1 Antigua and Barbuda, Brazil, Canada, Chile, Costa Rica, Guyana, Haiti, Jamaica, Mexico, Trinidad and Tobago, United States of America, Uruguay and Venezuela.
2 Antigua and Barbuda, Brazil, Canada, Chile, Grenada, Honduras, Mexico, United States of America, Uruguay and Venezuela.
3 World Resources Institute, *Data Tables: Part II* (EarthTrends, 2005), at 208.

60 percent of the continent and are home for over 45 percent of the population.[4] All states in South America are part of an international basin in at least 50 percent of their respective territories, with that ratio rising to 95 percent in the case of Bolivia.[5] The degree to which each state depends on water resources flowing from other countries varies: Argentina and Paraguay rely on their neighbors for 72 percent and 66 percent, respectively, of their external water flows; Uruguay is not far behind, with 58 percent of a dependency ratio; Colombia and Chile barely depend on river flows coming from neighboring countries (one percent and four percent, respectively), while Ecuador and Guyana, not at all.[6]

The Amazon River Basin at a glance

The Amazon River is the second longest in the world, stretching over 7000 km. The basin discharges on average 220,000 m³/second of water into the Atlantic Ocean, and up to 300,000 m³/second in the rainy season.[7] This represents 51 percent of South America's surface water discharge and over 15 percent of global freshwater flow,[8] which is 40,594 km³/year.[9] While recognized as the largest watershed in terms of land area on the planet,[10] the exact area of the Amazon Basin remains subject to debate. Depending on the definition of the basin and the source, the basin covers anywhere from 5,147,970 km² to 8,187,965 km², in eight countries and a small part of a French territory.[11] This chapter uses the figures compiled by a consortium of actors,[12] which defines the above numbers as 'Lesser' and 'Greater' Amazonia.[13] Given this range of possible areas of the

4 'Transboundary Freshwater Dispute Database' (Oregon State University). Available online at www.transboundarywaters.orst.edu/database (accessed March 24, 2013).
5 Wolf, A., Natharius, J. A., Danielson, J. J., Ward, B. S. and Pender, J. K. 'International River Basins of the World', *International Journal of Water Resources Development*, 1999; 15(4): 387–427.
6 World Resources Institute (note 3), at 208.
7 UNEP, Amazon Cooperation Treaty Organization (ACTO) and Universidad del Pacífico, *GEO Amazonia: Environment Outlook Amazonia* (UNEP and ACTO, 2009), at 34.
8 The discharge of the Amazon River is 6162 km³/year. 'Transboundary Freshwater Dispute Database' (note 4). In turn, South America's total discharge is 12,198 km³/year. World Resources Institute (note 3).
9 World Resources Institute (note 3).
10 'Transboundary Freshwater Dispute Database' (note 4).
11 UNEP, ACTO and Universidad del Pacífico (note 7); see also 'Transboundary Freshwater Dispute Database' (note 4); Organization of American States, Office for Sustainable Development and Environment, *Amazon River Basin: Integrated and Sustainable Management of Transboundary Resources in the Amazon River Basin* (OAS, October 2005) Water Project Series No. 8, at 1. Available online at www.oas.org/dsd/events/english/documents/osde_8amazon.pdf (accessed March 24, 2013); IUCN and others, *Watersheds of the World* (IUCN, IWMI, Ramsar Convention Bureau and WRI, 2003); World Resources Institute (note 3); 'Transboundary Freshwater Dispute Database' (note 4).
12 UNEP, ACTO and Universidad del Pacífico (note 7).
13 Ibid., at 40. The definition of 'Greater Amazonia' is the 'maximum extension of the Amazonian area based on at least one of the criteria (hydrographical, ecological or political administrative)'. 'Lesser Amazonia' is described as 'the minimum extension of the Amazonian area based on the three criteria taken together'.

Amazon Basin, the basin represents between 29 percent and 46 percent of South America's landmass.[14]

The value of the basin goes beyond just water resources, with the greatest amount of biodiversity in the world and a valuable indigenous heritage. The basin also acts as a major carbon sink.

Almost 63 percent of the Amazon Basin is in Brazil (Table 14.1). The Amazon River itself only drains the territories of Peru, Colombia and Brazil; in turn, the basin's tributaries flow through Bolivia, Ecuador, Venezuela and Guyana.[15]

Table 14.1 Amazon River Basin area by country (Greater Amazonia)

Country	Area of basin in country (%)
Brazil	61
Peru	12
Bolivia	9
Colombia	6
Venezuela	5
Ecuador	2
Guyana	3
Suriname	2
French Guiana[a]	0
Total	100

Notes: [a] 'Transboundary Freshwater Dispute Database' (Oregon State University). The Database states that a small amount of the Amazon River Basin resides within the political border of French Guiana (less than 0.01%)

Source: Adapted from UNEP, Amazon Cooperation Treaty Organization (ACTO) and Universidad del Pacífico, *GEO Amazonia: Environment Outlook Amazonia* (UNEP and ACTO, 2009 at 41.

As for the basin's groundwater, the Amazon transboundary aquifer system (ATAS)[16] is shared among six of the eight Amazon Basin states, and is estimated to extend over 3,950,000 km^2.[17] This aquifer system is thought to interact with both the Amazon and Orinoco River Basins. The aquifer's water resources are used mainly for irrigation, industry and rural population centers.[18]

The lack of agreement on the basin's borders makes it difficult to determine its exact population: Lesser Amazonia has an estimated population of 11 million

14   These percentages are based on the total landmass of South America being 17,800,000 km^2. Rand McNally, *The New International Atlas* (Rand McNally, 1996).
15   García. B. *The Amazon from an International Law Perspective* (Cambridge University Press, 2011), at 23–27.
16   da Franca Ribeiro dos Anjos, N., Miletto, M., Donoso, M. C., Aureli, A., Puri, S., Van der Gun, J. *Sistemas Acuíferos Transfronterizos en las Américas: Evaluación Preliminar* (UNESCO, 2007), at 124.
17   Ibid. Bolivia, Brazil, Colombia, Ecuador, Peru and Venezuela.
18   Ibid. The Orinoco is shared between Colombia and Venezuela.

Table 14.2 Amazon River Basin population by country (Greater Amazonia, 2007)

Country	Population (n)
Brazil	24,970,600
Peru	4,361,858
Bolivia[a]	6,541,020
Colombia	960,239
Ecuador	629,373
Venezuela	70,464
Guyana	751,223
Suriname	492,823
French Guiana	–
Total	38,777,600

Notes: [a] Bolivia's population was not included in the tables in the GEO resource and the reason for this was not explained. The number included for Bolivia was taken adding the other countries' populations and determining what was the remainder from the 38,777,600 population.

Source: Adapted from UNEP, Amazon Cooperation Treaty Organization (ACTO) and Universidad del Pacífico, *GEO Amazonia: Environment Outlook Amazonia* (UNEP and ACTO, 2009, at 67

inhabitants, and Greater Amazonia of almost 39 million. The population density of Greater Amazonia (4.74 inhabitants/km²) is more than double that of Lesser Amazonia (2.14 inhabitants/km²) – still far below the world average of 47.83 inhabitants/km².[19]

Despite this low population density, the Amazon Basin has been subject to increasing human pressure associated with large- and small-scale mining, forest burning and deforestation, farming,[20] and dam construction. With often significant impacts on water quality, quantity and timing, those activities may pose a threat to both the environment and society within the basin.

Rapid hydropower development within the basin is particularly relevant in this context: there are 11 dams either planned or currently under construction, and many have been subject to protest on social and environmental grounds.[21] An example of the controversial nature of these dams is the recently started construction of the Belo Monte Dam on the Xingu River, in Brazil. Indigenous and environmental groups have strongly protested the dam because of the alleged potential for significant social and environmental impacts.[22] In recent years, another hydropower complex in the Madeira River has also been the cause of controversy,

---

19   UNEP, ACTO and Universidad del Pacífico (note 7), at 67.
20   García (note 15). There is evidence that pesticides are widely used in the basin, but the resulting impacts on the environment are little known.
21   World Resources Institute (note 3), at 208.
22   Phillips, T. 'Belo Monte Hydroelectric Dam Construction Work Begins' *The Guardian* (London, March 10, 2011). Available online at www.guardian.co.uk/environment/2011/mar/10/belo-monte-hydroelectric-work (accessed March 24, 2013).

which included protests by indigenous communities and a protracted disagreement between Brazil and Bolivia.[23]

An overview of the Amazon River Basin governance regime

Various treaties govern the Amazon Basin. The handful of agreements adopted up to 1978 are all bilateral and of limited scope, addressing issues of navigation, hydropower generation and border demarcation.[24] Broader bilateral environmental treaties related to, but not specifically focused on, water management began to emerge following the UN Conference on the Human Environment in 1972.[25]

The Organization of American States (OAS) first suggested basin-wide agreements to Latin American states in the mid-1960s. This, relatively quickly, led to the adoption of the La Plata River Basin Treaty, in 1969.[26] The Amazon Basin states followed suit and began negotiations on a treaty in 1977, with the Brazilian Ministry of Foreign Affairs proposing a draft text after consultations with its neighbors. In 1978, the Amazon Cooperation Treaty (ACT) was adopted, entering into force two years later.[27]

The ACT aims to foster, among the basin states,

'joint [interstate] actions and efforts to promote the harmonious development of their respective Amazonian territories in such a way that these actions produce equitable and mutually beneficial results and achieve also the

23 See, e.g. 'Bolivia Enviará Misión a Brazil para Hablar sobre Posible Daño Ambiental Represas' *Terra Networks México* (Mexico City, July 24, 2007). Available online at www.terra.com.mx/articulo.aspx?articuloid=358375 (accessed March 24, 2013); 'Fórum dos Povos Indígenas da Amazônia Encerra com Protesto contra hidrelétricas no Rio Madeira e Reafirma Lutas em Defesa dos Direitos Indígenas' (Povos Indíginas no Brasil, November 30, 2007). Available online at http://pib.socioambiental.org/pt/noticias?id=51101&id_pov=216 (accessed March 24, 2013); 'Lula Vai Discutir Hidrelétricas do Madeira com Evo Morales' *Rondonia Dinamica* (Porto Velho, January 14, 2009). Available online at www.rondoniadinamica.com/arquivo/lula-vai-discutir-hidreletricas-do-madeira-com-evo-morales,1950.shtml (accessed March 24, 2013).
24 See, e.g. Declaration and Exchange of Notes concerning the Termination of the Process of Demarcation of the Peruvian–Ecuadorean Frontier (Peru–Ecuador) (May 22, 1944); Agreement concerning the Cachuela Esperanza Hydroelectric Plant, supplementary to the Agreement on Economic and Technical Co-operation between the Government of the Federative Republic of Brazil and the Government of the Republic of Bolivia (Brazil–Bolivia) (February 8, 1984); Exchange of Notes constituting an Agreement for the Construction of a Hydroelectric Plant in Cachuela Esperanza, supplementary to the Agreement on Economic and Technical Cooperation (Brazil–Bolivia) (adopted August 2, 1988) 1513 UNTS 6; Agreement on Fluvial Transportation (Brazil–Peru) (November 5, 1976); Technical and Economic Cooperation Agreement for the Improvement of Navigation and Fluvial Transportation in the River Madre de Díos and its Tributaries (Bolivia–Peru) (October 18, 1989), in García (note 15), at 153.
25 García (note 15), at 67.
26 Ibid., at 71.
27 Ibid., at 82–5

preservation of the environment, and the conservation and rational utilization of the natural resources of those territories'.[28]

The treaty itself barely mentions water, with the few exceptions referring mostly to navigation:

- Article III allows for 'complete freedom of commercial navigation on the Amazon and other international Amazonian rivers'.
- Under Article VI, parties may 'undertake national, bilateral or multilateral measures aimed at improving and making said rivers navigable', in order to promote the Amazon as an 'effective communication link'.
- Article V refers to other water uses: 'Taking account of the importance and multiplicity of the functions which the Amazonian rivers have in the process of economic and social development of the region, the *Contracting Parties shall make efforts aimed at achieving rational utilization of water resources*'.
- Other treaty provisions that indirectly relate to water include those dealing with health, research coordination, infrastructure and tourism (Articles VIII–X and XIII).
- Article XV, dealing with data sharing, does not refer specifically to water, but is important for sound water management: it requires states to 'seek to maintain a permanent exchange of information and cooperation among themselves and with the agencies for Latin American cooperation in the areas pertaining to matters covered by this Treaty'.

In 1998, an amendment to the ACT created the Organización del Tratado de Cooperación Amazónica [Amazon Cooperation Treaty Organization (ACTO)].[29] In March 2003, the ACTO was established in Brasilia, Brazil,[30] under the leadership of the respective member states' ministries of foreign affairs. All ACTO decisions require consensus.[31] The ACTO has, as its mission, to 'achieve a balanced distribution of benefits generated [from the basin], thus providing to create synergies that will increase the capacity of negotiation by the Country Members at a global level, in light of topics that relate to the Amazon'.[32] This goes along with the vision of the organization as being a recognized international forum for the integration and sustainable development of the region.[33]

---

28 Amazon Cooperation Treaty (Brazil-Bolivia-Colombia-Ecuador-Guyana-Peru-Suriname-Venezuela) (adopted July 3, 1978). Available online at www.otca.info/en/institucional/index.php?id=29 (accessed March 24, 2013).

29 Amazon Cooperation Treaty Amendment Protocol (ACT Protocol) (Brazil-Bolivia-Colombia-Ecuador-Guyana-Peru-Suriname-Venezuela) (adopted December 14, 1998) at Article I. Available online at www.otca.info/en/institucional/index.php?id=30 (accessed March 24, 2013).

30 Amazon Cooperation Treaty Organization (ACTO), *Strategic Plan: 2004–2012* (ACTO, 2004), at 15.

31 ACT Protocol (note 28), at Article II(3).

32 ACTO (note 29), 19.

33 Ibid.

Article 1 of the ACT Protocol delineates administrative and functional guidelines for the ACTO and empowers the organization to enter 'into agreements with Contracting Parties, non-member States and other international organizations'. In this sense, the ACTO has entered into several bilateral agreements with other organizations to assist in furthering its mission: these include the Coordinating Body for the Indigenous Organizations of the Amazon Basin, the Andean Community, the CIC-La Plata Basin, the Pan American Health Organization (PAHO) and the Inter-American Development Bank (IADB).[34]

Gaps in the Amazon River Basin governance regime

Most transboundary watersheds around the world are not under any governance regime, whether that includes a treaty and/or a river basin organization.[35] The existence of the ACTO and a governing treaty, therefore, is a significant achievement, placing the Amazon ahead of such other basins. This is especially true considering that the Amazon Basin is a large, vaguely defined region, with a significant amount of water resources, and shared among eight developing countries – countries at different development stages, and which have not always had good relationships with one another. Moreover, while the ACT is over 30 years old now, the ACTO has been in place for less than a decade, and significant funds in support of its work have only been coming in for a little over five years. Yet, projects exist, the ACTO Secretariat functions and the member states meet regularly at the Ministerial level.[36]

As discussed below, however, the international regime governing the Amazon Basin is not perfect and should be strengthened. Within the ACT itself, the provisions promoting the rational utilization of water resources and requiring permanent information exchange are important, but insufficient to ensure the sustainable and cooperative management of the basin's freshwater resources. There is thus much room for improvement to address the gaps and weaknesses identified and which might fail to foster, or even hamper, cooperation between riparian states.

Lack of dispute resolution mechanisms

The ACT has inadequate provisions governing disputes between the member states. This is a problem on several levels, as the ACT fails to offer an avenue or platform for the resolution of an interstate disagreement.

Taking a closer look at the ACT, Article XXI(1) of the treaty goes only as far as to charge the Amazon Cooperation Council with 'ensuring that the aims and

---

34   See, for example, Organización del Tratado de Cooperación Amazónica, 'Cooperación Internacional' [International Cooperation] (ACTO, 2013). Available online at www.otca.info/portal/cooperacao-internacional.php?p=otca (accessed March 24, 2013).

35   See, for example, Loures, F., Rieu-Clarke, A. and Vercambre, M. L. *Everything You Need to Know About the UN Watercourses Convention* (WWF International, 2010), at 5.

36   See Organización del Tratado de Cooperación Amazónica. Available online at www.otca.info/portal/index.php?p=index (accessed March 24, 2013)

objectives of the Treaty are complied with'. The treaty at hand does not offer any alternative means of dispute settlement in the event that the Council fails to provide for an equitable resolution of the case. This is likely to occur in every situation, since all Council decisions require unanimous vote by all parties, as per Article XXV. At the same time, since decision-making requires consensus, any state may have a veto power over an action with which it does not agree,[37] in line with the principle of sovereignty, frequently-stated in the ACT.

Apart from that provision, Article XXIV of the ACT establishes that parties may agree to create special commissions, as necessary to tackle problems, or topics related to the application of the treaty. However, there is no obligation for the implementing state to agree to the creation of such commissions upon a request by a potentially affected neighbor.

Outside the ACT, diplomatic channels always exist, but may lead to delays, instability and uncertainty, in the absence of clear rules, procedures and timeframes to govern negotiations. Then, there is the fallback to the international law regime (International Court of Justice), Permanent Court of Arbitration, etc.). However, these are costly routes to take, in terms of time, money, resources and even potential outright loss, with sometimes detrimental impacts on interstate relations and future cooperation. Moreover, since the ACT does not explicitly recognize the jurisdiction of those courts as compulsory, ipso facto, submitting a dispute thereto would require prior special agreement between the states concerned, on a case-by-case basis. Therefore, incorporating within the ACT more detailed provisions on dispute resolution, including those involving resort to third-parties, would help to mitigate conflicts that arise in the future and would even prevent the escalation of a dispute that might be resolved at the technical level.

Lack of provisions on equitable and reasonable use and harm prevention

In international water law, there are two key principles that govern the relationships between basin states: equitable and reasonable utilization and harm prevention.[38] This chapter does not attempt to determine which of these principles has priority, as is sometimes a topic of debate, but only points out that the ACT mentions neither.

Hence, the ACT lacks legal 'sticks' to dissuade a riparian from using the water resource inequitably or causing significant transboundary harm. Amazon River waters may be in abundance, but they are not immune to global changes taking place, such as climate change, urbanization, migration, etc. Such pressures on water resources, even in the Amazon, will require countries to protect the livelihoods of their citizens and cooperate more closely with their neighbors, including towards

37 UNDP-GEF International Waters Project, *International Waters: Review of Legal and Institutional Frameworks: Good Practice and Portfolio Learning in GEF Transboundary Freshwater and Marine Legal and Institutional Frameworks* (UNDP, White and Case, Global Environment Facility and UBC, 2011), at 14.

38 Ibid.

protecting shared freshwater ecosystems and the vital services they provide. Addressing and clarifying the issues of equitable utilization and significant harm could thus benefit all the basin states.

Lack of guidance on 'rational utilization' and sustainable water management

Besides navigation, the only specific mention of water in the ACT, in Article V, is for states to 'achieve the rational utilization of water resources'. This lofty and vague statement may offer some flexibility for parties to be innovative in their solutions. At the same time, however, the provision has no 'teeth' to encourage basin states to work towards this goal; or, for that matter, to go beyond rational utilization in terms of other water management objectives, such as the preservation of high-value conservation areas within the basin or the maintenance of overall river connectivity. On its own, therefore, 'rational utilization' is a vague concept and should have been defined more clearly in the ACT.

Insufficient provisions on information exchange

The provision for the exchange of information between states in the ACT is relevant for water resources management, as stated above. However, such a provision lacks detail in terms of the content, scope and extent of the obligations to which it may give rise.

Moreover, Article XV appears to focus the duty to exchange data on research and information in general. That provision lacks specific requirements for notification and dialogue over unilateral development projects, and in the case of emergency situations. There are no rules or procedures about how countries should act if such circumstances were to arise. It is thus no surprise that there has not been a history of practice among the ACTO member states of consulting one another before making unilateral decisions with regard to major planned measures.[39] The point is raised as another weakness in the treaty that prevents it from serving as an adequate conflict mitigation tool before future disputes affect diplomatic relations or unilateral measures lead to injury across borders.

Potential value added by the UN Watercourses Convention in the Amazon Basin

The above gaps in the water governance regime of the Amazon River Basin may appear to be inconsequential in today's context; apart from the aforementioned disagreement over a dam on the Madeira River between Brazil and Bolivia, there have not been serious disputes between the Amazon Basin states to date.

Still, a key factor in having a solid and improved governance regime is to be able to adapt to changes in the future; i.e. to have a robust enough system in place so as

---

39   García (note 15), at 123.

to be resilient to absorb those changes, and to mitigate potential interstate conflict. Filling the above gaps thus makes sense from a legal standpoint. From a political point of view, however, such changes may not be feasible. Despite being a downstream riparian, Brazil is the hegemon among the Amazon Basin states, and insists on adhering to the principle of sovereignty concerning resources within its own borders. The political will does not seem to exist to strengthen the ACT and/or to bolster ACTO's powers much more than how they exist today.

On the legal side, there are potential options for improvement and that is where the UNWC comes into play. The Convention is not designed to replace existing bilateral and multilateral treaties between states in transboundary basins, but to complement and supplement them, as per its Article 3. The aforementioned gaps could be addressed by the Convention in the following manner.

Lack of dispute resolution mechanisms

Article 33 of the UNWC presents detailed options for procedural steps that can be taken when a dispute arises between two or more parties. This provision would provide a clearly defined alternative avenue for Amazon Basin states to turn to in the event that the relevant procedures in the ACT, examined above, failed to lead to a solution.

Lack of provisions on equitable and reasonable use and harm prevention

Articles 5 and 7 of the UNWC, respectively, require states to develop and protect their international watercourses in an 'equitable and reasonable manner' and 'take all appropriate measures to prevent the causing of significant harm to other watercourse States'. The UNWC contains further guidance in the application of the principles at hand, in Article 6. As mentioned above, the ACT does not mention either of those principles, despite their wide acceptance in international water law. Recognizing that they exist, guiding their application, and defining how they relate to each other, would further clarify the rights and duties of the Amazon states in the management of their common water resources. Amending the ACT in that direction would also bring it up to speed with current international water law.

Lack of guidance on 'rational utilization' and sustainable water management

As mentioned above, the only mention of water in the ACT besides navigation is in Article V, which contains a general and abstract statement on the pursuance of 'the rational utilization of water resources'. The provision in question does not elaborate further on what 'rational utilization' means and how states are required to cooperate on water management and ecosystem protection more generally.

While the UNWC is not complete on the topic of water management, the Convention goes further in clarifying what the 'equitable and reasonable use' of international watercourses might entail. Article 5 refers to 'optimal and sustainable utilization' of watercourses 'consistent with [their] adequate protection'. Articles

20–26 expand beyond simple 'rational utilization' to address broader issues pertaining to water management. These include the protection and preservation of ecosystems, as well as of the marine environment; pollution; introduction of alien/new species; management and regulation of river flows; and the maintenance of installations. The Convention, therefore, offers more guidance for states to consider when managing international watercourses than the ACT.

Insufficient provisions on information exchange

The UNWC is more explicit and robust in terms of data and information sharing between basin states than the ACT. This is evident in Articles 9 (information and data exchange), 11–19 (notification, consultation and negotiation of planned uses), and 28 (emergency situations).[40] The Convention adds important dimensions to the lines of communication between riparian states, which is purportedly one of the goals of the ACT.

Conclusion

The Amazon River Basin is one of the most important basins in the world, yet its governance structure is not robust and comprehensive enough to assure conflict mitigation and sustainable water management across borders. While efforts have been made over the past several decades to improve this, and countries have in general cooperated in the context of specific projects, most notably since ACTO's establishment in 2003, political will to strengthen the existing governance regime itself does not seem to exist.

At the same time, the above analysis of the ACT and its deficiencies demonstrates the complementary role of the UNWC, and thus the potential benefits for the Amazon Basin states from acceding to that global instrument.

In this context, there are two options at the disposal of the basin states to create a more robust regime for the Amazon Basin. Such options are not mutually exclusive and should be preferably carried out together: (1) continue to develop the ACT and the ACTO, by interested basin states and the international community; and (2) ratify the UNWC. While the Convention is not a final solution to the problems that exist, there are provisions therein that can strengthen the governance regime in the basin and help mitigate future disputes as global changes continue to influence the region.

The likelihood that all ACTO member states would ratify the UNWC in the short term is slim. This does not mean, however, that, if some countries ratified, the others would not be held to that standard. Some of the Convention's most important provisions are widely accepted as customary international law, including equitable and reasonable use and harm prevention, as well as those obligations deriving therefrom, such as data sharing. Yet, ratification remains important.

---

40   UNWC, at Articles 9, 11–19 and 28.

Arguably, the more states that ratify the Convention, the more robust international water law will become, thereby leveling the playing field in interstate negotiations. In the absence of applicable international and regional governance structures, states' interactions over water are influenced by economics, geography and military might. The UNWC offers the opportunity to strengthen international water law as potentially a mitigating factor of other influential circumstances and an enabling tool of transboundary water cooperation.

15 Mekong Basin

Bennett Bearden, Alistair Rieu-Clarke and Sokhem Pech

Cooperation concerning the development and management of the Mekong River Basin dates back to 1957, when Cambodia, Laos, Thailand and Viet Nam (then known as South Viet Nam) created the Committee for Coordination of Investigations of the Lower Mekong, under the auspices of the UN.[1] However, it was not until the 1990s, and the end of the Cold War era, that significant cooperation reemerged in earnest. A major milestone was reached in 1995, with the adoption of the Agreement on the Cooperation for the Sustainable Development of the Mekong River Basin (1995 Mekong Agreement).[2]

The need for an effective cooperative mechanism by which basin states can protect, manage and develop the Mekong in an equitable and sustainable manner is evident. Most notable in recent years have been the opportunities and threats related to hydropower – although major irrigation schemes have also raised significant issues. The sudden groundswell of hydropower development by private power producers and private financing, along with the revitalization of the once abandoned hydropower dam projects along the Mekong mainstream in the Lower Mekong River Basin, starting from 2008, has taken many observers – including major regional organizations such as the Mekong River Commission (MRC) and other international financing institutions – by surprise.[3]

China, the most upstream state on the Mekong, and a non-party to the 1995 Mekong Agreement, is in the midst of developing a cascade of eight dams on the upper Mekong stretch in Yunnan Province, with a total installed capacity of 15,450 megawatts.[4] Contributing 55 percent of the total annual flow, the tributaries of the Lower Mekong have also been earmarked for their hydropower potential. However, earlier developments, such as Viet Nam's Yali Falls project, have resulted

1 Bearden, B. L. 'The Legal Regime of the Mekong River: A Look Back and Some Proposals for the Way Ahead', *Water Policy*, 2010; 12: 798–821.
2 Agreement on the Cooperation for the Sustainable Development of the Mekong River Basin (adopted April 5, 1995) (1995) 34 ILM 864 (1995 Mekong Agreement).
3 Pech, S. 'Cambodian and Mekong Water Resources Governance' in Sato, J. (ed.), *Transboundary Resources and Environment in Mainland Southeast Asia* (Institute of Advanced Studies in Asia, 2010).
4 Hirsch, P. 'Cascade Effect' (China Dialogue, February 8, 2011). Available online at www.chinadialogue.net/article/show/single/en/4093-Cascade-effect (accessed March 24, 2013).

in significant transboundary concerns.[5] Also within the Lower Mekong River Basin, 12 dams are being studied and planned on the mainstream, with the prior notification and consultation of the Xayabury Mainstream Dam, initiated by Laos, proving a major test for the Mekong treaty regime.[6]

Given these and more major current and future challenges facing the states of the Mekong, this chapter considers whether the 1997 UN Watercourses Convention (UNWC) might constitute an effective mechanism – among others – by which to strengthen the existing governance arrangements within the basin. The chapter therefore firstly examines the key issues and challenges faced by the Mekong Basin states in sharing its waters, and secondly, examines the similarities and differences between the existing Mekong legal framework and the text of the UNWC.

The Mekong River and its tributaries – challenges and opportunities

Basin characteristics

The Mekong River Basin is one of the world's largest river basins, with a length of 4,800 km and a catchment area of 795,000 km². About 20 percent of the Basin lies within China, three percent in Myanmar, 25 percent in Laos, 20 percent in Thailand, 19 percent in Cambodia and eight percent in Viet Nam. The upper catchment of the Mekong is characterized as steep and narrow. Consequently, soil erosion is a major issue, with approximately 50 percent of the sediment in the river coming from the Upper Mekong River Basin. Snow-melt floods from May to July dominate the upper Mekong (Lancang), and rain-induced floods from the summer monsoon (July to October) cause floods from tributaries to the lower Mekong, especially within Laos and the Central Highlands of Viet Nam. The Tonlé Sap, the largest freshwater lake in Southeast Asia and a designated UNESCO Biosphere Reserve, is situated in the lower part of the Mekong in Cambodia.[7] The lake

5 SWECO Grøner in association with Norwegian Institute for Water Research, ENVIRO-DEV, and ENS Council, *Electricity of Vietnam: Final Report – Environmental Impact Assessment on the Cambodian Part of the Se San River Due to Hydropower Development in Vietnam* (SWECO Grøner, 2006). Available online at www.ngoforum.org.kh/docs/publications/HCRP_FinalEIAReportofSeSan.pdf (accessed March 25, 2013); Rieu-Clarke, A. and Gooch, G. 'Governing the Tributaries of the Mekong: The Contribution of International Law and Institutions to Enhancing Equitable Cooperation over the Sesan', *Pacific McGeorge Global Business and Development Law Journal*, 2010; 22 (2): 193–224.

6 See Mekong River Commission Secretariat, *Procedures for Notification, Prior Consultation and Agreement (PNPCA) Proposed Xayaburi Dam Project – Mekong River: Prior Consultation Review Report* (MRC, March 24, 2011). Available online at www.mrcmekong.org/assets/Publications/Reports/PC-Proj-Review-Report-Xaiyaburi-24-3-11.pdf (accessed March 25, 2013).

7 UNESCO, 'Tonle Sap Biosphere Reserve' (UNESCO Office in Phnom Penh, 2011). Available online at www.unesco.org/new/en/phnompenh/natural-sciences/biosphere-reserves/tonle-sap-biosphere-reserve/ (accessed March 25, 2013).

changes flow direction twice a year, and the portion that forms the lake expands and shrinks dramatically with the season. The latter dynamics, along with high annual sediment and nutrient fluxes, make the lake one of the most productive inland fisheries in the world, and a source of rich biodiversity (Table 15.1).

Table 15.1 Key hydrological characteristics of the Mekong River Basin

	Yunnan	Myanmar	Laos	Thailand	Cambodia	Viet Nam	Total
Catchments (km²)	165,000	24,000	202,000	184,000	155,000	65,000	795,000
MRB (%)	22	3	25	23	19	8	100
Country area (%)	38	4	97	36	86	20	
Average rainfall (mm/year)	1,561		2,400	1,400	1,600	1,500	1,750
Average runoff (m³/second)	2,414	300	5,270	2,560	2,860	1,660	15,060
Average runoff (MCM/year)	76,128	9,461	166,195	80,732	90,193	52,350	474,932
In dry season	19,032	1,419	24,929	12,110	13,529	7,852	78,871
Average runoff (% of total)	16	2	35	17	19	11	100
In dry season	24.1	1.8	31.6	15.4	17.2	9.9	
Population (million)	10	0.5	4.9	24.6	10.8	21	71.8

Sources: MRC, *Overview of the Hydrology of the Mekong Basin* (MRC, 2005); Snidvongs, A. and Teng, S. K. *Global International Waters Assessment Mekong River: GIWA Regional Assessment 55* (UN Environment Programme, 2006).

National interests

Reconciling seemingly competing national interests will be key to addressing transboundary challenges within the Mekong River Basin. Most notably, China plays a major role in the geopolitical landscape of the Mekong – not only as a powerful upstream state, but also by providing loans for major infrastructure projects downstream.[8] Such investments result in reluctance from its trade partners to

8   Pech, S. and Sunada, K. 'Modern Upstream Myth: Is A Sharing and Caring Mekong Region Possible? Institutional Capacity Assessment for Sustainable Policy Scenarios', in Kummu, M., Keskinen, M. and Varis, O. (eds), *Modern Myths of the Mekong* (Water and Development Publications, 2008), at 135; Pech, S. 'Cambodian and Mekong Water Resources Governance', in Sato, J. (ed.), *Transboundary Resources and Environment in Mainland Southeast Asia* (Institute for Advanced Studies on Asia, University of Tokyo, 2010).

openly challenge upstream activities on the Mekong.[9] As far as regional coopera-tion frameworks are concerned, China favors the Association of Southeast Nations (ASEAN), the Greater Mekong Subregion (GMS), and bilateral cooperation for securing cross-border trade and investment, rather than legally binding watercourse agreements.[10]

In terms of the other riparian states, Myanmar, with only three percent of the total basin area of the Mekong, plays a limited role in Mekong Basin dynamics, having only recently begun to overcome international isolation through participa-tion in ASEAN and GMS.[11] The landlocked country of Laos, downstream of China and Myanmar on the Mekong, shares a 1,400-km common border with Thailand, 900 km of which is demarcated by the Mekong River. The Mekong and its trib-utaries have been recognized by the Lao government as offering tremendous potential for hydropower generation, as a main source for export earnings.[12] However, Laos is also concerned with the competition for hydropower markets within the basin. Plans for large energy and hydropower developments in Myanmar and China may produce surplus electricity for sale in other countries, particularly Thailand and Viet Nam.[13]

As a downstream country to China, Thai water and energy agencies view the Chinese dam development with hope that it would generate surplus water during dry season and cheaper energy for Thailand's fast growing economy, although it has some concerns about that development's negative impact on fisheries.[14] Thailand is also keen on improving cross-border trade and tourism along the Mekong River.

Cambodia – downstream of China, Myanmar, Laos, Thailand and Viet Nam (Central Highlands) – is one of the least developed countries in the Mekong. Most of Cambodia's territory (85 percent) is in the basin, and Cambodia's level of dependency on, and vulnerability from, basin developments is high.[15] The most

9   Pech, S. and Sunada, K. 'Management of Water Disputes in the Mekong River Basin: Roles of Law, Institution and Technology' (Proceedings of 3rd APHW Conference on Wise Water Resources Management Towards Sustainable Growth and Poverty Reduction, Bangkok, Thailand, October 16–18, 2006).

10 Pech, S. and Sunada, K. 'Managing Transboundary Rivers: The Case of the Mekong River Basin', *Water International*, 2007; 32 (4): at 503–23.

11 Hirsch, P. and Cheong, G. *Natural Resources Management in the Mekong River Basin: Perspectives for Australian Development Cooperation* (AusAID, 1996). Available online at http://sydney.edu.au/mekong/documents/report_mekongbasin1996.pdf (accessed March 25, 2013).

12 International Centre for Environmental Management, *Mekong River Commission Strategic Environmental Assessment for Hydropower on the Mekong Mainstream: Impacts Assessment (Opportunities and Risks) Vol. II: Main Report* (ICEM, 2010). Available online at www.icem.com.au/documents/envassessment/mrc_sea_hp/3.%20impacts/reports/pdf/SEA_IAR_vol2.pdf (accessed March 25, 2013).

13 Ratner, B. D. 'The Politics of Regional Governance in the Mekong River Basin', *Global Change, Peace and Security*, 2003; 15 (1): 59–76.

14 Hirsch and Cheong (note 11), at 41.

15 Browder, G. 'An Analysis of the Negotiations for the 1995 Mekong Agreement', *International Negotiation*, 2000; 5: 237–61; Browder, G. and Ortolano, L. 'The Evolution of an International Water Resources Management Regime in the Mekong River Basin', *Natural Resources Journal*, 2000; 40: 499–531.

recent MRC strategic environmental assessment found that Cambodia would be the biggest loser from the uncoordinated rapid development of hydropower dams on the Mekong.[16]

Finally, for Viet Nam, the Mekong Basin, and especially the Mekong delta and Central Highlands, plays a major role in agricultural and hydropower development – both existing and potential.[17] Additionally, the Mekong delta is crucial to Viet Nam as its 'rice bowl' – the richest agricultural zone in the country. Also, Viet Nam – as the most downstream country – currently uses almost 'the entire dry season flow of the Mekong River'.[18]

The 1995 Mekong Agreement and the UN Watercourses Convention

The 1995 Mekong Agreement includes a framework for regional cooperation in the Lower Mekong River Basin that is often lauded as a model in best practices for sustainable development of a transboundary watercourse.[19] However, several provisions in the treaty fall short of the guiding principles set forth in 'the most influential articulation of existing and emerging customary international law at the global level', the UNWC.[20] More than 15 years on, the current framework of the 1995 Mekong Agreement increasingly reflects a treaty with deficiencies and a flawed vision that signals a departure from the principles of customary law enshrined in the Convention.

The 1995 Mekong Agreement contains six chapters, the most important of which are chapters III (Objectives and Principles of Cooperation), IV (Institutional Framework), and V (Addressing Differences and Disputes). Among the international environmental law concepts embodied in the 1995 Mekong Agreement, the most significant is sustainable development, in Article 1, whereby the parties agree '[t]o cooperate in all fields of sustainable development, utilization, management and conservation of the water and related resources of the Mekong River Basin'. Drawing on such examples as the ongoing dispute between Viet Nam and Cambodia over hydropower development on the Sesan River at the Yali Falls dam,[21] and the controversy over the so-called 'Thai Water Grid' – the grandiose

---

16   ICEM, *Mekong River Commission Strategic Environmental Assessment for Hydropower on the Mekong Mainstream: Summary of the Final Report* (ICEM, 2010), at 17. Available online at www.icem.com.au/documents/envassessment/mrc_sea_hp/SEA_Final_Report_summary_Oct_2 010.pdf (accessed March 25, 2013).
17   Ibid., at 9.
18   Browder (note 15), at 239; Pech, S. and Sunada, K. 'Population Growth and Natural Resources Pressures in the Mekong River Basin', *Ambio*, 2008; 37: 219–24.
19   MRC, 'The Story of Mekong Cooperation' (MRC). Available online at www.mrcmekong.org/ about-the-mrc/history/ (accessed March 25, 2013).
20   Rieu-Clarke and Gooch (note 5), at 194.
21   Rieu-Clarke and Gooch (note 5), at 194; Lerner, M. 'Dangerous Waters: Violations of International Law and Hydropower Development Along the Sesan River' (unpublished manuscript); Rutkow, E., Crider, C. and Tyler, G. *Down River: The Consequences of Vietnam's Se San River Dams on Life in Cambodia and Their Meaning in International Law* (NGO Forum on Cambodia, 2005). Available online at www.law.harvard.edu/programs/hrp/documents/down_river.pdf (accessed March 25, 2013).

proposed water transfers from tributaries in Cambodia and the Lao PDR to support irrigation in north-eastern Thailand[22] – it is apparent that development concerns trump environmental ones in the Lower Mekong River Basin. Molle and others therefore maintain that 'megaprojects' on tributaries are transforming the Mekong into a basin of 'contested waterscapes'.[23]

Although it is fully binding as a treaty, the 1995 Mekong Agreement's broad, basic framework is, in effect, an agreement to agree and is essentially hortatory. The Lower Mekong River Basin countries set aside controversial issues such as water allocation (Article 5) and flow maintenance (Article 6) during the negotiations, with a view to developing further details at a later date. Under the MRC's Water Utilization Programme, which ran from 2000–08, the following procedures were developed: Procedures for Data and Information Exchange and Sharing, approved in 2001; Procedures for Water Use Monitoring, approved in 2003; Procedures for Notification, Prior Consultation and Agreement, approved in 2003; Procedures for Maintenance of Flows on the Mainstream, approved in 2006; and Procedures for Water Quality, approved by the MRC Council on January 26, 2011.[24] These procedures are referred to in Articles 5 and 6, and are mandated by Article 26 of the 1995 Mekong Agreement. Noteworthy, however, is that *all* the abovementioned 'procedures' are external to the Treaty, and it is questionable whether they are legally binding.

In addition to the abovementioned comments related to the UNWC and the 1995 Mekong Agreement, a number of clear differences can be found between the texts of the Treaties.

The UNWC defines 'watercourse' in Article 2(a) to include both surface waters (including tributaries) and connected groundwater. The UNWC then incorporates a holistic and conjunctive approach to freshwater management. Further, Article 3(4) of the Convention requires parties to define the waters that are the subject of an agreement. The 1995 Mekong Agreement refers to the Mekong 'Basin', as well as the terms 'river basin', 'basin level', 'basin-wide', and 'river system' throughout, but the Agreement fails to define those terms or the term 'tributary', which is defined only in external 'procedures', and in all likelihood is not legally binding on the Lower Mekong River Basin states. Additionally, the 1995 Mekong Agreement does not include China, the regional hydro-hegemon, that controls the headwaters in the Upper Mekong River Basin, and Myanmar, and accordingly, the governance approach of the Agreement is not basin-wide.

---

22 Molle, F. and Floch, P. 'Water, Poverty and the Governance of Mega Projects: The Thai "Water Grid"', M-POWER Working Paper MP-2008-08. (Unit for Social and Environmental Research, Chiang Mai University, 2007). Available online at www.mpl.ird.fr/ur199/resultats/textes%20PDF/ Water%20grid.pdf (accessed March 25, 2013); Molle, F., Floch, P., Promphaking, B. and Blake, D. J. H. 'The "Greening of Isan": Politics, Ideology, and Irrigation Development in the Northeast of Thailand', in Molle, F., Foran, T. and Kakonen, M. (eds), *Contested Waterscapes in the Mekong Region: Hydropower, Livelihoods and Governance* (Earthscan, 2009), at 253.

23 Molle and Floch (note 22); Molle *et al.* (note 22).

24 'Policies, Procedures and Guidelines' (MRC). Available online at www.mrcmekong.org/ publications/policies-procedures-and-guidelines/ (accessed March 25, 2013).

The UNWC, in Article 12, envisions that states will provide the results of any environmental impact assessments for all planned measures that may have a significant adverse effect upon other watercourse states. A recent ruling of the International Court of Justice, which maintains that there is now a customary international law obligation to conduct environmental impact assessments where there is a likelihood of transboundary harm, bolsters this provision.[25] The Mekong legal regime does not legally require such assessments.

In terms of substantive provisions, the UNWC includes factors relevant to equitable and reasonable utilization (Articles 5 and 6) but the Mekong legal regime does not. Additionally, the UNWC includes an obligation to take all appropriate measures not to cause significant harm (Article 7). The 1995 Mekong Agreement does not address significant harm, but rather Article 7 refers to 'harmful effects' and requires the victim state to provide evidence of 'substantial damage', both undefined terms in the current legal regime.

The UNWC does not differentiate between inter-basin and intra-basin diversions, nor does it differentiate regarding exchange of information and notification of planned measures related to such diversions.[26] Article 5(A) of the 1995 Mekong Agreement has a complex, convoluted matrix for reasonable and equitable use, although water is not allocated per se. This water use matrix emphatically marginalizes tributaries, treating them disparately compared with the mainstream, requiring merely notice to the MRC Joint Council before conducting planned measures. Passive regulation of tributaries, which allows the Lower Mekong River Basin states unilaterally to divert and transfer water without approval from other basin states, is perhaps the most glaring substantive deficiency in the 1995 Mekong Agreement. This regime passivity is especially important now because 'by 2010, more than 120 tributary projects [were] projected, under construction or operating on lower Mekong tributaries'.[27] With its 'marginalised tributaries'[28] and 'lopped headwaters',[29] the Mekong is the archetypal disservered river basin.[30]

The UNWC includes an obligation for protection of water quality and preservation of ecosystems, control of alien species and prevention, reduction and control of pollution.[31] In Article 3 of the 1995 Mekong Agreement, the Lower Mekong River Basin states merely agree to protect the environment, natural resources,

25   *Case Concerning Pulp Mills on the River Uruguay (Argentina v Uruguay)* (Judgment) [2010] ICJ Rep 14 at 82–83 (*Pulp Mills*).

26   UNWC, at Articles 5 and 11–14.

27   Hirsch, P. 'The Changing Political Dynamics of Dam Building on the Mekong', *Water Alternatives*, 2010; 3(2): 312–23 at 317.

28   Carrard, N. 'Mainstream or Marginal?' (8th International River Symposium: Water and Food Security – Rivers in the Global Context, Brisbane, September 2005).

29   Lebel, L. and Garden, P. 'Deliberation, Negotiation and Scale in the Governance of Water Resources in the Mekong Region', in Pahl-Wostl, C., Kabat, P. and Möltgen, J. (eds), *Adaptive and Integrated Water Management* (Springer, 2008), at 216.

30   Bearden, B. L. 'Following the Proper Channels: Tributaries in the Mekong Legal Regime' (JSD dissertation, McGeorge School of Law, University of the Pacific, 2011) (on file with author), at 263.

31   UNWC, at Articles 7 and 20–22.

aquatic life and conditions, and ecological balance of the basin from pollution or other harmful effects from development plans and uses of water and related resources. As previously noted, Article 7 refers to harmful effects and substantial damage, two undefined terms in the Mekong legal regime.

The UNWC, per Article 17(2), requires parties to negotiate in good faith and pay reasonable regard to the rights and legitimate interests of other states. The controversy surrounding the Yali Falls dam and continuing hydropower development on the Sesan River[32] irrefutably supplies the best evidence for the need of such requirements, which are conspicuously missing in the Mekong legal regime. Further, the 1995 Mekong Agreement has no provisions for stakeholder participation and access to information (transparency).

Article 27 of the UNWC requires prevention and mitigation of harmful conditions whether resulting from natural causes or human conduct, such as floods, ice conditions, erosion, siltation, drought, waterborne diseases, desertification or saltwater intrusion. There are no equivalent requirements to mitigate specific harmful conditions mentioned in the Mekong legal regime.

The UNWC's procedural framework includes detailed dispute settlement provisions in Article 33 for constructive resolution of emerging environmental challenges at the intersection of law and policy, emphasizing strategic counseling, crisis management and creative dispute resolution, including the use of good offices, mediation, fact finding, conciliation, arbitration and reference to the International Court of Justice. The dispute resolution procedure in the Mekong legal regime is convoluted and often protracted, as illustrated in the aftermath of the Yali Falls dam.[33]

Conclusion

Although the Mekong legal regime is a viable institutional framework evidencing over five decades of interstate cooperation during periods of conflict, it suffers from a lack of: (1) a basin-wide scope (China, which controls the headwaters in the Upper Mekong River Basin, and Myanmar are not parties); (2) specific legal principles and mechanisms regulating development on tributaries (Article 5(A)); (3) procedural elements of sustainable water development (environmental impact assessments, a legal framework for stakeholder participation and access to information); and (4) detailed dispute resolution provisions. These differences point to a broader scope, approach and framework in the UNWC compared with the 1995 Mekong Agreement. This does not auger well for future sustainable water development if the soft law[34] of the Mekong legal regime continues to hold sway.

---

32 Rieu-Clarke and Gooch (note 5) at 200–7.
33 Rieu-Clarke and Gooch (note 5), at 204–7.
34 For a critique of the soft-law approach to the evolution of the Mekong Treaty Regime, see Bearden (note 1), at 798.

All four Lower Mekong River Basin states voted in favor of the Convention, but none has signed it. The states should consider harmonizing and reconciling the current Mekong legal regime with the UNWC. Until they do, a recurring question is whether sustainable development in the Lower Mekong River Basin is a principle at odds with itself.

16 Ethiopia

Musa Mohammed Abseno

This chapter begins by examining conflict and cooperation in the Nile, laying emphasis on Ethiopia. It also looks at Ethiopia's role in the process of the work of the International Law Commission (ILC) and the UN General Assembly (UNGA) in the codification of the UN Watercourses Convention (UNWC). The role of domestic laws in supporting the principles of international water law is briefly assessed, followed by observations on the advantages and disadvantages of ratification of the UNWC by Ethiopia. The chapter concludes with an assessment reflecting varying possibilities on the question of ratification.

Ethiopia

Ethiopia has a large amount of surface-water resources and an appropriate climate that bestows a relatively higher amount of rainfall in many parts of the country. The annual fresh surface run-off from its 12 basins is estimated at 122 billion cubic meters/year, while the availability of ground water can be as much as 26.1 billion cubic meters.[1] However, amid such plenty, the country is considered water scarce.[2] Owing to an uneven spatial and temporal availability and distribution, access to water for micro and major economic activities and livelihoods of millions remains a long way off.[3] Large supplies of water, an estimated 80–90 percent, exist in the western and south-western parts of the country, where only 30–40 percent of the population live.[4] The challenges of development of Ethiopia's transboundary water resources are further complicated by sustained riparian conflict and lack of cooperation in the region.

1 World Water Assessment Programme, *National Water Development Report for Ethiopia* (UNESCO, 2004), at 62. Available online at http://unesdoc.unesco.org/images/0014/001459/145926e.pdf (accessed March 25, 2013).
2 Ohlsson, L. and Appelgren, B. 'Water and Social Resource Scarcity' (*The Water Page*, March 1998). Available online at www.africanwater.org/SoicalResourceScarcity.htm (accessed March 25, 2013).
3 See Ministry of Water Resources, Ethiopian Water Resources Management Policy (The Federal Democratic Republic of Ethiopia 1999) (on file with author).
4 World Water Assessment Programme (note 1).

Conflict and cooperation

More than 90 percent of the country's rivers are transboundary.[5] Ethiopia is the source of many rivers that flow to the Horn and North Africa: from the Wabi-Shebele and Dawa, shared with Somalia and Kenya, to the Abbay (the Blue Nile), Tekeze (Atbara) and Baro-Akobo (Sobat), the most important subsystems to the great Nile River. Despite its immense contribution to transboundary waters and the huge potential for domestic utilization for irrigation and hydropower, it has been unable to utilize its transboundary water resources for centuries – the reasons for this are social, economic, political and legal difficulties caused as a result of unequal development and utilization in the basin.

Ethiopia's economy is predominantly dependent upon rural agriculture, an important sector which sustains 85 percent of the population and which provides 90 percent of its export earnings.[6] The distribution of the rainfall in terms of volume, space and time is highly uneven, and usually entails serious crop failures and shortage of water supply; as a result, the country continues to suffer from cyclical droughts and famine.[7] Achieving food security for an ever-increasing population is a double challenge for a capacity- and finance-strapped country like Ethiopia. According to the 2010 census, Ethiopia's population exceeds 82 million, which makes the question of food security the most critical issue in the country's economic policy.[8]

In the first ten years of the twenty-first century, Ethiopia's transboundary rivers have become the most important strategic resource in national economic development and food security. Every effort is being made to concretize this potential through master plan studies, and to adopt water resources management policy, strategy and action programmes for the realization of safe drinking water and sanitation, self-sufficiency in crop production through irrigated agriculture and hydroelectric power. Ethiopia has 3.7 million hectares of irrigable land, of which, by 2002, only 179,000–250,000 hectares had come under irrigation.[9] Official

---

5   See Ministry of Water Resources (note 3).
6   Country Programme Action Plan (2007–2011) between the Government of the Federal Republic of Ethiopia and the UN Development Programme (adopted February 27, 2007). (on file with author).
7   Ministry of Water and Energy, 'Ethiopian Country Paper, Proceedings' (Vth Nile 2002 Conference, Addis Ababa, Ethiopia, February 24–28, 1999).
8   Central Statistical Agency of Ethiopia, 'National Statistics 2010: Population' (CSA, 2012). Available online at www.csa.gov.et/index.php?option=com_rubberdoc&view=category&id=75&Itemid=561 (accessed March 25, 2013).
9   Awulachew, S. B., Yilma, A. D., Loulseged, M., Loiskandl, W., Ayana, M. and Alamirew, T. *Water Resources and Irrigation Development in Ethiopia*, Working Paper 123 (International Water Management Institute, 2007), at 2. Available online at www.iwmi.cgiar.org/publications/ Working_Papers/working/WP123.pdf (accessed March 25, 2013). See also Awulachew, S. B., Merrey, D. J., Kamara, A. B., Van Koppen, B., Penning de Vries, F. and Boelee, E. *Experiences and Opportunities for Promoting Small-Scale/Micro Irrigation and Rain Water Harvesting for Food Security in Ethiopia*, Working Paper 98 (International Water Management Institute, 2005), at 8. Available online at www.iwmi.cgiar.org/Publications/Working_Papers/working/WOR98.pdf (accessed March 25, 2013).

sources indicate that Ethiopia has an estimated 13,000–30,000 megawatts (MW) of hydropower potential, which is recognized as readily available, affordable and a clean source of energy.[10]

As a national economic imperative to overcome the scourge of poverty and the realization of food security, the country has embarked on the construction of a series of dams. In the first ten years of the twenty-first century alone, more than three dams were constructed over the Tekeze (Atbara), Beles (Blue Nile), and Omo-Gibe rivers. Currently, a plan to build the Grand Renaissance Dam, a mega-dam project on the Abay River (Blue Nile), which has an installed capacity of 5,250 MW and 15,128 gigawatt hours (GWh) annual energy generation, drew opposition from Egypt.[11] Ethiopia stresses that the project causes no harm to its downstream neighbors, Egypt and Sudan, and sees Egypt's opposition as a claim that is tantamount to a veto power over developments upstream on the Nile waters.[12]

The current status quo based on existing agreements has been challenged by Ethiopia, which is keen to see that it is replaced by a legal framework agreement that establishes equitable entitlement to all the basin states. Ethiopia has therefore been instrumental in pursuing a unanimous adoption by the Technical Cooperation Committee for the Promotion of the Development and Environmental Protection of the Nile Basin on the elaboration of a framework for cooperation in 1999 and the establishment of the Panel of Experts for negotiating the current Cooperative Framework Agreement (CFA).[13] Its commanding influence in the signature and future ratification of CFA, and its attempt to put facts on the ground through hydraulic structures, has brought the relationship between Ethiopia and Egypt to its lowest level since the establishment of the Nile Basin Initiative (NBI).[14]

---

10 Ministry of Water Resources, *Ethiopian Water Sector Strategy* (Federal Democratic Republic of Ethiopia, 2001).

11 New Business Ethiopia, 'Meles Launches Millennium Dam Construction on Nile River' (New Business Ethiopia, April 2, 2013). Available online at http://newbusinessethiopia.com/index.php?option=com_content&view=article&id=466:meles-launches-millennium-dam-construction-on-nile-river&catid=35:trade&Itemid=12 (accessed April 2, 2013).

12 The two notable Nile agreements are: 'Exchange of Notes between His Majesty's Government in the United Kingdom and the Egyptian Government in Regard to the Use of the Waters of the River Nile for Irrigation Purposes' (adopted May 7, 1929). Available online at http://ocid.nacse.org/tfdd/tfdddocs/92ENG.pdf (accessed March 25, 2013) (1929 Nile Agreement); 'Agreement between the Republic of the Sudan and the United Arab Republic of Egypt for the Full Utilization of the Nile Waters' (adopted November 8, 1959, entered into force December 12, 1959) 453 UNTS 66 (1959 Nile Agreement).

13 Waterbury, J. *The Nile Basin: National Determinants of Collective Action* (Yale University Press, 2002), at 79. See also International Water Law Project, *Agreement on the Nile River Basin Cooperative Framework* (opened for signature May 14, 2010). Available online at www.internationalwaterlaw.org/documents/regionaldocs/Nile_River_Basin_Cooperative_Framework_2010.pdf (accessed March 25, 2013).

14 Extraordinary Meeting of Nile Council of Ministers (Dar es Salaam, Tanzania, February 22, 1999), 'Agreed Minutes' (1999).

Egypt has officially requested to be provided with all information related to the proposed Renaissance Dam, stating that its final position will be determined on the effect on its water quota.[15] Following a diplomatic visit by Egyptian delegation, the Ethiopian Prime Minister has agreed to a request by Egypt for an expert examination of the impacts of Ethiopian projects on the Nile waters.[16]

The current conflict between Egypt and Ethiopia over the planned measures on the Blue Nile coincides with disagreement over the CFA, which is signed by six countries, passing the threshold for its entry into force once ratified or acceded to. The CFA still divides two sets of basin states, the downstream countries Egypt and Sudan on the one hand, and the rest of the riparian states, mainly over the interpretation of the provision on 'water security' in Article 1(f). Apprehensive of the unfolding reality of a new legal regime, which is a danger to its 'historical rights' under the 1929 and 1959 Agreements, Egypt has undertaken a diplomatic campaign to make every effort to delay the process of ratification. Accordingly, when its diplomatic mission visited Ethiopia in April 2011, it was able to secure Ethiopia's agreement to postpone the ratification until such time that a new election is made and a new government formed in Egypt.[17]

Although some of the above measures seemed to have some effect in mitigating the heightened tensions over the CFA and the new dam in Ethiopia, the issue of conflict and cooperation continue to remain as critical as ever, unless an amicable solution is found to basic legal issues splitting downstream Egypt and Ethiopia, as well as the rest of the basin states.

Ethiopia and the UNWC

Ethiopia was actively involved in the work of the ILC and the debates in the 6th Committee (Legal) of the UNGA in the law of the non-navigational uses of international watercourses. It was among three other Nile Basin countries, Egypt, Rwanda and Tanzania, that abstained during the vote on adoption of the UNWC. Among Ethiopia's reasons for its abstention was that, 'the Convention was not balanced, particularly with respect to safeguarding the interest of upper riparian states', and as a result 'put an onerous burden on upper riparian states'.[18]

Regarding existing agreements, it argued in favor of adjustment of specific

---

15 'Egypt Looks Closer Ties with Ethiopia' (*Ahramonline*, April 21, 2011). Available online at http://english.ahram.org.eg/News/10503.aspx (accessed March 25, 2013).

16 'Political Relations: Exchanged Visits' (Egypt State Information Service, April 2, 2013). Available online at http://new.sis.gov.eg/En/Templates/Articles/tmpArticles.aspx?CatID=1231 (accessed April 2, 2013).

17 Yohanes, R. and Jemal, N. 'Egyptian Public Diplomatic delegation meets Prime Minister Meles Zenawi' (Ethiopian Radio and Television Agency, April 3, 2011). Available online at www.ertagov.com/en/component/content/article/34-top1-news/339-egyptian-public-diplomatic-delegation-meets-pm-meles-.html (accessed March 25, 2013).

18 UNGA, 'General Assembly Adopts Convention on Law of Non-Navigational Uses of International Watercourses' (Press Release GA/9248, May 21, 1997). Available online at www.un.org/News/Press/docs/1997/19970521.ga9248.html (accessed March 25, 2013).

watercourse arrangements to the UNWC, so that the implementation of its provisions to the characteristics of a particular watercourse could not be undermined.[19]

However, Ethiopia's abstention and the reasons for it should not be construed to mean a lack of support for the UNWC. On the contrary, in its statement after the vote it declared that 'the Convention might encourage negotiations to ensure equitable utilisation and promote cooperation'.[20] Moreover, its positive contribution in the adoption of the UNWC's substantive and procedural rules by the CFA, and the incorporation of some of these principles into its national laws and policies, convey affirmative support for the UNWC.

The role of Ethiopian laws and policies

The ancient Ethiopian law, the Fetha Negest, or the Law of the Kings, compiled around 1240, combined rules administering spiritual and temporal matters and set a number of rules governing water use.[21] Among them are rules requiring upstream users not to cause harm to downstream users, the right of compensation to upstream inhabitants for the fertile soil provided to the downstream users, the right to draw water (*aqua haustus*), the right to an access point to water animals and continued dominion over eroded soils.[22]

The Ethiopian Civil Code provides specific provisions regarding ownership and use of water. The provisions under the Ethiopian Civil Code constitute the first set of comprehensive rules on water resources management. Government, as the competent authority, is responsible under the Civil Code for the control and preservation of all surface and ground water.[23] The Civil Code, per Article 1229(2), also provides priority of water use to the community, while allowing private ownership of water collected in a man-made reservoir, basin or cistern from which it does not flow naturally.

The power to resolve disputes in water use is given to the courts. The law protects ownership rights by imposing compensation for their infringement, although the same could be set aside by the law.[24] The Ethiopian Constitution exclusively vests the right of ownership of rural and urban land and all natural resources, including water, in the state.[25] Under Article 51(11) of the current Constitution, the power to determine and administer the utilization of the waters or rivers and lakes linking two or more states or crossing the boundaries of the national territorial jurisdiction belongs to the Federal Government.

---

19 Ibid.
20 Ibid.
21 Tzadua, A. P. (trans.), Strauss, P. L. (ed.), *The Fetha Nagast: The Law of the Kings* (Carolina Academic Press, 2009).
22 Arsano, Y. *Ethiopia and the Nile: Dilemmas of National and Regional Hydropolitics* (Center for Security Studies, 2007), at 110–11.
23 Civil Code of the Empire of Ethiopia, Article 1228(2), Negarit Gazeta, 1960, Vol. 165, No. 2.
24 Ibid.
25 Constitution of the Federal Republic of Ethiopia, Article 40(3), Proclamation No 1/1995, Federal Negarit Gazeta, 1995.

The 1999 Ethiopian Water Resources Management Policy establishes equitable, sustainable and efficient development of the country's water resources as its main guiding principle.[26] The policy affirms water as a natural endowment commonly owned by all the peoples of Ethiopia, where as far as conditions permit every citizen shall have access to sufficient water of acceptable quality to satisfy basic human needs. Water is also considered an economic and social good. Social equity, economic efficiency, and stakeholder participation are part of policy principles.[27]

On transboundary waters, the policy principles promote the establishment of an integrated framework for joint utilization, equitable cooperation and agreement. It also ascertains and promotes Ethiopia's entitlement and use of transboundary waters based on accepted international norms and conventions endorsed by Ethiopia.[28] The policy promotes joint basin agreements for a joint and efficient use of transboundary waters, based on the 'equitable and reasonable' use principles. The implementation aspect of the policy provides for compliance with those international covenants adopted by Ethiopia in the management of transboundary waters.[29]

Should Ethiopia ratify the Convention?

The question of Ethiopia's ratification shall be viewed within the ongoing global initiative for the world's governments to join and implement the UNWC, as part of a call for its entry into force.[30] To date, 30 contracting states have become parties to the UNWC, only five short of the required number for its entry into force.[31]

Ethiopia's advantage of ratification envisaged from the perspective of the value of the entry into force of the UNWC is to further strengthen the commitments under the CFA, most of which are consistent with the UNWC and endorsed by Ethiopia. Externally, Ethiopia's status can be enhanced, as well as the relevance and global recognition of the Convention, and it will be a contribution for the benefit of future agreements.[32]

A 2012 assessment by a group of researchers, including the author, shows that the majority of opinion on whether Ethiopia should support the UNWC stressed the need for a clear understanding of whether the UNWC promotes Ethiopia's interest on the use of its transboundary waters.[33] In particular, there is a need for a

26 Ministry of Water Resources (note 3), at 1.1.
27 Ibid, at 1.2–1.3.
28 Ibid, at 2.1.2.
29 Ibid.
30 See WWF Global, 'UN Watercourses Convention' (WWF). Available online at http://wwf.panda.org/what_we_do/how_we_work/policy/conventions/water_conventions/un_watercourses_convention/ (accessed March 25, 2013).
31 Ibid.
32 Rieu-Clarke, A. and Loures, F. R. 'Still Not in Force: Should States Support the 1997 UN Watercourses Convention?', *Review of European Community and International Environmental Law*, 2009; 18: 185–97.
33 IHP-HELP Centre for Water Law, Policy and Science, 'UN Watercourses Convention (UNWC) Global Initiative' (University of Dundee). Available online at www.dundee.ac.uk/water/research/unwcglobalinitiative/ (accessed March 25, 2013).

better understanding of the content of the provisions of the Convention, and the implications of their implementation in light of the current developments in the Nile, in particular the construction of the Renaissance Dam in the country and the entry into force of the CFA. The relevance of the UNWC in consolidating Ethiopia's position, therefore, requires further deep awareness creation and capacity building for the implementation of the CFA.

Conclusion

Ethiopia had actively participated in the process of codification and development throughout the work of the ILC and the UNGA. Despite its state position during the adoption of the UNWC, its participation, in particular its comments and observations in the ILC, the UNGA and the 6th Committee on issues relating to non-navigational uses on international watercourses offered vital inputs, and thus formed the basis towards the codification and progressive development of the UNWC. On the other hand, its valuable contribution at all stages of the process had a profound impact in dealing with a number of difficult issues in the CFA. Consequently, the study of these lessons and experiences has shaped the process of the negotiating the CFA, which incorporates a number of principles of the UNWC, and can be utilized as a basis for Ethiopia's future ratification of the UNWC. Getting on board countries such as Ethiopia can be considered an important breakthrough, not only from a point of view of the Nile legal discourse, but from the wider implication on upstream positions, which seem less enthusiastic about the UNWC.

17 El Salvador

Meg Patterson and Alexander López

This chapter looks at the water resources of El Salvador and related governance frameworks, as a basis to consider the specific role and relevance of the UN Watercourses Convention (UNWC) in El Salvador. After an overview of the available water resources and current problems affecting them, the chapter examines the political dynamics between El Salvador and its neighbors. Next, the chapter looks at domestic and international legal instruments governing fresh water in El Salvador. Finally, the chapter assesses how the UNWC might fit into this complicated picture of transboundary water management.

Water resources and related challenges in El Salvador

El Salvador has considerable freshwater resources, with ten river basins and numerous aquifers. Annual precipitation in the country is well above the world average, varying from 140 cm in the coastal zone to more than 240 cm in the mountainous interior. Internal renewable water resources in El Salvador are estimated at 17.8 km³.[1] Among Central American countries, however, El Salvador has the lowest availability of water, per capita, in absolute terms, as well as the highest rate of extraction.[2] Furthermore, most rain falls between May and October, with frequent droughts during the winter months. Seasonal concentration of precipitation contributes to high runoff rates and erosion. The uneven distribution of water resources across El Salvador's territory is another challenge. Projections indicate that, by 2030, 13.2 percent of the population will be under water stress – a number that could jump to almost 30 percent by 2050.[3]

The demographics of El Salvador pose another challenge to freshwater resources. El Salvador is the most densely populated country in Central America. Every year, its population grows by 100,000. Although 88 percent of its arable land

1 UN Food and Agriculture Organization, 'El Salvador' (Aquastat, 2000). Available online at www.fao.org/nr/water/aquastat/countries_regions/el_salvador/indexesp.stm (accessed March 25, 2013).
2 López, A. 'Política de la Subregión Centroamérica Hacia el V Foro Mundial del Agua' (Comité Regional de Recursos Hidráulico del Sistema de la Integración Centroamericana, 2009), at 3.
3 López, A. and Porta, M. A. 'Estudio Línea Base Convención de la ONU Sobre los Cursos de Aguas Internacionales: El Salvador' (2011), at 3 (on file).

is under cultivation, El Salvador is a net food importer. As its population continues to grow, El Salvador will have to find ways to improve agricultural efficiency. Currently, irrigation covers just 22 percent of all irrigable land, with considerable losses due to poor maintenance. More irrigation could help improve food productivity, but would also require more water. Agriculture is already the largest water user by sector, accounting for 46 percent of extractions. Yet, agriculture only employs 19 percent of the population and accounts for 11.3 percent of gross domestic product.[4]

Water quality is a serious concern in El Salvador, with most surface water unsuitable for consumption. Much of El Salvador's industrial and domestic effluent is discharged into rivers and coastal areas without any treatment. The Acelhuate River, which drains the capital, San Salvador, is so highly contaminated with heavy metals and domestic and industrial waste that it is considered a biohazard.[5]

While El Salvador's plentiful groundwater resources have thus far been sufficient to meet the country's freshwater needs, pollution is beginning to threaten aquifers, too. In addition, deforestation has negatively impacted recharge rates, and saltwater intrusion has begun to affect coastal aquifers.[6]

El Salvador's reservoirs, currently numbering four, and all located on the Lempa River, are also at risk. The Cerron Grande reservoir, for example, contains significant amounts of industrial and domestic waste, as well as high levels of sedimentation, which threaten the reservoir's health and sustainability.

To complicate matters, the future of El Salvador's water resources depends to a large extent on its neighbors. El Salvador relies heavily on its three international watercourses: the Lempa, Paz and Goascorán Basins. These basins extend over almost 62 percent of El Salvador's territory and account for 34 percent of the country's available freshwater.[7] The Lempa is El Salvador's most important river, spanning almost half of its territory. The basin is home to 48 percent of the population and provides water for irrigation, fishing, industry and domestic consumption.[8] The four hydropower facilities on the Lempa River provide close to 40 percent of the country's electricity, though only 21 percent of the basin's hydropower potential has been developed within El Salvador.[9] The two water treatment plants within the Lempa Basin provide water to San Salvador.

El Salvador shares the Lempa Basin with Guatemala and Honduras. As the most downstream riparian, El Salvador is the most vulnerable to significant changes in river flows through unilateral developments upstream. Honduras produces 48 percent of the sediment in the Lempa Basin, with Guatemala contributing another

4   FAO (note 1).
5   Buckalew, J. O. *Water Resources Assessment of El Salvador* (US Army Corps of Engineers, 1998), at 11.
6   López, A. *Environmental Conflicts and Regional Cooperation in the Lempa River Basin: The Role of Central America's Plan Trifinio* (Adelphi Research, and others, 2004), at 10. Available online at www.ce.utexas.edu/prof/mckinney/ce397/Topics/El%20Salvador/Lopez_2004.pdf (accessed March 25, 2013).
7   See FAO (note 1).
8   López (note 6), at 11.
9   López and Porta (note 3), at 8.

13 percent. Deforestation in the Honduran and Guatemalan reaches of the Lempa Basin and, to a lesser extent, within El Salvador, has reduced land infiltration capacity and caused erosion, leading to sedimentation of reservoirs and increased flooding in the coastal plains. Pollution from Honduras and Guatemala is also a concern – 16 of 19 Guatemalan districts in the Lempa Basin do not treat their sewage.

Therefore, as the case of El Salvador makes clear, state sovereignty can no longer dominate foreign policy. Successfully addressing the challenges faced by the Lempa and other transboundary basins requires international cooperation. The 'internationalization' of environmental problems challenges traditional notions of state sovereignty. Delineating borders on a map and defending them is no longer a straightforward process. Pollution and other environmental problems do not respect international boundaries. Maintaining the integrity of state borders and a healthy environment within border zones now requires not just military and political force, but also communication and cooperation. International law has evolved to acknowledge this, and the time has come for Central American countries to embrace the change.

International watercourses have not yet been a source of major disputes between El Salvador and its neighbors. The possibility, however, is there: the levels of freshwater degradation are high, the sources of the degradation are known, and El Salvador has a high reliance on the impacted resource and is particularly vulnerable as a downstream riparian.[10] As shown above, pollution upstream and changes in river flows are already having a considerable impact on El Salvador. Hence, cooperation towards protecting the basin from further degradation is critical for the country's socioeconomic development and political stability in the region.

The poor status of legal and institutional water governance, both at the domestic and international levels, heightens the likelihood of El Salvador and its co-riparians becoming involved in disputes over water resources, and any disagreement quickly escalating, as examined in the next sections.

The international governance of El Salvador's transboundary waters

Until recently, water cooperation between El Salvador and its neighbors would have seemed unlikely. El Salvador and Honduras have had a tense relationship since at least 1969, when the two countries went to war over disputed border regions and the emigration of large numbers of Salvadorans to Honduras. The 'Soccer War', as it was known, only lasted several days, but the resulting tensions continued for decades and contributed to serious social conflict in both countries.[11] The border issues were not resolved until after El Salvador's civil war, following a 1992 ruling by the International Court of Justice (ICJ) and intervention by the Organization

---

10   López (note 6), at 11.
11   López and Porta (note 3), at 8.

of American States.[12] That ruling gave much of the disputed territory to Honduras, but did not completely settle the boundary issues between the two neighbors – in 2006, a small island in the Gulf of Fonseca was the source of another border friction.[13]

Despite this, the two countries, along with Guatemala, have a history of cooperation under the Trifinio regime. Such efforts were launched in 1986, when the three states signed a technical cooperation agreement to foster regional integration.[14] That Agreement lays out the principles, conditions and terms of reference for the elaboration of a joint plan to foster the integrated socioeconomic development of the border area, as well as the protection and sustainable management of its natural resources.

Among other obligations, Article 11(A) requires the three states to share all the information available that may be relevant for carrying out the Agreement. This includes the governments' goals and planned and ongoing projects for the development of the border area in their respective territories. Under the provisions of Article 13, any dispute regarding the application or interpretation of the Agreement is to be resolved through negotiation, or other agreed means.

To inform and guide the cooperation process, the Annex to the 1986 Trifinio Plan Agreement refers to the development of studies to understand the natural resource base of the area and the problems that hinder its development and the formulation of a strategy and a plan for border zone development, including a subprogramme for the management of fresh water and other renewable natural resources.

In 1988, the parties to that agreement concluded the first version of the Trifinio plan, outlining 29 projects. Included among these projects was one devoted to the joint planning and development of the upper Lempa Basin, including studies of present and potential water uses and the drafting of a water management proposal for submission to the three governments.[15]

In 1997, El Salvador, Guatemala and Honduras concluded the Treaty for the Execution of the Trifinio Plan,[16] as updated and approved in 1993, to further cooperation in the region. The 1997 Treaty lays out the institutional structure for

12 *Land, Island and Maritime Frontier Case (El Salvador/Honduras, Nicaragua intervening)* (Judgement) [1992] ICJ Rep 351.
13 López and Porta (note 3), at 8.
14 Acuerdo de Cooperación Técnica de los Gobiernos de las Republicas de Guatemala, Honduras y El Salvador con la Secretaria General de la Organización de los Estados Americanos y el Instituto Interamericano de Cooperación para la Agricultura para la Formulación de un Plan de Desarrollo Integral en la Región Fronteriza de los Tres Países (adopted November 12, 1986) (1986 Trifinio Plan Agreement); López (note 6), at 16.
15 Plan Trifinio, *Plan de Desarrollo Regional Fronterizo Trinacional Trifinio* (September 1988), at 133, 135. Available online at www.sica.int/busqueda/busqueda_archivo.aspx?Archivo=info_4929_1_30112005.pdf (accessed March 25, 2013) (1988 Trifinio Plan).
16 Tratado entre las Republicas de El Salvador, Guatemala y Honduras para la Ejecución del Plan Trifinio (adopted October 31, 1997, entered into force May 28, 1998). Available online at www.sica.int/busqueda/busqueda_archivo.aspx?Archivo=trat_1302_3_28042011.pdf (accessed March 25, 2013).

ensuring the implementation of the Plan and its regular updating and defining the roles and responsibilities of the several bodies involved – all operating under the umbrella of the Trinational Commission. Apart from establishing an institutional framework, the 1997 Treaty does not incorporate any specific principles, rules or procedures of international water law. With regard to disputes, Article 22 determines that parties shall resort to negotiations and, in the absence of agreement, to other peaceful means, including the Central American Court of Justice.

Together, the above agreements and arrangements form the cooperation regime for the Trifinio region, including its water resources. Although the Trifinio regime is much broader than water management per se, it does not overlook the need to foster the cooperative management of the shared Lempa Basin in a sustainable manner. Overall, the Trifinio regime offers an important precedent in Central America – it has built interstate confidence, facilitated post-conflict high-level dialogue and fostered cooperation among border communities.

The Trifinio regime, however, focuses on the border region, and only encompasses the upper 19 percent of the Lempa Basin. Moreover, environmental efforts so far have been limited to reforestation. While that provides innumerable benefits for the watershed, reforestation alone will not address all of the management challenges facing water resources and ecosystems in the border area and the larger catchment.

Beyond the Trifinio regime, El Salvador and Honduras established a bi-national commission for the sustainable management of the Goascorán Basin in 2006. The Commission coordinates transboundary activities and joint management of the shared resource, with the involvement of local and national governments, civil society and other institutions. This Commission also plays a role in improving the living standards of communities in the basin.[17]

El Salvador and Guatemala have also concluded bilateral agreements. A 1938 border agreement[18] addresses their shared waters. Article 2 of the 1938 Agreement grants each government the right to utilize half the volume of water in shared rivers for agricultural or industrial purposes. Within the framework of that Agreement, those states later created the International Border and Waters Commission (CILA, based on its Spanish initials) to conserve the border line fixed in the 1938 Treaty.[19] CILA is also charged with studying river flows and developing works and projects for the beneficial use and exploitation of fresh waters, such as for defense against floods. In planning such measures, CILA must safeguard the rights of each party under the Agreement. In the case of disputes, questions and problems relating to international watercourses must be dealt with according to the norms and principles of recognized international law.

---

17 López and Porta (note 3), at 15.
18 Tratado de Límites Territoriales entre Guatemala y El Salvador (adopted April 9, 1938, entered into force May 24, 1938) 4390 LNTS 295.
19 See Comisión Internacional de Límites y Aguas entre Guatemala y El Salvador, DO No 2, T 261, 22 Nov 1978, at 8 (El Salvador).

In 2000, through a Memorandum of Understanding,[20] Guatemala and El Salvador decided to establish a Binational Commission with a broader mandate than CILA. This new Commission serves as a forum for bilateral coordination, consultation and negotiation on a broad range of topics related to cooperation between El Salvador and Guatemala. The Commission's mandate includes cooperation on economic, commercial, financial, educational, scientific, technical, cultural, touristic, energy generation, transport, security, organized crime, natural disasters and border development issues. Article 2 of the Memorandum directs the Commission to study and monitor the application of cooperation agreements concluded or under consideration between the two countries, identifies new possibilities for cooperation and proposes the relevant recommendations and facilitates the exchange of information and bilateral consultations.

These commissions are a step in the right direction, but have not fully addressed the need for integrated river basin management. Overall, the commissions in questions have operated largely without a proper legal basis determining the rights and obligations of the riparian states in the use, development and protection of international watercourses and their ecosystems.

El Salvador's domestic governance of water resources

The Constitution of El Salvador gives the state the right to administer water, and grants it expropriation rights for the public good.[21] The Constitution also addresses natural resources, including water, and affirms the public interest in protecting, restoring and rationally using such resources.[22]

However, the diffusion of competencies among numerous agencies does little to further that goal – there are 25 agencies with responsibility for various aspects of water resources, with little coordination among them. The agency in charge of providing 'water and sewers' for the population also regulates water extractions.[23] The country's hydroelectric agency has the responsibility to develop and utilize the hydropower potential. The Environment and Natural Resources Ministry is in charge of preventing and controlling pollution.[24]

The laws governing El Salvador's water resources are also disparate. According to one of the relevant statutes, the state owns surface and underground water resources, water is a national good and the state must prioritize allocation of water for human consumption.[25]

---

20 Memorandum de Entendimiento para la Creación de la Comisión Binacional Salvadoreña-Guatemalteca [Memorandum of Understanding for the Creation of the Salvadoran-Guatemalan Binational Commission] (August 19, 2000) at Article 1 (on file with authors).
21 Constitución Política de le Republica de El Salvador (1983), at Article 106.
22 Ibid., at Article 117.
23 Ley de la Administración Nacional de Acueductos y Alcantarillados, DL No. 341, October 17, 1961, at Article 2 (El Salvador); López and Porta (note 3), at 15.
24 Ley de la Comisión Ejecutiva Hidroeléctrica del Río Lempa, DL No. 45, September 18, 1948 (El Salvador); Ley de Medio Ambiente, DL No. 233, May 4, 1998 (El Salvador).
25 Ley de Riego y Avenamiento del Ministerio de Agricultura y Ganadería, DL No. 153, 1970, at Article 3 (El Salvador).

In an attempt to better systematize water governance, El Salvador passed a new water law in 1981 – the first of its kind in Central America and, in many ways, ahead of its time. The law: a) acknowledges that water, wherever it is located and in whatever physical state, is always part of the hydrological cycle, and should be regulated under one framework; and b) creates a new office in charge of developing a national water plan and coordinating with relevant entities for its execution and evaluation of results, to avoid duplication and conflict over water.[26] The law neither establishes principles to assist the newly created water office in developing a national water policy, nor creates mechanisms for coordination and integration among ministries.

To address these remaining gaps, a general water law draft bill is currently going through the legislative process. Submitted by the Environment and Natural Resources Ministry under the previous administration, the bill focuses on the integrated management of water resources at the basin level. The bill recognizes the fundamental right to water, and considers water to be a strategic but limited and vulnerable resource.[27] This bill, however, is unlikely to be approved under the current administration – the Environment and Natural Resources Minister has already voiced his opposition to the bill.

In the international arena, since the 1990s, El Salvador has been an active participant in global conventions. The most relevant agreements to which El Salvador is a party are the Ramsar Convention, the UN Framework Convention on Climate Change, the UN Convention on Biological Diversity and the UN Convention to Combat Desertification.[28] El Salvador has also joined regional agreements on biodiversity, climate change, forests and hazardous wastes.[29]

---

26 Ley sobre Gestión Integrada de los Recursos Hídricos, DO Decreto No. 886, December 2, 1981, at Preamble, Recital 1 and Articles 1–2 (El Salvador).

27 See, e.g. Llewellyn, R. O. 'El Salvador Mulls Total Ban on Mining' (*Mongabay.com*, October 22, 2012). Available online at http://news.mongabay.com/2012/1022-mining-el-salvador-llewellyn.html (accessed March 25, 2013).

28 The Convention on Wetlands of International Importance especially as Waterfowl Habitat (adopted February 2, 1971, entered into force December 21, 1975) 996 UNTS 245 (Ramsar Convention); UN Framework Convention on Climate Change (adopted May 9, 1992, entered into force March 21, 1994) 1771 UNTS 107; UN Convention on Biological Diversity (adopted June 5, 1992, entered into force December 29, 1993) 1760 UNTS 79; UN Convention to Combat Desertification in Those Countries Experiencing Serious Drought and/or Desertification, Particularly in Africa (adopted June 17, 1994, entered into force December 26, 1996) 1954 UNTS 3.

29 Convenio para la Conservación de la Biodiversidad y Protección de Areas Silvestres Prioritarias en América Central (Costa Rica-El Salvador-Guatemala-Honduras-Nicaragua-Panamá) (adopted 5 Jun 1992). Available online at www.inbio.ac.cr/estrategia/Leyes/convenio_C_A.HTML (accessed March 25, 2013); Convenio Regional sobre Cambios Climáticos (Costa Rica-El Salvador-Guatemala-Honduras-Nicaragua-Panamá) (adopted October 29, 1993). Available online at www.sica.int/busqueda/busqueda_archivo.aspx?Archivo=conv_1255_2_16062005.htm (accessed March 25, 2013); Convenio Regional para el Manejo y la Conservación de los Ecosistemas Naturales Forestales y el Desarrollo de Plantaciones Forestales (Costa Rica-El Salvador-Guatemala-Honduras-Nicaragua-Panamá) (adopted October 29, 1993). Available online at www.inbio.ac.cr/estrategia/coabio/Conv_Fores.html (accessed March 25, 2013); Acuerdo Regional sobre Movimiento Transfronterizo de Desechos Peligrosos (Costa Rica-El Salvador-Guatemala-Honduras-Nicaragua-Panamá) (adopted December 11, 1992). Available online at http://ban.org/library/centroamerica.html (accessed March 25, 2013).

El Salvador and the UN Watercourses Convention

El Salvador did not take part in the drafting process of the UNWC, and was absent from the final vote. This is not surprising given the political turmoil in the country during much of the negotiations.

More recently, El Salvador's active participation in negotiations on the Draft Articles on the Law of Transboundary Aquifers suggests some interest in the topic of shared freshwater resources.[30] During such discussions, El Salvador underscored 'that States had sovereignty over aquifers located within their territory, but that such sovereignty must be exercised in accordance with all the obligations laid down in the draft articles and in international law'.[31] This indicates that El Salvador does not consider itself to have absolute control over its water resources, but rather accepts the limitations that emerge from international law. El Salvador also noted the importance of preventing environmental harm, particularly 'given the irreversible nature of certain processes, such as the damage caused to water resources by excessive pollution'.[32] Although El Salvador made these comments in the context of transboundary aquifers, the fact that 90 percent of its surface waters are too polluted for consumptive uses suggests that it might have said similar things during the UNWC negotiations, had it participated.

Within this context, the absence of clearer, more detailed watercourse agreements spelling out the rights and duties of El Salvador and its co-riparians under international water law makes the UNWC an attractive legal instrument to manage their shared basins. The UNWC could fill in legal gaps, strengthen existing cooperation arrangements and provide a solid legal basis for informing the work of the joint bodies mentioned above.

For example, the focus of the Trifinio regime is on the border area, not on the international waters that drain the region. Therefore, the regime does not cover the entire Lempa Basin, but only its upper portions. Adopting the UNWC would allow the parts of the basins already covered to remain under the Trifinio regime, while extending some legal protection to those sections not covered by that regime.

In addition, the Trifinio regime lacks specific principles, rules and procedures for the cooperative, equitable and sustainable development, management and protection of international watercourses. In this sense, the UNWC's better developed normative content could supplement the Trifinio regime, without affecting its continued validity, as per Article 3(1) of the Convention. This will be particularly important when each state decides to undertake unilateral development projects in the border area or as climate change begins to demand better coordinated responses across international boundaries.

---

30 See, e.g. Report of the Secretary General, 'The Law of Transboundary Aquifers' (June 29, 2011) UN Doc A/66/116, at 8–10 (2011 Country Comments); UNGA 6th Committee (66th Session), 'The Law of Transboundary Aquifers' (October 18, 2011) (on file).

31 See, e.g. 2011 Country Comments (note 30).

32 Ibid.

For example, the Trifinio regime has a strong focus on regional cooperation and joint actions, including carrying out joint studies and sharing existing information. However, those actions apply specifically to information related to projects carried out under the Trifinio Plan itself. Article 9 of the UNWC can supplement the regime by encouraging information sharing on a broader range of topics of relevance for the sound management of shared water resources. Articles 11–19, on planned measures, and 27–28, on harmful conditions and emergencies, provide further guidance on data sharing.

The UNWC's dispute resolution procedures could also provide useful additional guidance in this context. Given the history of tension among the countries in this region, a more detailed dispute resolution procedure, like the one spelled out in the UNWC, might help to avoid conflicts or resolve disputes before they escalate.

The UNWC could also play an important role beyond the Trifinio regime and the Lempa Basin. For example, CILA is charged with carrying out projects to benefit both states. The Commission, however, does not have detailed legal guidance on how to do that, what factors to consider when providing advice on water uses or what the consequences may be if injury through an international watercourse occurs.

Generally speaking, El Salvador's current agreements do not specifically address ecosystem protection, pollution or mitigation measures for causing transboundary harm. Without such measures, El Salvador has to rely on the goodwill of its neighbors to protect its interests. This is far from ideal, especially given El Salvador's reliance on its shared waters and its history of strained international relations. The current level of contamination in its surface waters underscores the vulnerability of El Salvador's watercourses. In this sense, the UNWC offers guidance: Article 7 deals with transboundary harm, Article 20 governs the protection of the ecosystems of international watercourses and Article 21 contains specific obligations pertaining to the prevention, reduction and control of water pollution.

El Salvador has indicated its willingness to enter into and implement agreements on water resources with its neighbors. The UNWC could guide negotiations on new treaties and supplement existing watercourse agreements with detailed principles, rules and procedures that codify and develop international water law, making the rights and obligations of all states clear at the outset.

Conclusion

El Salvador is already a party to several multilateral environmental agreements. El Salvador has also taken a proactive role in developing and implementing a trilateral agreement covering the upper Lempa Basin. Acceding to and implementing the UNWC would further demonstrate the country's support for international law as an enabling tool for effective and sustainable transboundary water cooperation.

Part 4

The UN Watercourses Convention, multilateral environmental conventions and international water and environmental policy goals

18 Convention on Climate Change

Flavia Rocha Loures, Christian Behrmann and Ashok Swain

Climate change is expected to exacerbate existing pressures over transboundary water systems – pressures that have caused environmental degradation and fomented disagreements among watercourse states over the years. However, international law can aid in preventing interstate disputes and resolving disagreements over freshwater resources before such quarrels escalate into conflict.

The exact impact of climate change is not yet known. Yet, climate change will have clear bearing upon access to shared water resources, in that it affects the hydrological cycle at geographical scales from global to local. Some regions will become much drier; others, wetter. Variations in precipitation are already leading to more frequent, widespread and severe droughts and floods; changes in the rates of groundwater recharge; high evaporation from freshwater systems; and alteration in river flow regimes. Under a changing climate, an increasing number of high and untimely floods will threaten the safety of dams and other water infrastructure, and severe droughts will drastically reduce water supplies, including for irrigation and hydropower generation. As Arnell argues, 'climate change may affect the demand side of the balance as well as the supply side'.[1] Climate change will not only reduce water availability; it may also increase water demand for various uses. Hence, the projected impacts of global climate change over freshwater resources will be huge and dramatic.

Climate change is set to add another layer of complexity to the challenge of transboundary water cooperation. For example, in certain basins, some parts will experience higher flows, with lower flows dominating other portions of the system. All these factors are likely to place significant strain on existing agreements and institutions for the management of shared water resources. As a global legal framework, the UN Watercourses Convention (UNWC) comes into this picture when we consider whether existing watercourse agreements – and, for that matter, formal cooperation at the basin level alone – offer sufficient legal basis for riparian states to respond cooperatively to the added challenge posed by climate change on water resources.

---

1 Arnell, N. W. 'Climate Change and Global Water Resources', Global Environmental Change, 1999; 9: 31–49.

Watercourse agreements are important to govern the equitable and collaborative management and use of transboundary waters. When sufficiently comprehensive and flexible, and adequately enforceable, watercourse treaties may offer a solid foundation for evolving water-sharing regimes capable of, in situations of crisis and water scarcity, reconciling conflicting interests and maintaining peaceful interstate relations. Such a legal foundation for transboundary water cooperation is vital for supporting sustainable climate change adaptation and mitigation with respect to international watercourses.

Although numerous watercourse agreements exist, most of the world's transboundary water resources still lack adequate legal protection. In most basins, no management agreement is in place, existing agreements are inadequate, or not all states within a basin are parties to existing treaties.[2] These regulatory weaknesses and gaps are significant when they prevent the law, as a pillar of good governance, from fully contributing to cooperation-building over water in the face of climate variability and change.

Therefore, climate change adds uncertainties to the smooth functioning – or even survival – of existing transbourday water regimes and, more broadly, to the cooperation process between co-riparian states. The long-term changes to water availability resulting from climate change are likely to require water agreements and institutions to be flexible and robust enough to cope with greater uncertainties, more frequent emergencies and widespread harmful conditions.[3] As the effects of climate change grow more severe and widespread, therefore, the need for a global legal framework for freshwater to supplement water agreements and spur cooperation at the regional, basin and sub-basin levels becomes ever more evident.

At the global level, the UN Convention on Climate Change (UNFCCC)[4] was adopted in 1992 to govern both mitigation and adaptation to climate change. The UNFCCC, however, lacks a specific requirement for basin states to cooperate with each other in developing and protecting international watercourses against climate change and variability, such as the effects of droughts and floods. Moreover, the UNFCCC provides little guidance with regard to how such collaboration should take place.

This chapter thus explores the potential role of the UNWC in sustaining and improving cooperation between watercourse states in the context of a changing, increasingly unpredictable climate. We ask whether and to what extent the UNWC may supplement regional and watercourse agreements, as well as advance the goals pursued by the UNFCCC, by guiding and fostering cooperative, sustainable and informed mitigation and adaptation strategies among co-basin states. Our conclusion is that an effective UNWC can serve as a unified, widely accepted,

2 UN Environment Programme, *Challenges to International Waters: Regional Assessments in a Global Perspective* (UNEP, 2006), at 35.
3 McCaffrey, S. C. 'The Need for Flexibility in Freshwater Treaty Regimes', *Natural Resources Forum*, 2003; 27: 156–162.
4 UN Framework Convention on Climate Change (adopted May 9, 1992, entered into force March 21, 1994) 1771 UNTS 107 (UNFCCC).

authoritative and comprehensive global legal framework with minimum standards of cooperation for transboundary watersheds in the face of climate variability and change. As such, the UNWC can provide the legal global foundation needed to supplement the UNFCCC and water treaties at various levels, and can draw states' attention to the issues at the core of climate change mitigation and adaptation with regard to international watercourses.

Water-related adaptation and mitigation governance under the Climate Change Convention

There are four aspects of the UNFCCC that are relevant for this chapter: a) the notion of *cooperative* climate change adaptation; b) the regional programmes to facilitate adaptation; c) the preparation of national adaptation programmes of action (NAPAs); and d) the minimization of adverse effects from adaptation and mitigation measures.

Article 4(1)(e) of the UNFCCC contains a general requirement for parties to *cooperate* in preparing for climate change adaptation, and to develop integrated plans for managing coastal zones, water resources, droughts and floods. In Paragraph 1(b), Article 4 acknowledges that climate change adaptation might have regional dimensions. It does so by requiring parties to 'formulate, implement, publish and regularly update national and, where appropriate, *regional* programmes containing... measures to facilitate adequate adaptation to climate change'. Transboundary water cooperation among basin states could be interpreted as a specific application of these provisions. Yet, apart from the idea of regional programmes, Article 4(1)(b) and (e) does not establish any substantive or procedural rights and obligations to guide states through the cooperation process.

Still with regard to adaptation planning, the NAPAs offer an approach to address the adaptation needs of the least-developed countries through enhancing their adaptive capacity to climate variability:

> The NAPA takes into account existing coping strategies at the grassroots level, and builds upon that to identify priority activities, rather than focusing on scenario-based modeling to assess future vulnerability and long-term policy at state level... NAPAs provide a process for LDCs to identify priority activities that respond to their urgent and immediate needs with regard to adaptation to climate change.[5]

Hence, NAPAs focus on local level adaptation strategies, *rather than on long-term policy reform and enforcement mechanisms,* even less so on joint or coordinated inter-state action. Working at the local level is crucial, including for addressing the needs

---

5 UNFCCC, 'Chronological Evolution of LDC Work Programme and Introduction to the Concept of NAPA' (UNFCCC, May 29, 2012). Available online at http://unfccc.int/cooperation_support/ least_developed_countries_portal/ldc_work_programme_and_napa/items/4722.php (accessed March 25, 2013).

of local communities. But will it be sufficient? As Bangladesh's NAPA explains, '92%... of [the country's] annual runoff enters the country from outside its borders... Trans-boundary inflow in the dry season has decreased due to upstream development, and withdrawal of water for irrigation and other purposes'.[6] In the context of climate change, therefore, conditions of high interdependency among co-riparian countries can exacerbate uncertainties about annual and seasonal water availability and variability. Bangladesh is not alone in such a dramatic picture. India is also likely to be hit with more frequent heavy rainfall events and flash floods. Such extreme events might in the future be aggravated or mitigated by increased Chinese control upstream of the waters the two countries share.[7]

Therefore, basin-wide/cross-border thinking is needed to supplement the scope of NAPAs. When climate change gives rise to transboundary water issues, mitigation and adaptation strategies should involve an ongoing dialogue among basin states to be successful. In fact, greater cooperation between co-riparian states may be desired even during the development stage of the NAPAs. Such collaboration at the outset would allow the states concerned to identify priorities on a basin scale that could support coping strategies at the grassroots level. The process would also enable states to explore opportunities for the sharing of benefits and costs from joint planning and implementation of adaptation and mitigation measures within transboundary watersheds.

Article 4(1)(f) of the UNFCCC covers the last of the aspects listed above. That provision requires parties to act diligently with a view to *minimizing* adverse effects from the implementation of climate change adaptation and mitigation measures. The provision in question refers to 'appropriate methods... formulated and determined *nationally*'. Hence, the UNFCCC does not: a) refer specifically to harm prevention; or b) spell out how the less demanding obligation of harm *minimization* is to be implemented in the case of transboundary river basins, lakes and aquifers. In other words, the UNFCCC fails to foresee the possibility that such methods of harm minimization might have to be agreed upon in a transboundary context, taking into account the applicable principles and rules of international water law. As a result, the UNFCCC fails to provide for a duty on basin states to cooperate in the prevention and minimization of transboundary harm potentially arising from activities they undertake to mitigate or adapt to climate change.

In this context, we examine below the relevant provisions of the UNWC that could supplement and strengthen the regulatory framework under the UNFCCC. We look first at provisions of value to cooperative adaptation, including through regional programmes or plans, and then at those relevant to the general duty of transboundary harm prevention and minimization.

---

6   Bangladesh Ministry of Environment and Forest, *National Adaptation Programme of Action (NAPA): Final Report* (UN Development Programme and Bangladesh Ministry of Environment and Forest, 2005), at 5. Available online at http://unfccc.int/resource/docs/napa/ban01.pdf (accessed March 25, 2013).

7   Khadka, N. S. 'Concerns over India Rivers Order', *BBC News* (London, March 30, 2012). Available online at www.bbc.co.uk/news/science-environment-17555918 (accessed March 25, 2013).

Cooperative management

Watercourse agreements

Articles 3–4 of the UNWC govern the adoption of watercourse agreements in a manner that aims to ensure that such agreements take into account the interests of all states within a basin. Article 3(4) and (6) deals with partial agreements; i.e. agreements not involving all states within a basin. Such agreements cannot be significantly detrimental to the riparians not parties thereto, and do not affect the rights and obligations of those third states under the UNWC. Article 4 then secures the right of all basin countries to participate in consultations and negotiations for, and to become a party to, agreements applicable to the entire watercourse. The same applies to consultations and, where appropriate, negotiations of new agreements that only refer to a portion of the basin or to a specific project, programme or use, *to the extent that such an agreement could affect the rights of any of the basin states.*

To illustrate how important the above provisions are, let us consider an example from the Ganges Basin: India and Nepal have entered into water treaties without involving Bangladesh; and Bangladesh has bilateral water agreements with India that do not necessarily consider changes to the basin further upstream in Nepal,[8] even though 'the rivers of Nepal contribute more than 45% of the total flow of the Ganges'.[9] Bangladesh depends on a long-term solution to increase flows in the Ganges during the dry season. In addressing the issue, Bangladesh has proposed a series of storage dams in the river's upper regions as part of its bilateral dialogue with India. While Nepal has traditionally not been involved in these discussions, it is there that such developments would have the most significant impacts with benefits accrued mainly by India and Bangladesh. Arguably, therefore, the development of the Bangladeshi proposal should have been part of a trilateral planning process leading up to a mutually agreed, equitable sharing of costs and benefits[10] – a process that would have given full consideration to Nepal's concerns pertaining to '[s]ubmergence of land, displacement of people upstream and downstream benefits resulting from flood control, flow regulation and hydropower generation'.[11]

---

8 See, e.g. Agreement between the Government of India and the Government of Nepal on the Kosi Project (adopted April 25, 1954). Available online at http://ocid.nacse.org/tfdd/tfdddocs/177ENG.pdf (accessed March 25, 2013); Agreement between His Majesty's Government of Nepal and the Government of India on the Gandak Irrigation and Power Project (adopted April 12, 1959). Available online at http://ocid.nacse.org/tfdd/tfdddocs/233ENG.pdf (accessed March 25, 2013); Statute of the Indo-Bangladesh Joint Rivers Commission (adopted November 24, 1972). Available online at http://ocid.nacse.org/tfdd/tfdddocs/353ENG.pdf (accessed March 25, 2013); Treaty between the Government of the Republic of India and the Government of the People's Republic of Bangladesh on Sharing of the Ganga/Ganges Waters at Farakka (adopted December 12, 1996) (1996) 36 ILM 519.
9 Rahaman, M. M. 'Integrated Ganges Basin Management: Conflict and Hope for Regional Development', *Water Policy*, 2009; 11: 168–90, at 169.
10 Rahaman (note 8), at 176–77, 181 and 186.
11 Ibid., at 186.

Without entering the merits of ongoing disputes in the region, the example above shows that all riparian states need to be involved in discussions regarding droughts and floods, in line with Articles 3–4 of the UNWC.

Basic principles

Equitable and reasonable use (and participation) is the fundamental principle of international water law; it calls for an ongoing process of evaluation of all relevant factors and circumstances, and aims to maintain a balance among the rights of basin states to a reasonable and equitable share of the uses and ecosystem services (and related costs/benefits) of an international watercourse. As formulated in the UNWC, the ultimate goal of equitable and reasonable utilization and participation is sustainable development, consistent with the watercourse's adequate protection.

Article 6(1)(a) of the UNWC makes equitable and reasonable utilization relevant in the context of climate change and variability by including climatic factors among the issues to be considered in the application of that principle. One must interpret that provision in combination with the requirement under Article 6(2), which reads: 'In the application of [equitable and reasonable use and participation], watercourse States concerned shall, when the need arises, enter into consultations in a spirit of cooperation'. These mandatory good-faith negotiations imply a duty on basin states parties to the UNWC to respond to requests by their neighbors to establish or revisit water allocation standards and requirements in the face of climate change and variability, based on the Convention's provisions and other applicable agreements. In so doing, the Convention creates a forum for ongoing interstate dialogue, thereby providing some flexibility to the legal regimes to which co-basin states are bound.

Water allocation readjustments may include measures to, e.g. reallocate existing water uses, conserve water and temporarily or permanently forego certain uses. Arguably, equitable and reasonable use might also entail assessing and securing (or restoring) environmental flows, where necessary for the preservation and protection of the ecosystems of international watercourses, as mandated by Article 20 of the UNWC.

Approaching water rights allocation and the sharing of benefits as part of an ongoing dialogue process, as envisioned under Article 6(2) of the UNWC, will become vital in a changing climate, with the expected impacts on the timing, quantity and quality of river flows. Such impacts will affect how much water can be captured, by which state and when. This, in turn, will call for the periodical revision of current allocations under a flexible apportionment process that addresses annual and seasonal water surpluses or shortages, and considers short- and long-term changes to water availability. The framework of duties and obligations under the UNWC offers the basic principles and rules to be considered in the development and implementation of such a process.

Article 9 of the UNWC codifies and details another basic principle, although of a more procedural nature: the regular exchange of information. When clarifying the nature of the data to be shared, that provision expressly refers to *data and information of a meteorological nature and related forecasts.* In so doing, Article 9 encourages states to assess and evaluate freshwater and meteorological conditions *throughout a river basin*. This basin-wide approach is needed to enable the monitoring of climate change and variability, and to forecast extreme events across larger regions and watersheds.

The idea is to share the information necessary to predict and prepare for climate variability, and determine how and to what extent climate change is affecting the basin and/or parts thereof. This would enable states to distinguish between: a) *long-term climate-related changes*, requiring overall adjustments in water use and allocations, such as permanent cutbacks in water consumption and their apportionment among riparians; and b) *ordinary weather variability*, which can be addressed with extraordinary temporary measures. Such measures may include, e.g. the exploitation of recharging aquifers above replenishment levels during dry periods, followed by compensation measures for the overdraft once the emergency ceases.

Water management

Article 24 of the UNWC deals with the cooperative management of international watercourses. Under this provision, the Convention requires watercourse states to enter into consultations on water management issues, including the establishment of joint governance bodies, upon unilateral request from any co-riparian. Further specifying the content of the above provision, Article 25 of the UNWC governs the regulation of international watercourses, which is of significant importance in the context of climate change and variability. The article requires states to respond cooperatively to needs and opportunities to regulate river flows, and to share the related costs and benefits in an equitable manner.

The Convention also governs harmful conditions and emergencies, which 'differ from each other according to the degree of seriousness and of the imminence of the danger of harm being caused that they, respectively, address. Therefore, the same factual situation, being susceptible to escalation, may, in different points in time',[12] qualify as either one or the other. With climate change, droughts and floods are expected to become more frequent, widespread and intense, making water availability increasingly unpredictable. In such a scenario, neighboring countries will have to agree on basic duties and procedures for addressing those issues collectively or, at a minimum, in a coordinated manner.

As per Article 27 of the UNWC, 'harmful conditions' are those 'related to an international watercourse that *may be* harmful to other watercourse States, resulting from natural causes or human conduct'. They may include, e.g. flood or ice conditions, water-borne diseases, saltwater intrusion, drought or desertification. The specific reference to saltwater intrusion makes the provision relevant for the preservation of deltas, estuaries and coastal aquifers – all ecosystems highly vulnerable to climate change and sea-level rise. The Convention establishes a due diligence duty for riparian states to prevent and/or mitigate harmful conditions, through individual

---

12 Tanzi, A. and Arcari, M. *The UN Convention on the Law of International Watercourses* (Kluwer Law International, 2001), at 223.

and, where appropriate, joint measures. An example of such measures is the Lower Danube Green Corridor – a joint initiative by Bulgaria, Romania, Moldova and Ukraine to restore floodplains along the Danube River and, among other goals, improve the retention and safe release of floodwaters.[13]

In its turn, Article 28(1) of the UNWC defines an emergency as 'a situation that causes, or poses an imminent threat of causing, serious harm to watercourse States or other States and that results suddenly from natural causes, such as floods, the breaking up of ice, landslides or earthquakes, or from human conduct, such as industrial accidents'. The key elements of an emergency, therefore, are transboundary harm (actual or imminent) and the suddenness of the situation, regardless of whether the cause is natural or results from human conduct. In the context of climate change, the breaking up of ice, floods and landslides – all mentioned in Article 28 for illustration purposes – are of particular importance.

The UNWC creates a framework within which states can tackle emergencies cooperatively. Under Article 28(2)–(3) of the Convention, the state where an emergency originates is required to a) notify its neighbors and relevant international organizations, without delay and by the most expeditious means available; and b) immediately take all practicable measures needed to prevent, mitigate and eliminate the emergency's harmful effects, in cooperation with the states and water governance bodies concerned. These obligations apply regardless of liability 'on the part of a watercourse State for the harmful effects in another watercourse State of an emergency originating in the former and *resulting entirely from natural causes*'.[14] As climate change leads to more frequent, widespread and severe floods and other water-related emergencies, increased levels of cooperation between states, in accordance with the above provisions, will become increasingly important.

Article 28(4) spells out an additional requirement, but one of an *anticipatory* nature. The provision in question mandates co-riparians, 'when necessary... [to] jointly develop contingency plans for responding to emergencies, in cooperation, where appropriate, with other potentially affected States and competent international organizations'. A duty to develop *joint* contingency plans for responding to

---

13 Facilitated by WWF, the Lower Danube Green Corridor Agreement was signed in 2000 by the governments of Romania, Bulgaria, Ukraine and Moldova, recognizing a need and shared responsibility to protect and manage in a sustainable way one of the most outstanding biodiversity regions in the world... In 2010... the level of achievement was much higher than expected, with some 1.4 million ha brought under protection... [T]he countries [are] slightly more than a quarter of the way to their target of restoring 224,000 ha of former wetlands.' WWF Global, 'Lower Danube Green Corridor' (WWF). Available online at http://wwf.panda.org/what_we_do/where_we_work/black_sea_basin/danube_carpathian/our_solutions/freshwater/floodplains/lower_danube_and_danube_delta/ (accessed March 25, 2013).

14 ILC, 'Draft Articles on the Law of the Non-Navigational Uses of International Watercourses and Commentaries thereto and Resolution on Transboundary Confined Groundwater', in *ILC, Report of the International Law Commission on the work of its forty-sixth session, 2 May–22 July 1994, Official Records of the General Assembly, Forty-ninth session, Supplement No. 10* (May 2–July 22, 1994) UN Doc A/49/10 (1994). Extract from the Yearbook of the ILC, 1994, Vol. II(2), at 130. Available online at http://untreaty.un.org/ilc/documentation/english/A_49_10.pdf (accessed March 25, 2013). (1994 ILC Report).

possible emergencies is crucial in the context of climate variability and change, and their water-related effects. Such a duty has been incorporated into other treaties, too.[15] Under the Convention, 'whether such plans would be necessary would depend, for example, upon whether the characteristics of the natural environment of the watercourse, and the uses made of the watercourse and adjacent land areas, would indicate that it was possible for emergencies to arise'.[16] Yet, with climate change, the need for such plans – underlying the co-obligation on watercourse states, under Article 28(4), to develop them – will likely become ever more evident in most transboundary basins.

Therefore, the entry into force of the UNWC would bind parties to a clear set of obligations pertaining to harmful conditions and emergency situations. At the same time, arguably, an effective UNWC would contribute to the clarification of the content of those obligations under customary law. Once in force, the Convention would serve as a more compelling policy incentive for states, whether parties to it or not, to incorporate similar provisions into existing and future watercourse agreements; as well as to comply with their obligations under customary law, as codified in the UNWC.

Environmental protection

Climate change is expected to exacerbate existing pressures on aquatic ecosystems, including impacts on the quality, quantity and timing of river flows and the recharge of aquifers. With climate change, therefore, it will become increasingly important for co-riparians to cooperate on matters of ecosystem protection and cross-border pollution control.

In this sense, the UNWC contains an entire section devoted to the environmental protection of international watercourses. In Article 20, the Convention requires states to protect and preserve the ecosystems of international watercourses. Arguably, this duty includes the maintenance of environmental flows where necessary to ensure the protection of aquatic ecosystems. The provision is also unique in that it requires states to act individually and, where appropriate to ensure compliance with this duty, jointly. *Joint* action could involve, e.g. joint assessments of areas of high conservation value within a river system, followed by a collective commitment to keep such areas free from infrastructure development to ensure river connectivity and the overall sustainability of the basin.

Article 21 deals with pollution, defining it, in Paragraph 1, as 'any detrimental alteration in the composition or quality of the waters of an international watercourse which results directly or *indirectly* from human conduct'. A question here that arguably remains open is whether pollution caused or aggravated by climate

---

15 For example, the UN Convention on the Law of the Sea requires states to 'jointly develop and promote contingency plans for responding to pollution incidents in the marine environment'. UN Convention on the Law of the Sea (adopted December 10, 1982, entered into force November 16, 1994) 1833 UNTS 3, Article 199.

16 1994 ILC Report (note 14), at 130.

change (e.g. changes in water temperature) could be considered as resulting indirectly from human conduct. As in Article 20, this provision requires not only individual but also joint action, where needed, to prevent, reduce and control the pollution of an international watercourse that may cause significant harm to other riparian states or their environment. As per Article 21(2), this may include 'harm to human health or safety, to the use of the waters for any beneficial purpose or to the living resources of the watercourse'. The Convention then requires states to: a) take steps to harmonize their policies in this connection; and b) upon request, consult on joint activities to address pollution (e.g. the establishment of quality standards and lists of substances whose release into the environment is to be prohibited, limited, investigated or monitored).

Harm prevention and minimization

This subsection focuses on provisions related to the prevention and minimization of transboundary harm associated with climate change mitigation and adaptation measures and which may affect international watercourses. This is important because, in their intent to reduce carbon dioxide emissions (mitigation) and increase water security (adaptation), states are expected to build more dams and reservoirs – the type of infrastructure development that may have major impacts on freshwater resources and related transboundary repercussions. Another important trend in relation to climate change mitigation is the expansion in biofuel production, which should further increase water demand and the pressure on already stressed supplies.

In this context, Article 7(1) of the UNWC codifies the duty of states to, 'in utilizing an international watercourse in their territories, take all appropriate measures to prevent the causing of significant harm to other watercourse States'. The provision incorporates a due diligence duty of prevention with respect to harm that is transboundary in nature and of significance. This excludes mere inconveniences or minor disturbances states are expected to tolerate from one another, in conformity with the rule of good neighborliness.[17]

The UNWC makes this substantive duty operational through a well-developed body of procedural rules. The Convention requires states to exchange information, consult each other and, if necessary, negotiate on the possible effects of planned measures on an international watercourse. The Convention establishes detailed procedures and timelines applicable before the implementation of actions with potentially significant transboundary effects. Such actions could include, e.g. those capable of significantly altering river flows, thereby increasing vulnerability to climate change in neighboring nations.

17 UN General Assembly 6th Committee (51st Session), 'Report of the Working Group on the Convention on the Non-Navigational Uses of International Watercourses' (April 11, 1997) UN Doc A/51/869.

The process applicable to planned measures is to be triggered with a *timely* notification from the state of origin to all states that may be affected by the measure in question. The state of origin must attach to the notification all available information and existing environmental impact assessments. This documentation is needed for the recipient country to assess and evaluate the effects that may result from a project. The notified country then has six months to reply; silence on the part of the latter means tacit consent and the implementing state may proceed with the proposed project; on the other hand, a finding of inconsistencies with the UNWC must be formalized with a documented explanation. This reply triggers a process of consultations and negotiations between the parties involved.

In the course of these procedures, countries would investigate and discuss the possible effects of adaptation and mitigation activities on an international watercourse or related ecosystems. During this period, measures to prevent, mitigate and/or compensate for significant transboundary harm may be proposed and adopted. Interstate consultations on planned measures are also an opportunity for countries to assess the potential for equitable benefit- and cost-sharing from the measures proposed.

During the consultation period, implementation actions are to remain suspended, as per Articles 14(b) and 17(3) of the UNWC. If no agreement is reached between the state of origin and the potentially affected state upon completion of the time-bound consultation and negotiation procedures, the former can proceed with implementation. Therefore, the Convention does not provide for a veto power. The obligation imposed on the state implementing the measure has a procedural nature: to consult and negotiate in good faith. There is no requirement of mutual agreement. In proceeding unilaterally, i.e. without the consent of an affected country, the state of origin must not interfere with equitable and reasonable uses and benefits enjoyed by its neighbors. In this case, the state implementing the measure risks being held responsible if it fails to comply with its due diligence duty to take all appropriate preventive measures against significant and inequitable harm to other basin states.

Article 16(2) of the UNWC is part of the package of procedural rules on planned measures and aims to ensure a balance in the consultation procedures, in a spirit of good-faith cooperation. The provision at hand determines that claims to compensation by a notified state that has failed to reply within the preset period may be offset by the costs incurred by the notifying state for actions undertaken after the expiration of that period and which would not have been undertaken if the notified state had objected in a timely manner.

If, once the above procedures are concluded, the parties to a dispute fail to reach an agreement, the UNWC's dispute settlement mechanisms apply.

Article 33 establishes guidance and requirements for the peaceful resolution of water disputes between parties, which are applicable in the absence of related provisions in watercourse agreements. Negotiation is the primary means for resolving disputes, with either party entitled to request negotiations with any co-riparian. If states do not reach an agreement through negotiations, they may agree, as a second step, to submitting the dispute to the good offices of, or mediation or conciliation by, a third party; or to resorting to joint bodies, arbitration or the International Court of Justice. *If*, after six months from the time of the request for negotiations, parties have not been able to reach an agreement, any party may unilaterally submit the controversy to impartial fact-finding, unless they agree otherwise. A commission is then created to analyze the case and issue a final report, with findings, motivations and, if appropriate, recommendations. The idea is to prevent the dispute from escalating into a real conflict, while achieving an equitable solution, based on an accurate knowledge and impartial assessment of the facts.

These provisions are of particular importance here, since experts already predict an increase in conflicts between watercourse states in a rapidly changing climate. Through its impacts on freshwater resources, climate change will exacerbate other correlated potential drivers of conflict, such as crop failure and food shortages, energy insecurity and economic stagnation, border disputes among local users, humanitarian crises and population movements. Climate change will do no less than turn parts of the world uninhabitable, thereby forcing a mass migratory movement that is expected to alter ethnic composition and population distribution within and between states. Such changes are likely to cause regional instability and increase domestic and international tensions in affected areas, as a greater number of water users and sectors compete over dwindling resources. Adding up to such physical threats, most watercourse agreements lack adequate dispute prevention and settlement mechanisms, which makes it harder for states to resolve their disputes in a peaceful and equitable manner.

In a basin as important as the Amazon, for example, the existing mechanisms for dealing with disagreements, under the existing multilateral treaty involving all basin states,[18] are inadequate and insufficient. The Amazon Cooperation Council is charged with ensuring compliance with the aims and objectives of the treaty. Solving disputes, however, is not clearly included among the Council's functions. Even if it were, there are no procedures for the Council to follow. And what happens when the Council fails to solve a dispute? While this is likely to happen, since Council decisions must be adopted by consensus, the treaty does not contain any provisions allowing for third-party intervention in dispute settlement. Finally, the treaty envisions the creation of special commissions to solve problems, without establishing any related conditions, procedures or requirements, or defining criteria for a state's refusal to participate in such commissions.

Finally, the environmental provisions of the UNWC, examined in the previous subsection, are also important for purposes of harm prevention and minimization. After all, countries would be bound by those obligations in implementing mitigation and adaptation measures.

---

18   Treaty for Amazonian Cooperation (Brazil-Bolivia-Colombia-Ecuador-Guyana-Peru-Suriname-Venezuela) (adopted July 3, 1978) (1978) 17 ILM 1045.

Conclusion

Through its direct and indirect impacts on water resources, climate change will aggravate the global water crisis, and is likely to increase the risk for disputes between and within riparian states. Climate change has operated and will increasingly do so as a threat multiplier of unprecedented proportions, magnitude and consequences. Although in most cases freshwater has been a catalyst for interstate cooperation, this trend may change as freshwater resources dwindle worldwide, their availability becomes more unpredictable and extreme events grow in intensity and frequency, as a result of climate change. For example, stakeholders might resist, potentially through violence, necessary readjustments in resource use due to reduced water availability. Ongoing tensions in the Middle East illustrate the strong links between water, peace and security.[19] Tensions have also been growing among the Mekong Basin countries, where the planned Xayaburi dam, in Laos, brings to the forefront the water-food-energy nexus and potential security implications thereof.[20]

How climate change might affect peace and security is only now beginning to be assessed, just as the role for interstate cooperation on climate change adaptation and mitigation in a transboundary context needs to be further explored. But it seems clear that the impacts of climate change on freshwater will interconnect us all through the hydrological cycle and through the long-range effects of political instability and failing states; an increasing number of environmental refugees from flooding, disease and famine; mass migration across international borders and even between continents; and multiple natural disasters taking place simultaneously around the globe.

For all these reasons, legal experts must lead the task of developing and testing the effectiveness of international law principles, procedures and regimes for sharing water between states and across sectors, reconciling conflicting interests in situations of water scarcity and maintaining peaceful interstate relations in moments of crisis. As seen above, the UNFCCC does not offer specific guidance on transboundary cooperation at the basin level, and existing watercourse agreements are unlikely to, on their own, offer an adequate and sufficient legal response to the transboundary dimensions of climate change in relation to water resources. This is where the multi-level governance of transboundary basins comes to play, with the UNWC as a unified, widely accepted, authoritative and comprehensive global code. As such, the Convention would ensure legal (and so, to some extent, political) stability in the relations among watercourse states, either by informing the negotiation of new or revised agreements, by supplementing or supporting the application of the UNFCCC and existing watercourse agreements and by governing directly interstate relations in the absence of applicable agreements.

19 See, e.g. Zecchini, L. 'Is Water Being Used as a Weapon in the West Bank?' (*World News Australia*, March 15, 2012). Available online at www.sbs.com.au/news/article/1634580/Is-water-being-used-as-a-weapon-in-the-West-Bank- (accessed March 25, 2013).
20 See Prak Chan Thul, 'Cambodian Villagers Protest Controversial Laos Dam' (*AlertNet*, June 29, 2012). Available online at www.trust.org/alertnet/news/cambodian-villagers-protest-controversial-laos-dam/ (accessed March 25, 2013).

In Magsig's words, an effective UNWC would:

> lift the profile of international water law, increase its enforceability, and accelerate the process of codification of customary law[, as well as] provide for consistency among all states in the management of shared water resources, since all watercourse states would have to implement the provisions in their entirety (including procedural norms)... At its current stage, customary international water law is providing the needed flexibility, but lacks in predictability and stability. Watercourse states try to address these shortcomings by concluding freshwater treaty regimes, which are more precise, but also much more rigid... To overcome these obstacles, it would be helpful to have one global framework in place... The [UNWC] could provide such a comprehensive framework, once it has entered into force.[21]

The UNWC is not the final or only answer to the problems of cooperative climate change adaptation and mitigation in international watercourses, but the Convention can be a big part of the solution. Although states will still need to agree on more detailed mechanisms within watercourse agreements, the UNWC can serve as a general framework always in place and sufficiently flexible to adapt to rapidly evolving conditions.

21 Magsig, B. O. 'International Water Law and Climate Change: How Can International Freshwater Regimes Cope with Increasing Water Stress?' (LLM thesis, University of Dundee, 2008), at 13, 21 (on file).

19 Convention to Combat Desertification

Christian Behrmann, Ashok Swain and Flavia Rocha Loures

The UN Convention to Combat Desertification in Countries Experiencing Serious Drought and/or Desertification, Particularly in Africa (UNCCD)[1] is the main international legal instrument dealing with the urgent global problem of land degradation in arid, semi-arid and dry sub-humid areas. This treaty is sometimes described as the world's 'sustainable development convention', in that it frames desertification as both an environmental and a development issue.[2]

Drylands cover more than 40 percent of the Earth's land surface and are home to nearly 40 percent of the total global population. Land degradation, drought and desertification represent major threats to those biomes.[3] Climate change, excessive cultivation, overgrazing, faulty irrigation management and deforestation are considered to be directly responsible for such threats. However, land degradation, drought and desertification are often exacerbated by a number of underlying social, economic and political realities in the drylands.[4]

Desertification was first conceptualized as an issue in need of global political attention following the severe drought and associated famine in the Sudano-Sahel region of Africa, between 1968 and 1974.[5] In 1977, the UN Conference on Desertification (UNCOD) prescribed a Plan of Action to Combat Desertification, but its implementation was never up to the mark. In 1992, due to increasing desertification, particularly in Africa, the UN Conference on Environment and Development (UNCED) called upon the UN General Assembly (UNGA) to establish an Intergovernmental Negotiating Committee to prepare a convention on desertification. Such a convention was to recognize that 'desertification is a major

---

1   UN Convention to Combat Desertification in Those Countries Experiencing Serious Drought and/or Desertification, Particularly in Africa (adopted June 17, 1994, entered into force December 26, 1996) 1954 UNTS 3 (UNCCD).
2   Stringer, L. C. 'Reviewing the International Year of Deserts and Desertification 2006: What Contribution towards Combating Global Desertification and Implementing the UN Convention to Combat Desertification?', *Journal of Arid Environments*, 2008; 72: 2065–74.
3   Reynolds, J. F. 'Global Desertification: Building a Science for Dryland Development', *Science*, 2007; 316: 847–51.
4   Chasek, P. S. 'The Convention to Combat Desertification: Lessons Learned for Sustainable Development', *Journal of Environment Development*, 1997; 6: 147–69.
5   Thomas, D. S. G. and Middleton, N. J. *Desertification: Exploding the Myth* (Wiley, 1994).

economic, social and environmental problem of concern to many countries in all regions of the world'.[6] The UNCCD was adopted in Paris in 1994 and entered into force in 1996.

The UNCCD enjoys quasi-universal ratification[7] and aims to facilitate effective action through innovative local programmes and supportive international partnerships. In combating desertification, the UNCCD promotes activities to mitigate drought and desertification in the context of the integrated and sustainable development of drylands. As per Article 1(b) and (d) of the UNCCD, such activities are to: (i) prevent and/or reduce land degradation; (ii) rehabilitate partly degraded land; (iii) reclaim desertified land; (iv) aid in the prediction of drought; and (v) reduce the vulnerability of society and natural systems to drought as it relates to combating desertification.

The obligations under the UNCCD vary depending on whether a party is affected by desertification or not:

• Parties affected by desertification in Africa, Asia, Latin America and the Caribbean and the Northern Mediterranean undertake to prepare national action programmes and to cooperate at the regional and subregional levels. Other affected parties have the option of preparing action programmes following UNCCD guidelines or, more generally, establishing strategies and priorities for combating desertification.

• Developed country parties unaffected by desertification have specific obligations to support affected countries (particularly, but not exclusively, affected developing countries) by providing financial resources and facilitating access to appropriate technology, knowledge and know-how.

• The common obligation of all parties, whether affected or not by desertification, relates to international cooperation in implementing the UNCCD through: (i) the collection, analysis and exchange of information; (ii) research and technology transfer; (iii) capacity- and awareness-building; (iv) the promotion of an integrated approach in developing national strategies to combat desertification; and (v) assistance in ensuring that adequate financial resources are available for programmes to combat desertification and mitigate the effects of drought.

Overall, the UNCCD supports countries in sustainably managing land and water resources in collaboration with their neighbors. Like the UN Watercourses

6   UNCCD, 'History' (UNCCD, 2012). Available online at www.unccd.int/en/about-the-convention/history/Pages/default.aspx (accessed March 25, 2013).

7   As of April 20, 2012, 195 states and one regional economic integration organization are parties to the UNCCD. See UN Treaty Collection, 'Chapter XXVII: Environment. 10. UN Convention to Combat Desertification in Countries Experiencing Serious Drought and/or Desertification, Particularly in Africa' (Paris, October 14, 1994) (UN, 2012). Available online at http://treaties.un.org/pages/ViewDetails.aspx?src=TREATY&mtdsg_no=XXVII-10&chapter=27&lang=en (accessed March 25, 2013).

Convention (UNWC), the UNCCD is a framework convention, rooted in the understanding that its objectives can only be reached on a case-by-case basis and as part of a comprehensive balancing process – a process that must be flexible enough to allow for tailor-made solutions in each specific case. Under the UNWC, every legal assessment is subject to the principle of equitable and reasonable utilization and participation; i.e. the ongoing balancing process of all significant aspects of a given situation by means of cooperation. In its turn, Article 10(1) of the UNCCD calls for the establishment of national, subregional and regional action plans to identify the contributing factors and the practical measures necessary to combat desertification and mitigate the effects of drought. In doing so, the UNCCD avoids – similar to the UNWC – the temptation of a one-size-fits-all approach. Rather, both Conventions create a systematic process for identifying and evaluating the factors pertaining to the specific circumstances of an affected area, with a view to reaching sustainable use and development of scarce and/or degraded water and land resources.

Against this background, this subchapter examines if and to what extent the UNWC can support the implementation of the UNCCD through improved cooperation between states in international watercourses. We do so by assessing the relationship between the two Conventions, with focus on the common goals they share and potential areas of synergies in which the UNWC may contribute to strengthening the regulatory framework under the UNCCD with respect to transboundary river basins. In particular, we conclude that the regular sharing of hydrological, meteorological and other relevant data and information among countries within the ambit of the UNWC will enable them to address drought and desertification challenges jointly and more effectively.

Thematic overlap between the UNCCD and the UNWC

The UNCCD applies to the degradation of 'land'. Article 1(e) defines that term broadly: 'the terrestrial bio-productive system that comprises soil, vegetation, other biota, and the ecological and hydrological processes that operate within the system'. This definition leads to a large overlap *ratione materiae* with the UNWC – and rightly so, as the key challenges addressed by each Convention are intimately interrelated.[8] Any attempt to successfully protect, preserve and manage international watercourses cannot ignore that desertification, land degradation and drought – resulting from various factors, including climatic variations and human activities – pose a major threat to the availability, quantity and quality of water resources. At the same time, increasing long-term imbalances between available water resources and demand can trigger and intensify the effects of desertification, including the drying up of freshwater resources, an increased frequency in droughts

---

8 The UNCCD's Preamble makes explicit reference to the relationship between desertification and other environmental problems of global dimension. In turn, the 23rd Recital pays tribute to the contribution that combating desertification can make in support of the objectives of related environmental conventions. See UNCCD (note 1), at Preamble, 22nd and 23rd Recitals.

and sand and dust storms, and a greater occurrence of flooding due to inadequate drainage or poor irrigation practices. The loss of vegetation cover may, in turn, lead to further land and water degradation, through, e.g. water pollution and salinization, as well as erosion, siltation and alkalization of soils.

In this sense, the UNWC and the UNCCD are responses to interconnected, urgent and growing issues of global dimension: the international water crisis is caused by poor water management and governance, coupled with growing water demand, the decrease in availability of water of the required quality, the unequal distribution of available water resources across different regions and climate variability and change; in turn, land degradation affects and is affected by the timing, quantity and quality of freshwater flows,[9] as well as management practices that fail to integrate land and water resources. Moreover, land degradation is a leading source of land-based pollution, as polluted sediments and water are washed away. At the same time, integrated water resources management in critical watersheds can contribute to the fight against desertification, land degradation and drought.

In particular, international watercourses bear the potential for conflict over competing water uses, raising political tensions in many parts of the world. About 1,400 million people inhabit these shared basins, many of which suffer from water stress. National and international politics further complicate the challenge of transboundary river basin management. As land degradation, drought and desertification take their toll, in the context of climate change and variability, water disputes can be expected to raise political tensions in drylands, contributing to international conflicts where water resources straddle or delineate country borders.

For these reasons, the UNCCD and the UNWC address the environmental degradation of water and land resources in an integrated manner. Both Conventions are crisis prevention and mitigation mechanisms, dealing each with natural resources management issues that are closely interrelated and should be tackled in a coordinated manner.

Article 2(1) of the UNCCD promotes an 'integrated approach' to the Convention's implementation. In Paragraph 2, Article 2 calls for 'long-term integrated strategies that focus simultaneously, in affected areas, on improved productivity of land, and the rehabilitation, conservation and sustainable management of land and water resources'. Furthermore, in pursuing its objectives, Article 4(2)(a) of the UNCCD suggests that parties follow 'an integrated approach addressing the physical, biological and socio-economic aspects of the processes of desertification and drought'. As a more specific example of integration, Article 11(4) of the UNCCD suggests that, where appropriate, national action programmes should include measures for strengthening assessment and systematic observation capabilities, including hydrological and meteorological services.

Within this framework, Article 8(1) of the UNCCD anticipates the potential for horizontal synergies and encourages the coordination of activities carried out under the UNCCD 'and other relevant international agreements... in order to

---

9   Ibid., at Preamble, 4th Recital.

derive maximum benefit from activities under each agreement while avoiding duplication of effort'.

In turn, as per Article 2(a), the UNWC applies to 'a system of surface waters and ground waters constituting by virtue of their physical relationship a unitary whole and normally flowing into a common terminus'. Yet, commentators widely agree that the Convention is also concerned with land-based activities that may affect an international watercourse. This understanding draws from Article 1(2), which determines that the UNWC applies 'to uses of international watercourses and of their waters for purposes other than navigation and to measures of protection, preservation and management related to the uses of those watercourses and their waters'. In addition, Article 20 requires states to protect and preserve the ecosystems of international watercourses. In order to comply with this provision, states will necessarily have to tackle land management issues that may affect such ecosystems. Finally, Article 27, dealing with the prevention and mitigation of harmful conditions, makes express reference to drought and desertification.

Therefore, both the UNCCD and the UNWC follow an integrated approach to land and water resources management. However, the UNCCD does not lay out the principles and procedures to guide cooperation between states where combating land degradation may require joint or coordinated action between co-riparian states. The next section thus examines further the potential role that an effective UNWC could play as a natural counterpart to the UNCCD towards fostering and enabling integrated land and water management in international watercourses.

Sustainable development as a common objective

Article 2 of the UNCCD defines the achievement of sustainable development in affected areas as its ultimate aspiration.[10] To achieve this objective, Article 3(c) requires parties to 'develop, in a spirit of partnership, cooperation among all levels of government, communities, non-governmental organizations (NGOs) and land-holders to establish a better understanding of the nature and value of land and scarce water resources in affected areas and to work towards their sustainable use'.

As a consequence, Article 10(3)(e)–(4) of the UNCCD includes the 'development of sustainable irrigation programmes' and the 'sustainable use of natural resources' among the measures that the national action programmes may incorporate. Such programmes are expected to be a central and integral part of each country's strategic planning framework for sustainable development, as per Articles 5(b), 9(1) and 10(2)(a). Furthermore, Article 11 of the UNCCD explicitly suggests including 'agreed joint programmes for the sustainable management of transboundary natural resources, scientific and technical cooperation and strengthening of relevant institutions' into subregional and regional action plans. As part of the suggested technical and scientific cooperation, Article 17(1)(a) and (f) determines that research activities should focus on the 'sustainable use and management of

10   Ibid., at Preamble, 8th, 9th, 11th and 14th Recitals, Articles 1(b) and 4(2)(b).

resources' and 'improved, affordable and accessible technologies for sustainable development'.

Moreover, Article 18(1) of the UNCCD promotes the transfer and development of environmentally sound, economically viable and socially acceptable technologies, in support of sustainable development in affected areas. Article 19(1)(c) then recognizes the significance of institution-building and training and development of relevant local and national capacities for the conservation and sustainable use of natural resources. Finally, Article 19(3)(e) regards as crucial the support for public awareness and educational programmes on the conservation and sustainable use and management of the affected areas' natural resources.

The UNWC also places sustainable development at its center. Already the Preamble, in its fifth Recital, underscores the Convention's role in ensuring 'the utilization, development, conservation, management and protection of international watercourses and the promotion of the optimal and sustainable utilization thereof for present and future generations'. Moreover, Article 5 of the UNWC makes clear that sustainable utilization is the ultimate goal to be pursued through the development, management and protection of international watercourses, both in the context of water allocation and benefit sharing.

Hence, although the UNCCD is richer in detail in terms of the recommended content of the national action programmes, both treaties at hand have sustainable development as a core objective. In supplementing regional plans or programmes to combat desertification, substantive and procedural rules under the UNWC may guide parties to the UNCCD in cooperating towards the sustainable and integrated management of land and water resources. Similarly, joint programmes developed under the UNCCD may serve as useful management tools to advance transboundary water cooperation on the ground.

Common environmental focus

The UNCCD and the UNWC reinforce each other in their common recognition of the necessity to protect and preserve ecosystems, thriving to integrate the ecological and economic dimensions of sustainable development.

Article 2(2) of the UNCCD calls for the 'rehabilitation, conservation and sustainable management of land and water resources'. Article 3(c) highlights the need for 'a better understanding of the nature and value of land and scarce water resources in affected areas'. Based on those principles, Article 4(2)(d) promotes 'cooperation among affected country Parties in the fields of environmental protection and the conservation of land and water resources, as they relate to desertification and drought'.

This correlates with the obligations under the UNWC – particularly in Part IV, on protection, preservation and management of international watercourses, stipulating that an adequate protection of a watercourse has to take its bearings from the ecological equilibrium of the entire aquatic ecosystem. Part IV addresses the protection and preservation of aquatic ecosystems; the prevention, reduction and control of pollution; introduction of alien or new species; the protection and preservation of the marine environment; water management and regulation; and the maintenance and protection of installations. These provisions contain a series of substantive and procedural obligations on the issue of environmental protection, which could supplement and provide further detail to the duty under Article 4(2)(d) of the UNCCD.

Common focus on cooperation

Both the UNCCD and the UNWC recognize the importance of and need for international cooperation and partnership in reaching their respective objectives.[11] Articles 3(b)–(c), 4(2)(e) and 10(2)(e) of the UNCCD emphasize the 'spirit of international solidarity and partnership' to improve cooperation and coordination at subregional, regional and international levels, as well as cooperation among governments at all levels, local populations and community groups, NGOs and landholders. Particularly pertinent in the context of international watercourses is the importance that Article 12 of the UNCCD attributes to cooperation between affected country Parties and other Parties and the international community to ensure the promotion of an enabling environment for its implementation.

Articles 11–23 of the UNCCD also explicitly promote interstate cooperation through joint programmes for the sustainable management of transboundary natural resources. The UNCCD foresees a variety of detailed measures for such cooperation to materialize. As per Articles 10(3)(a) and 16–19, those measures may include, e.g: (i) the establishment and/or strengthening, as appropriate, of joint early warning systems at the subregional and regional levels; (ii) scientific and technical cooperation through (potentially joint) research and development; (iii) technology transfer; and (iv) capacity building, education and public awareness. Article 4(2)(g) of the UNCCD encourages the establishment of institutional mechanisms, provided they are appropriate and avoid duplication.

Similarly, parties to the UNWC must cooperate through, inter alia, the establishment of joint mechanisms or commissions and the regular exchange of data and information, including that of a meteorological nature and related forecasts. In particular, Article 27 requires states to, 'individually and, where appropriate, jointly, take all appropriate measures to prevent or mitigate conditions related to an international watercourse that may be harmful to other watercourse States . . . such as . . . drought or desertification'.

Hence, the treaties at hand are in harmony with each other in promoting cooperation between states that share natural resources. Still, the UNWC's provisions detailing the general duty to cooperate in Article 8 may also guide states dealing with drought and desertification with regard to the coordinated or joint management of affected river basins.

Finally, both treaties establish a dispute settlement mechanism. According to Article 28(6) of the UNCCD, if the parties to a dispute do not agree on, or cannot

---

11 See UNCCD, at Preamble, 17th and 25th Recitals, Article 2(1).

settle it through, negotiations or other peaceful means within a specified period, 'the dispute shall be submitted to conciliation at the request of any Party to the dispute, in accordance with procedures adopted by the Conference of the Parties in an annex as soon as practicable'. As of the date of writing, such an Annex had not been established and so parties to the UNCCD cannot resort to its unilateral conflict resolution mechanism.

In its turn, Article 33(3)–(9) of the UNWC establishes a similar dispute settlement system, in the form of an impartial fact-finding commission. The UNWC already regulates such a commission. Hence, upon entry into force, this will be a ready-made solution for parties to both Conventions to resort to in solving their disputes on the basis of facts and the recommendations of a neutral body.

Development focus of the UNCCD

The UNWC aims to promote optimal and sustainable water utilization. From its Preamble, the UNWC underscores 'the special situation and needs of developing countries'. This recognition then permeates the entire Convention. Expressions such as 'all appropriate measures' aim to, among other things, take into account the financial capabilities of each party. This applies, for example, to the requirement for states to 'take *all appropriate measures* to prevent or mitigate conditions related to an international watercourse that may be harmful to other watercourse States'. The UNWC thus integrates socioeconomic and ecological interests, enabling and obliging parties to take development-related aspects into account. In particular, this concern is inherent to the relevant factors for determining an equitable and reasonable utilization in Articles 5–6 and 10(2), which also require states to give special regard to vital human needs, and to the application of its Articles 11 et seq. on planned measures.

However, the UNCCD goes beyond the UNWC in this respect, with its strong focus on the links between development and desertification and droughts. As per the very title of the Convention and as corroborated in Articles 6(a) and 7, the UNCCD is especially concerned with Africa and the least developed countries. Article 1(b) of the UNCCD considers the fight against desertification and droughts as 'part of the integrated development of land in arid, semi-arid and dry sub-humid areas'.[12] In addition, Article 4(2)(c) considers poverty eradication as an integral part of the UNCCD's strategy to combat desertification and mitigate the effects of drought.[13]

As a consequence, Articles 5–6 of the UNCCD define distinct obligations for parties affected by desertification, on the one hand, and developed parties unaffected by desertification, on the other. In Article 4(2)(h), the UNCCD also explicitly promotes 'the use of existing bilateral and multilateral financial mechanisms and arrangements that mobilize and channel substantial financial resources to affected developing country Parties in combating desertification and mitigating the effects of drought'. In addition, '[a]ffected developing country Parties are eligible

12   Ibid., at Preamble, 8th–9th and 15th Recitals, Article 10(4).
13   Ibid., at Preamble, 7th–8th Recitals.

for assistance in the implementation' of the UNCCD, within the framework of Articles 4(3), 6(b)–(d), and 20–21.

Furthermore, under Articles 16-17 of the UNCCD, scientific and technical cooperation, as well as research and development, are explicitly subject to the 'respective capabilities' of the parties. Articles 18-19 – on technology transfer, capacity building, education and public awareness – confirm and are a function of the UNCCD being fundamentally development-driven.

Participation focus of the UNCCD

The UNCCD puts a particular emphasis on the participation of local populations and NGOs – to a degree that goes far beyond of what the UNWC provides for in the context of equitable and reasonable participation, in Article 5(2), and the non-discrimination clause, in Article 32.

According to Article 3(a) of the UNCCD, 'decisions on the design and implementation of programmes to combat desertification and/or mitigate the effects of drought are taken with the participation of populations and local communities'.[14] Article 5(d) goes on to require affected parties to 'promote awareness and facilitate the participation of local populations, particularly women and youth, with the support of [NGOs], in efforts to combat desertification and mitigate the effects of drought'.

In making public participation operational, national action programmes must specify the respective roles of government, local communities and land users and the resources available and needed to effectively implement the Convention. According to Article 10(2)(f), national action programmes must also:

> provide for effective participation at the local, national and regional levels of [NGOs] and local populations, both women and men, particularly resource users, including farmers and pastoralists and their representative organizations, in policy planning, decision-making, and implementation and review of national action programmes.

In addition, Articles 17(1)(f), 18(2)(b) and 19(3)(b) consider effective participation of local populations and communities to be a core element of joint research programmes for the development of improved, affordable and accessible technologies, technology transfer, as well as capacity building, education and public awareness activities.

The special role of NGOs and the far-reaching rights provided to them by the UNCCD are a particularity of this Convention.[15] As per Articles 5(d), 6(d), 20(2)(d) and (4) and 21(1)(c)–(d), (3) and (5)(c), NGOs play a key role in supporting the participation of affected populations and raising funds. NGOs also support, in

---

14 Ibid., at Preamble, 20th Recital.
15 Ibid., at Preamble, 21st Recital.

accordance with their mandates and capabilities, the elaboration, implementation and follow-up of action programmes, as per Articles 9(3), 10(2)(f) and 13(1)(b). Under Article 14(2), NGOs take part in the elaboration and operational implementation of coordination mechanisms. Moreover, within the framework of Articles 16(d), 18(1)–(2)(a) and 19(1)(a)–(4), NGOs play a role in scientific and technical cooperation, technology transfer and capacity building, education and public awareness. Article 22(2)(h) illustrates the institutional role of NGOs under the UNCCD, requiring the Conference of the Parties to seek, as appropriate, their cooperation and utilize their services. Finally, under Article 21(1)(b), NGOs can be involved in financial mechanisms, such as national desertification funds, to channel financial resources rapidly and efficiently to the local level in affected developing country parties.

Conclusion

The threats of water scarcity to arid, semi-arid and sub-humid areas are set to increase in magnitude and scope, due to climate change and mismanagement of natural resources. It is thus vital that states take into account the need to manage land, water and related natural resources in an integrated manner. Desertification and land degradation adversely affect the availability of water resources. According to the existing climate change models and scenarios, an increasing number of people are going to live in areas of high water stress. Appropriate management of water resources represents a key factor in mitigating the adverse effects of drought, as well as in combating land degradation. In this context, coherent policies for land and water resource management are needed to maintain a healthy environment and enable socioeconomic development. Both the UNCCD and the UNWC work in this direction by promoting interstate cooperation on the conservation, protection and rehabilitation of land and freshwater resources, as the natural basis for sustainable development.

The UNCCD supports two primary objectives of sustainable development: poverty eradication and sustainable production and consumption. Although the UNWC is not as explicit about its development objectives, its focus on equitable utilization of water resources and ecological sustainability paves the way in that direction. Neither Convention has a financing tool. However, consistent with Agenda 21, they are expected to support sustainable development through meaningful multi-level actions and through international cooperation and partnership arrangements. Both Conventions also mutually strengthen each other. In Bruch's words, 'they contain provisions that specify obligations and responsibilities for the riparian countries to use their international water resources reasonably and equitably while also promoting open and active participation and collaboration between riparian states'.[16] Widespread ratification and implementation of these Conventions would provide evidence of political will and shared spirit of cooperation among neighboring states to foster the sustainable management of precious internationally shared and mutually dependent natural resources.

16 Bruch, C., Jansky, L., Nakayama, M. and Salewicz, K. A. (eds), *Public Participation in the Governance of International Freshwater Resources* (UN University Press, 2005).

20 UN Economic Commission for Europe Water Convention

Attila Tanzi

The two conventions under review address the same subject matter. This is confirmed by their respective titles: UN Economic Commission for Europe (UNECE) Convention on the Protection and Use of Transboundary Watercourses and International Lakes (UNECE Water Convention),[1] and UN Convention on the Law of the Non-Navigational Uses of International Watercourses (UNWC).

The coincidence of the scope of the two instruments under consideration calls for a comparative analysis of their most relevant provisions from two standpoints. On the one hand, from the point of view of international customary law, if such an analysis were to bring to the conclusion that there is a high degree of conformity between the basic rules and principles as they have been spelled out in the two conventions, one could easily argue that they would jointly contribute to the consolidation of the international water law process. Conversely, were one to determine that there is incompatibility between the basic provisions of the two instruments, it could be concluded that they undermine the formation and consolidation of the general customary law of transboundary waters. On the other hand, a comparative analysis is required also from a treaty law point of view to be of practical assistance to states that would intend to ratify either of the conventions in point, or both of them. The relevance from a treaty law point of view of this assessment is increasingly gaining importance in view of the imminent entry into force of the UNWC, as well as of the amendments to the UNECE Water Convention allowing for the accession to it by non-UNECE member states.[2]

Despite the basic coincidence between the two Conventions in scope, some differences in content may remain. Among these, one may detect language suggesting that the primary focus of the UNECE Water Convention is on water quality

---

1 UNECE Convention on the Protection and Use of Transboundary Watercourses and International Lakes (adopted March 17, 1992, entered into force October 6, 1996) 1936 UNTS 269 (UNECE Water Convention).
2 UNECE Convention on the Protection and Use of Transboundary Watercourses and International Lakes, Amendments to Articles 25 and 26 of the Convention on the Protection and Use of Transboundary Watercourses and International Lakes (adopted November 28, 2003) ECE/MP.WAT/14. Available online at http://treaties.un.org/doc/Treaties/2004/02/20040217%2005-46%20AM/Ch_XXVII_05_bp.pdf (accessed March 27, 2013) (UNECE Water Convention Amendment).

issues, while the UNWC would be more concerned with problems of apportionment of water. However, it would be wrong to assume that the difference between the two conventions in point on this matter is more than one of emphasis. The above statement is supported by the physical interdependence between water quantity and water quality issues. Accordingly, from a legal standpoint, any regulation addressing either of the two aspects at issue would inevitably affect the other one. Indeed, a utilization that leads to a significant reduction in the water flow inevitably affects its quality as it diminishes its capacity to absorb pollutants; conversely, the significant pollution of a watercourse would reduce the availability of significant amounts of water otherwise allotted to important uses of the watercourse, even if the quantity of the water flow remained unaffected. One may think of irrigation for agriculture, or recreation, apart from drinking, among other uses of the watercourse. This had been emphasized in legal literature more than 30 years ago.[3]

Furthermore, Article 1(1) of the UNWC, in enunciating its scope, expressly indicates, next to 'non-navigational uses', 'measures of protection, preservation and management related to the uses of [the] watercourses and their waters'. This provision introduces an 'integrated' approach to the determination of the scope of the Convention. It provides the basis for the structural linkage between the core principles of equitable utilization and no-harm (Articles 5–7), on the one hand, and water quality issues – also encompassed in Articles 5–7 and further specified in Part IV, on protection, preservation and management – on the other. That is also to say, the provisions contained in Part IV of the UNWC cannot be deemed to address exclusively pollution, but are to be considered to extend their reach also to questions of water quantity, i.e. apportionment. Conversely, the rules on equitable utilization and no-harm do not govern only questions of water apportionment, but also cover problems of pollution.

Following the above introductory remarks, this section starts by addressing the respective scope of the two conventions. Consideration follows of the basic substantive principles of international water law as codified in each convention: the equitable utilization principle and the no-harm rule. Attention then focuses on the different approaches followed by each of the instruments in point, with respect to the general obligation of cooperation as the catalyst for the case-specific interpretation and application of the substantive principles of international water law as codified in both conventions. Finally, the relationship between them is analyzed in terms of their applicability to states that become parties to both instruments. Such an analysis shows that the complementary character between the UNECE Water Convention and the UNWC allows them to fall perfectly within the *harmonization*

---

3 See Gaja, G. 'River Pollution in International Law' in *Colloque of the Hague Academy of International Law on The Protection of the Environment and International Law* (Kluwer Law International, 1973), at 366, 371, 377ff [French]. See, more recently, Brunnée, J. and Toope, S. 'Environmental Security and Freshwater Resources: A Case for International Ecosystem Law', *Yearbook of International Environmental Law*, 1994; 5 (1): 41–76, at 47.

principle advocated by the International Law Commission (ILC) in its study on the issue of 'fragmentation' of international law. According to that study, 'when several norms bear on a single issue they should, to the extent possible, be interpreted so as to give rise to a single set of compatible obligations'.[4]

Physical scope

This section identifies the geographical, hydrological and geological entities falling within the reach of each convention, with special regard to: a) the areas on which activities are carried out that are subject to the evaluation as to whether they are equitable and reasonable or may cause transboundary impact; and b) those areas and physical entities that may be adversely affected by activities carried out outside them.

Article 2(a) defines the UNWC's hydrological and geographical scope with the term 'watercourse', intended as 'a system of surface waters and ground waters constituting by virtue of their physical relationship a unitary whole and normally flowing into a common terminus'. This focus on the watercourse and its waters may seem at odds with the outright adoption of the ecosystem approach by the UNECE Water Convention, as examined below, as well as with the drainage basin concept endorsed by the Institute of International Law (IIL),[5] the International Law Association (ILA),[6] as well as relevant conventional practice.[7] However, a broader interpretation of the UNWC's scope seems admissible – one that reads the Convention as also covering land-based activities within the larger river basin and which may affect the ecosystems of international watercourses, in line with an ecosystem or river basin approach.

In the first place, the watercourse system terminological approach under the UNWC goes far beyond the traditional definition of watercourse limited to the

---

4   ILC, 'Report on the Work of its Fifty-eighth Session' (May 1–June 9 and July 3–August 11, 2006) UN Doc A/61/10 (2006), at 408 (Report on Fragmentation).

5   See IIL, 'Utilization of Non-Maritime International Waters (Except for Navigation)', *Annuaire de l'Institut de Droit International*, 1961; 49(2): 371, at Article 1 (Salzburg Resolution).

6   Article II of the Helsinki Rules reads: '[a]n international drainage basin is a geographical area extending over two or more States determined by the watershed limits of the system of waters, including surface and underground waters flowing into a common terminus'. ILA, 'Helsinki Rules on the Uses of the Waters of International Rivers' in Report of the 52nd Conference (Helsinki, 1966) (1966 ILA Report). See also ILA, 'Resolution on the Relationship of International Water Resources with other Natural Resources and Environmental Elements' in Report of the 59th Conference (Belgrade 1980), Article 1.

7   See, e.g. Agreement on Action Plan for Environmentally Sound Management of the Common Zambezi River System (adopted May 28, 1987) (1988) 27 ILM 1109, at Article 1; Agreements on the Rivers Meuse and Scheldt (adopted April 26,1994) (1995) 34 ILM 851, at Articles 1 and 3; Agreement on the Cooperation for the Sustainable Development of the Mekong River Basin (adopted April 5, 1995) (1995) 34 ILM 864, Article 3; Convention on Cooperation for the Protection and Sustainable Use of the Danube River (adopted June 29, 1994, entered into force October 22, 1998) 994 EMuT 49, Article 3.

main arm of the river, in that it encompasses all tributaries, lakes, aquifers and other hydrological components connected to a major internationally shared river.[8]

More importantly, from a contextual interpretation of the term 'watercourse' in conjunction with other relevant provisions of the UNWC, one may reach the conclusion that the drainage basin area falls under the purview of its rules. The drainage basin area can come into play as *the area on which the activity causing harm is carried out, or as the area affected by activities carried out in a neighboring country*.

These assumptions find their textual ground in several provisions of the UNWC: Part IV deals with 'protection, preservation and management'. Under Article 20, in particular, states must protect and preserve the ecosystems of international watercourses. Then, Article 5(2), on equitable utilization, invokes Part IV when requiring states to utilize international watercourses in a manner *consistent with their adequate protection*. Moreover, Article 7 incorporates a duty of prevention of significant transboundary harm resulting from the uses of an international watercourse. Such harm may include detrimental effects to the watercourse itself or to the larger river basin area. Along the same lines, Article 21(2) requires states to 'prevent, reduce and control the pollution of an international watercourse that may cause significant harm to *other watercourse States or to their environment*'.

Hence, implicit in the above provisions, when read in an integrated manner, is the notion that an activity carried out in the drainage basin which pollutes an international watercourse or alters it to the extent that it may cause significant harm to other riparians falls under the scope of application of the UNWC.[9] The same is true for the uses of an international watercourse that affect the drainage area. This contextual interpretation of the UNWC brings it in line with the UNECE Water Convention.

In fact, the UNECE Water Convention merits a special place among the international instruments to which a contextual interpretative reference, leading to the application of the ecosystem approach[10] under the UNWC, should be made. In this sense, Article 31(3)(c) of the Vienna Convention on the Law of Treaties[11] establishes

---

8   As it encompasses 'a number of different components through which water flows, both on and under the surface of the land ... [including] rivers, lakes, aquifers, glaciers, reservoirs and canals'. ILC, 'Draft Articles on the Law of the Non-Navigational Uses of International Watercourses and Commentaries thereto and Resolution on Transboundary Confined Groundwater' in ILC, 'Report on the Work of its Forty-sixth Session' (May 2–July 22, 1994) UN Doc A/49/10(Supp) (1994), at 90 (1994 ILC Report).

9   This argument finds support in the 'authentic interpretation' provided by the former Special Rapporteur, Stephen McCaffrey, of the 1994 ILC Draft Articles, which, in their relevant parts for our purposes, have not been changed by later negotiations leading up to the adoption of the UNWC. McCaffrey, S. (Special Rapporteur), '7th Report on the Law of the Non-Navigational Uses of International Watercourses' (March 15, 1991) UN Doc A/CN.4/436, at 59; *Yearbook of the ILC*, 1991; 2 (I): 45–69. Available online at http://untreaty.un.org/ilc/documentation/english/a_cn4_436.pdf (accessed March 27, 2013).

10   For a synthetic inventory of such instruments, see particularly, Brunnee, J. and Toope, S. J. 'Environmental Security and Freshwater Resources: Ecosystem Regime Building', *American Journal of International Law*, 1997; 91: 26–59, at 50ff.

11   Vienna Convention on the Law of Treaties (adopted May 23, 1969, entered into force January 27, 1980) 1155 UNTS 331 (Vienna Convention).

that, in the interpretation of a treaty, '[t]here shall be taken into account, together with the context... any relevant rules of international law applicable in the relations between the parties'. In line with this rationale, the drafters of the UNWC recall, in Recital 9 of the Preamble, 'the existing bilateral and multilateral agreements regarding the non-navigational uses of international watercourses', among which is the UNECE Water Convention.

Besides spelling out that the promotion of 'the application of the ecosystems approach' is among its normative aims, the UNECE Water Convention[12] provides an articulated series of obligations on the prevention of 'transboundary impact', whereby,

> [t]ransboundary impact' means any significant adverse effect on the environment resulting from a change in the conditions of transboundary waters caused by a human activity, the physical origin of which is situated wholly or in part within an area under the jurisdiction of a party, within an area under the jurisdiction of another party.[13]

The analysis above corroborates the conclusion that the scope of both conventions under review goes beyond the concept of a 'watercourse' in that they also aim to prevent: a) harm to the water of a watercourse caused by activities that may take place outside the actual watercourse, provided a linkage of interdependence may be established between the aquatic ecosystem and the environment which is primarily affected, or on which the activity has been carried out; and b) transboundary harm caused by uses of the watercourse to elements of the environment different from the water of the watercourse.[14]

Groundwater

The inclusion of groundwater within the scope of the UNECE Water Convention is beyond question. In Article 1(1), the Convention defines 'transboundary waters' as 'any surface or groundwaters which mark, cross or are located on boundaries between two or more States'. According to this definition – differently from the UNWC, as we shall see below – the UNECE Water Convention addresses fossil aquifers, in addition to groundwater interacting, directly or indirectly, with transboundary surface waters. Hence, the principles and provisions of the Convention applicable to transboundary surface water also apply to groundwater systems,

---

12 Article 3(1) reads: 'To prevent, control and reduce transboundary impact, the Parties shall develop, adopt, implement and, as far as possible, render compatible relevant legal, administrative, economic, financial and technical measures, in order to ensure, inter alia, that: ... (i) Sustainable water-resources management, including the application of the ecosystems approach, is promoted'.

13 Ibid., at Article 1(2).

14 Art 1(2) of the UNECE Water Convention goes so far as to specify that '[s]uch effects on the environment include effects on human health and safety, flora, fauna, soil, air, water, climate, landscape and historical monuments or other physical structures or the interaction among these factors; they also include effects on the cultural heritage or socio-economic conditions resulting from alterations to those factors'.

regardless of whether these are components of an international watercourse or not.

This interpretation has been corroborated by the UNECE *Guidelines on Monitoring and Assessment of Transboundary Groundwaters*,[15] and later confirmed by the *Guide to Implementing the Convention*, which stresses that, '[a]s for groundwaters, the Convention includes both confined and unconfined aquifers'.[16] The Legal Board of the Convention is currently studying the issue of the application of the principles of the Convention to ground waters upon request by the Meeting of the Parties, and has produced so far two working documents.[17] These documents may lead to the elaboration of a set of model provisions on transboundary ground waters to guide future negotiations on specific groundwater agreements.[18]

In turn, under Article 2(a) of the UNWC, for ground waters to be considered within its purview, they have to be connected with surface waters so as to constitute a 'unitary whole'.[19] Accordingly, unrelated ground waters, even if intersected by a boundary, seem to fall beyond the reach of the Convention. Be that as it may, in 2002, the ILC started anew its study of this topic and brought it to completion in 2008 with the adoption of the Draft Articles on the Law of Transboundary Aquifers.[20]

15   UNECE Task Force on Monitoring and Assessment, *Guidelines on Monitoring and Assessment of Transboundary Groundwaters* (UNECE, 2000). Available online at www.unece.org/env/water/publications/documents/guidelinesgroundwater.pdf (accessed March 27, 2013).

16   UNECE, 'Integrated Management of Water and Related Ecosystems: Draft Guide to Implementing the Convention' (August 31, 2009) ECE/MP.WAT/2009/L.2, at para. 73. Available online at www.unece.org/fileadmin/DAM/env/documents/2009/Wat/mp_wat/ECE_mp.wat_2009_L2_%20E.pdf (accessed March 27, 2013).

17   Ibid.

18   See UNECE, 'Application of the UNECE Water Convention to Groundwater: Explicatory Recognition of the Existing UNECE Regulatory Language' (February 24–25, 2011) ECE/WAT/LB/2011/INF.2. Available online at www.unece.org/fileadmin/DAM/env/documents/2011/wat/AC/LB_2011_Inf-2_E.pdf (accessed March 27, 2013); UNECE, 'Application of the UNECE Water Convention to Groundwater and Possible Developments' (April 15–16, 2012) LB/2010/INF.2. Available online at www.unece.org/fileadmin/DAM/env/water/meetings/legal_board/2010/Groundwater_discussion_paper_inf2.pdf (accessed March 27, 2013).

19   As per the ILC's commentary to draft Art 2(b) of the 1994 ILC Draft Articles, '[i]t . . . follows from the unity of the system that the term "watercourse" does not include "confined" groundwater, i.e. which is unrelated to any surface water'. 1994 ILC Report (note 8), at 201.

20   See ILC, 'Draft Articles on the Law of Transboundary Aquifers, with Commentaries' in ILC, 'Report on the Work of its Sixty-third Session' (May 5–June 6 and July 7–August 8, 2008) UN Doc A/63/10 (Supp) (2008). Available online at www.unece.org/env/water/meetings/legal_board/2010/annexes_groundwater_paper/Annex_VI_Draft_Articles_Law_Transboundary_Aquifers_ILC.pdf (accessed March 27, 2013). While completing a second reading its draft articles, the ILC adopted a resolution inviting states to apply to ground waters the same principles set forth in the articles. It also recommended 'States to consider entering into agreements with the other State or States in which the confined transboundary groundwater is located', and 'that, in the event of any dispute involving transboundary confined groundwater, the States concerned should consider resolving such a dispute in accordance with the provisions of article 33 of the draft articles, or in such manner as may be agreed upon'. 1994 ILC Report (note 8), at 326. See, on such a resolution, Idris, K. and Sinjela, M. 'The Law of the Non-Navigational Uses of International Watercourses: The ILC's Draft Articles – An Overview', *African Yearbook of International Law*, 1995; 3: 183–203, at 201–2; McCaffrey, S. C. and Rosenstock, R. 'The International Law Commission's Draft Articles on International Watercourses: An Overview and Commentary', *Review of European Community and International Environmental Law*, 1996; 5: 89–96, at 93.

Substantive principles

The UNWC may appear to devote more attention than the UNECE Water Convention to the substantive principles governing the utilization of international watercourses as opposed to the principle of cooperation, which has a procedural character. On closer scrutiny – particularly looking at the overall picture of the two texts – one gathers quite different indications. Articles 5–7 of the UNWC, on the equitable utilization principle and the no-harm rule, are the result of a compromise to resolve a mostly symbolic and rhetorical debate on whether the former has priority over the latter, or vice-versa. The debate has continued at the academic level and, while some scholars seem to argue that the UNWC gives priority to equitable and reasonable utilization,[21] it is the view of the present writer that the great merit of that Convention is precisely that of having carefully crafted the provisions codifying the two principles in point so as to place them on an equal footing, with the advantages from the legal and diplomatic viewpoint already stressed elsewhere.[22]

The UNECE Water Convention may seem at first to have followed a different approach on the point at issue. In fact, Article 2(1), opening up the 'General Provisions' of the UNECE Water Convention enunciates at the outset the no-harm rule providing that '[t]he Parties shall take all appropriate measures to prevent, control and reduce any transboundary impact'. However, it would seem sterile to claim that this drafting approach would, give priority to the no-harm rule. Likewise, it would serve little constructive purpose to maintain that the intention to give priority to the no-harm rule would be corroborated by Article 2(2)(c), which refers to the equitable utilization principle within the context of the harm prevention or that, on the contrary, the latter provision would render the no-harm rule subservient to the equitable and reasonable utilization principle. The drafting of the provisions of the UNECE Water Convention codifying the principles in hand rather corroborates the idea of one complex substantive customary normative setting of which both principles are part and parcel, being totally entangled with each other. This water law substantive normative setting appears to have been expressed in more concise terms in the UNECE Water Convention than in the UNWC. However, the UNECE Water Convention provides thorough guidelines for states to adopt individually and to adapt concretely to a specific watercourse in cooperation with their co-riparians. These guidelines are many more, and more detailed, than those set out in the UNWC, giving clearer substance to the general principles at issue. On account of the integration of the two general principles in

---

21   See, among others, Utton, A. E. 'Which Rule Should Prevail in International Water Disputes: That of Reasonableness or that of No Harm?', *Natural Resources Journal*, 1996; 16: 635–41, at 638ff; Subedi, S. P. 'Resolution of International Water Disputes: Challenges for the 21st Century' in Permanent Court of Arbitration International Bureau, *Resolution of International Water Disputes: Papers Emanating from the Sixth PCA International Law Seminar* (Kluwer Law International, 2003), at 33ff.

22   Tanzi, A. and Arcari, M. *The UN Convention on the Law of International Watercourses: A Framework for Sharing* (Kluwer Law International, 2000), at 172ff.

point into one normative setting, the concrete guidelines for the prevention of transboundary impact serve also for the determination of the equitable and reasonable utilization of an international watercourse. Against this background, the two conventions under review complement each other.

Furthermore, under both conventions, the no-harm rule is set out in terms of an obligation of due diligence.[23] While Article 7 of the UNWC does not expressly provide clues for the identification of 'all the appropriate measures' of prevention, the UNECE Water Convention does so. This is one of the many cases in which the latter Convention complements the former. The preparatory works of the UNWC offer some ground for this interpretative approach. Indeed, the ILC had indicated to have deduced the due diligence obligation of prevention 'as an objective standard . . . from treaties governing the utilization of international watercourses'[24] and the UNECE Water Convention is a key instrument among multilateral treaties on international watercourses. This reasoning is furthered by the fact that the wording of Article 7(1) of the UNWC largely coincides with that of Article 2(1) of the UNECE Water Convention.

Accordingly, the concrete determination of 'all appropriate measures' to be taken in a given case, i.e. the due diligence standard only abstractly announced by Article 7(1) of the UNWC, should be made in the light of the more specific guiding principles contained in the UNECE Water Convention. In this exercise, special regard should be given to the ecostandards consisting of the 'best available technology'[25] and the 'best environmental practices',[26] as well as to the 'previous environmental impact assessment' and the 'precautionary principle'.

Moreover, in the light of the interconnections between water quantity and water quality issues and the resulting indivisibility of the international regulation thereof, the concept of transboundary impact under both conventions covers harm caused by the amount of water flow, as well as harm caused by pollution. The extension of the reach of the obligation of prevention as to encompass water quality issues emerges with clarity in the definition of 'transboundary impact' in Article 1(2) of the UNECE Water Convention. The same conclusion can be reached with regard to the UNWC, through a contextual interpretation of its Articles 5–7 and Part IV, on protection, preservation and management.

As far as the threshold of non-permissible harm is concerned, in both conventions this is not so low as to include any degree of perceptible harm, but only that

---

23 See, generally, Barnidge, R. P. 'The Due Diligence Principle under International Law', *International Community Law Review*, 2006; 8: 81–121.

24 See 1994 ILC Report (note 8) at 237.

25 See Article 3(1)(f), which includes 'the application of the best available technology' among the '[a]ppropriate measures . . . to reduce nutrient inputs from industrial and municipal sources' as a specification of the obligation of prevention, control and reduction. See also Annex I, on the definition of best available technology.

26 See Article 3(1)(g), which provides for the development and implementation of 'appropriate measures and best environmental practices . . . for the reduction of inputs of nutrients and hazardous substances from diffuse sources, especially where the main sources are from agriculture'. See also Annex II, entitled 'Guidelines for Developing Best Environmental Practices'.

of a 'significant' nature.[27] However, whatever adjective is used, the term would provide little guidance for the assessment *in concreto* of the exact nature and extent of the harm to be prevented.[28] Hence, further specification of the 'significant' threshold should be reached through cooperation between the co-riparians concerned on a case by case basis, i.e. through agreements setting out more precise parameters. This conclusion emanates from the rationale of both conventions, which are geared towards the promotion of joint efforts among co-riparians in applying the two substantive water law principles in concrete terms.

Here, again, a complementary role with respect to the UNWC can be played by the UNECE Water Convention. Article 9 of the UNECE Water Convention sets out the obligation for co-riparians to enter into 'agreements or arrangements' for the establishment of joint bodies whose various tasks include '[t]hat to elaborate joint water-quality objectives and criteria'. Appendix III to the UNECE Water Convention then provides a number of guidelines to that end. In its turn, Article 21(3) of the UNWC, requires, in complementary terms, watercourse states to 'consult with a view to arriving at mutually agreeable measures and methods to prevent, reduce and control pollution of an international watercourse, such as [s]etting joint water quality objectives and criteria'.

Cooperation

However, the main difference between the two conventions bears on the way each of them addresses the issue of cooperation. Yet, under both instruments, the principle in point is given appropriate prominence and compulsory nature as the key catalyst for the concrete application of the substantive principles of water law. The UNECE Water Convention complements the general obligation of cooperation by *requiring* watercourse states to enter into agreements setting up joint bodies, and by providing a detailed set of functions and parameters for the operation of such bodies. Furthermore, under the UNECE Water Convention, cooperation is centered on the institutional setting of the Meeting of the Parties and its permanent subsidiary bodies, which may assist parties in the Convention's implementation. Neither of those features exists in the UNWC. Article 8(2) of the UNWC simply indicates that 'watercourse States may *consider* the establishment of joint mechanisms or commissions' as a means of cooperation. The provision in point has thus no normative force. The same applies to Article 24 of the UNWC, which refers to the possibility of establishing joint mechanisms for the management of an international watercourse.

27 See generally on the question at issue, Sachariew, K. 'The Definition of Thresholds of Tolerance for Transboundary Environmental Injury Under International Law: Development and Present Status', *Netherlands International Law Review*, 1990; 37: 193–206; Nollkaemper, A. *The Legal Regime for Transboundary Water Pollution: Between Discretion and Constraint* (Martinus Nijhoff/Graham and Trotman, 1993), at 35–39.

28 Much skepticism on the usefulness of adjectives qualifying the threshold of acceptable harm has been expressed by Zemanek, K. 'State Responsibility and Liability' in Lang, W., Neuhold, H. and Zemanek, K. (eds), *Environmental Protection and International Law* (Graham and Trotman, 1991), at 196.

The above does not prevent cooperation from being an essential feature of the UNWC. First and foremost, cooperation is part and parcel of the normative setting which contains the equitable utilization principle and the no-harm rule under Articles 5–7. Cooperation is also reflected and specified in Article 9, on exchange of data and information; in Part III, on notification, consultation and negotiation concerning planned measures; in most of Part IV, on protection, preservation and management; in Part V, on harmful conditions and emergency situations; in Article 30, on indirect procedures; as well as in Article 33, on dispute settlement.

Arguably, the UNWC provides a more extensive set of procedural obligations than the UNECE Water Convention, particularly on notification and consultation. It is so because such functions under the UNECE Water Convention are, to a large extent, expected to be performed by the joint water bodies which the Convention in question provides as compulsory.

The applicability of the conventions to states that are parties to both

As discussed above, overall, there is compatibility between the two conventions. In most cases, the two texts appear complementary to each other. As a result, those articles in the UNWC and the UNECE Water Convention, respectively, which provide for more detailed rules, offer important elements complementing the guideline and the prescriptive function of those in the other convention that are less stringent and/or detailed. It has been shown that, more often than not, it is the UNECE Water Convention that offers complementary guidelines for the application and implementation of the UNWC, with the exception of some procedural rules, especially those on planned measures, and, possibly, those on the consequences of the occurrence of 'lawful' harm.

Based on the above conclusions, this section examines how the two conventions would apply in the case of states that become parties to both instruments. In that regard, the Vienna Convention confirms that two or more treaties on the same subject matter may be applicable at the same time between the same parties to such treaties, provided there is mutual compatibility between their provisions.[29] This is so under Article 30(3),[30] on the application of successive treaties relating to the same subject matter, as well as under Article 59(1)(b),[31] on the termination or

29 See generally, Rocounas, E. 'Engagements Parallèles et Contradictoires' (1987-VI) 206 Recueil des Cours de l'Académie de Droit International 9; Czaplinsky, W. and Danilenko, G. 'Conflicts of Norms in International Law', *Netherlands Yearbook of International Law*, 1990; 21: 29ff; Mus, J. B. 'Conflicts between Treaties in International Law', *Netherlands International Law Review*, 1998; 45: 208–32; Vranes, E. 'Lex Superior, Lex Specialis, Lex Posterior: Zur Rechtsnatur der "Konfliktlösungsregeln"', *ZAöRV*, 2005; 65: 391–405. See also Report on Fragmentation (note 4).

30 Article 30(3) reads: '[w]hen all the parties to the earlier treaty are parties also to the later treaty but the earlier treaty is not terminated or suspended in operation under article 59, the earlier treaty applies only to the extent that its provisions are compatible with those of the later treaty'.

31 Para. 1(b) reads: '[a] treaty shall be considered as terminated if all the parties to it conclude a later treaty relating to the same subject-matter and ... the provisions of the later treaty are so far incompatible with those of the earlier one that the two treaties are not capable of being applied at the same time'.

suspension of the operation of a treaty implied by the conclusion of a later treaty. Despite the generally mutual compatibility, if not full complementarity, between the UNWC and the UNECE Water Convention, one might wonder about the legal situation deriving for the future states parties to both conventions from the few cases of difference between their respective provisions bearing on the same matter.

One would make the case that the more specific and articulated provisions of the earlier convention, i.e. the UNECE Water Convention, would not be derogated from the more general ones, or less stringent, contained in the later UNWC. Articles 30 and 59 of the Vienna Convention uphold the principle *lex posterior derogat priori* with regard to conflicting provisions contained in two treaties on the same subject matter as between the parties to both treaties. However, the case law of the International Court of Justice (ICJ) seems to give prevalence to the principle of the *lex specialis* over that of the *lex posterior*. Moreover, Articles 30 and 59 of the Vienna Convention leave the parties to international treaties free to regulate, on a case by case basis, the legal effects of such treaties between the parties with respect to pre-existing or future treaties on the same subject matter. This contractual freedom has been exercised by the drafters of both conventions in a way that rules out plain derogation of the provisions of the earlier convention by the latter.

In this sense, Article 9(1) of the UNECE Water Convention provides not only for the possibility, but for the parties' obligation to enter into bilateral or multilateral watercourse agreements setting out more specific rules, *without*, however, *deviating from the 'mother' convention*. This provides for the applicability to the parties of the UNECE Water Convention that become parties also to the UNWC of the few rules contained in the latter that provide for more detailed or stringent standards. At the same time, by implication, this provision rules out any derogatory effect on the rules of the UNECE Water Convention of the less stringent or detailed rules contained in the UNWC.

The above should be combined with the relevant rules on the issue contained in the UNWC. The key general provision on the point at issue is in Article 3(1), which reads: 'In the absence of an agreement to the contrary, nothing in the present Convention shall affect the rights or obligations of a watercourse State arising from agreements in force for it on the date on which it became a party to the present Convention'. This provision clearly preserves the normative force of the UNECE Water Convention for those parties to it that were to ratify also the UNWC. At the same time, by its reference to the admissibility of 'an agreement to the contrary', in combination with Article 9(1) of the UNECE Water Convention, the provision quoted above also ensures the applicability to the parties to the UNECE Water Convention of those provisions of the UNWC that were to be more specific *and* in line with its basic principles.

Conclusion

The relationship between most of the provisions of the two conventions under review is exemplary of what the ILC defined in its work on the fragmentation of international law a *relationship of interpretation* – as opposed to one *of conflict* – whereby 'one norm assists in the interpretation of another...for example as an application, clarification, updating, or modification of the latter. In such situation, both norms are applied in conjunction'.[32]

Indeed, where there is no full coincidence between the provisions of the two conventions under review, there is no conflict between them. Those of the UNECE Water Convention are generally more stringent than those of the UNWC. The UNECE Water Convention sets out more precise guidelines and advanced standards of conduct for the prevention of transboundary impact. On the other hand, one can derive more guidance from the UNWC on the consequences of the occurrence of harm, despite appropriate due diligence measures having been applied. As to the procedural rules, special emphasis has been placed on the mandatory character of institutional cooperation under the UNECE Water Convention, which is only hortatory under the UNWC. This explains the greater level of detail in the procedural rules as codified by the UNWC. Therefore, it would not be appropriate to consider the relation between the provisions of the two instruments that are not perfectly coincident as one of derogation from the former by the later convention, under the rule *lex posterior derogat priori*.

From the standpoint of treaty law, on the basis of Article 3(1) of the UNWC, whatever doubts were to arise as to the compatibility between specific provisions of the two conventions, the *lex posterior derogat priori* rule cannot operate invalidating the provisions of the UNECE Water Convention due to subsequent ratification, acceptance, approval or accession to the UNWC. It is only natural that a negotiation carried out at the universal level yields to common denominators much lower than those that may be agreed upon in a less heterogeneous regional context, such as that of the UNECE. Though, since the aim has been the same in both processes, namely, that of enhancing the common interest for the benefit of all parties involved, it would be inconsistent with the rationale of both instruments if the more detailed standards set out in a regional context were to be effaced from the rules governing the relations between states that are parties to that regional process just because they would take part in and promote a similar process at the universal level.

The treaty law perspective of the relationship between the two conventions gains special relevance at a time when the entry into force of the UNWC appears imminent, just like the entry into force of the amendments to the UNECE Water Convention that would open accession to it by non-UNECE countries.

---

32 Report on Fragmentation (note 4), at para. 2.

21 International development and environmental goals

Nicole Kranz, Lesha Witmer and Uschi Eid

This chapter[1] explores linkages between the UN Watercourses Convention (UNWC) and other internationally agreed development, equity, empowerment and environment policy goals. In particular, we examine whether certain development and environment goals find explicit mention or are somehow reflected in the UNWC. Secondly, we ask whether, regardless of any direct reference in the UNWC, the implementation of the Convention as such could support the attainment of certain development and environment goals, thus potentially leading to a positive resonance between the Convention and other policy initiatives.

With respect to the linkages between water and other development goals,[2] it is argued that the UNWC could serve as an effective guidance for developing countries and countries in transition in managing shared waters to reap benefits for development. In this regard, the chapter investigates in what way the UNWC supports attainment of the Millennium Development Goals (MDGs)[3] – with which MDGs do we see the most obvious interactions? Where could windfall profits occur for others? What are potential implications for peace building and overall human security? In addition, we look beyond the MDG debate to explore

---

1 Reviewed by Abby M. Onencan, Regional Manager, Nile Basin Discourse, and Susan Bazilli, LLB, LLM, Project Manager Good Practices and Portfolio Learning in GEF Transboundary Freshwater and Marine Legal and Institutional Frameworks University of British Columbia.

2 Sadoff, C. and Grey, D. 'Beyond the River: The Benefits of Cooperation on International Rivers', *Water Policy*, 2002; 4: 389–403.

3 '2015 Millennium Development Goals' (UN). Available online at www.un.org/millenniumgoals/bkgd.shtml (accessed March 27, 2013). The MDGs were developed in the UN context on occasion of the 2000 Millennium Summit. See UNGA, Res 55/2 'UN Millennium Declaration' (September 18, 2000) UN Doc A/Res/55/2. They seek to translate broadly agreed development and human security challenges at that time into clearly formulated and measurable goals. These goals were specified through sub-goals and indicators, which were to be reached and implemented by a 2015 deadline.

the importance of the UNWC for the currently emerging policy discourse in the run up to and following the Rio+20 Summit.[4]

Taking into consideration these potential benefits of the UNWC, the chapter reflects on the relevance of the Convention as an instrument for donor organizations and other actors to achieve additional benefits for development through sound transboundary water management.

The second part of the chapter deals with equity and participation, with a specific focus on gender equality as an increasingly important issue in the water governance discourse – does the UNWC allow for the strengthening of (gender) equality? And does it emphasize the requirement for enhanced consultation and participation of non-state actors, especially marginalized and vulnerable stakeholders, in transboundary water management and development? Can the Convention serve as a tool to enhance the leverage and empowerment of non-state actors in international processes?

Millennium Development Goals and Rio+20

The MDGs cover a broad range of singular aspects, ranging from the alleviation of poverty and hunger throughout the world – the share of those people living in extreme poverty and hunger is to be halved by 2015 – to educational and equity related goals (MDG 2, on access to basic education, and MDG 3, on equity and the empowerment of women). The health-related MDGs address child mortality, mothers' health and combating diseases, such as HIV/AIDS and malaria. MDG 7 comprises several environmental sustainability issues, e.g. the integration of sustainable development as a core element in national policies and programmes to reverse the loss of or damage to natural resources. Although energy is not specifically mentioned, there is large consensus that natural resources and basic services are closely interlinked and that water and energy are cross-cutting issues to achieve all the MDGs, as much as all MDGs are dependent on the attainment of the other Goals.

MDG 7 contains two directly water related sub-goals: sustainable access to drinking water and access to basic sanitation; and a habitat-related goal, calling for the improvement of living conditions in slum areas. MDG 8 addresses a collection of issues under the heading 'worldwide development cooperation', pertaining to the allocation of official development assistance, access to markets and debt relief for developing countries.

---

4   Marking the 20th anniversary of the inaugural 1992 UN Conference on Environment and Development, and the 10th anniversary of the 2002 World Summit on Sustainable Development in Johannesburg, the UN Conference on Sustainable Development took place on June 20–22, 2012 in Rio de Janeiro. The conference convened stakeholders from governments, the private sector and NGOs to advance poverty alleviation, while securing the protection of the environment and firmly establishing social standards. The conference focused on two distinct streams: the boosting of a green economy in the context of poverty alleviation and the redesign of the (global) institutional framework for sustainable development.

The success record for attainment of the MDGs has been mixed at best. The 2010 Review Summit and the Annual Report 2011 documented progress as well as stagnation.[5] The sanitation-related target, in particular, is lagging behind in implementation. And, even in the case of the target on access to water, experts have raised questions as to whether it has been adequately met.[6] Much criticism has also been voiced with regard to the rather insular and end-of-pipe approaches promoted by the Goals.[7] Nevertheless, there is a general consensus about the need to address the issue areas under the MDGs at the international level.[8]

Substantial support has been mobilized over the past ten years to move towards the realization of the Goals.[9] Currently, discussions on how to proceed with the Goals after the 2015 deadline are unfolding.[10] It is thus timely to review the relationship of the MDG set with other frameworks, such as the UNWC.

In addition, the MDGs have constituted an important element of the Rio Process and consequently featured prominently in the discourse leading up to Rio+20. While the outcome document, *The Future We Want*,[11] specifically discusses the establishment of the so-called Sustainable Development Goals, which are to be defined in the year to come, the MDGs remain an important reference for the follow-up process to that Conference. In the following section, we show how the UNWC supports water-related objectives in the context of sustainable development and poverty eradication, thereby enabling the placement of the MDGs and the future SDGs in a broader perspective.

5   Significant strides have been made towards the Millennium Development Goals. Yet, reaching all the goals by the 2015 deadline remains challenging, as the world's poorest are being left behind, according to the MDG Report 2011, the UN's annual progress report. 'The MDGs have helped lift millions of people out of poverty, save countless children's lives and ensure that they attend school', UN Secretary-General Ban Ki-Moon said. 'At the same time, we still have a long way to go in empowering women and girls, promoting sustainable development, and protecting the most vulnerable'.

6   For example, Payen observes that the target in question refers to access to improved drinking water sources, which is not the same as safe drinking water sources. According to that author, when considering this qualitative criterion, access to safe drinking water might in fact be very limited. Gérard Payen, 'Worldwide Needs for Safe Drinking Water Are Underestimated: Billions of People Are Impacted' in Smets *et al*, Le Droit à l'Eau Potable et à l'Assainissement, Sa Mise en Œuvre en Europe (Académie de l'Eau, 2011), at 45–63 [French]. Available online at www.circleofblue.org/waternews/wp-content/uploads/2013/02/Payen_DrinkingWaterNeeds UnderEstimate_EN_2011-11-09.pdf (accessed March 27, 2013).

7   Vandemoortele, J. 'The MDG Story: Intention Denied', *Development and Change*, 2011; 42: 1–21, at 2–21.

8   Bullock, A., Cosgrove, W., van der Hoek, W. and Winpenny, J. 'Getting Out of the Box: Linking Water to Decisions for Sustainable Development', in World Water Assessment Programme, *World Water Development Report 3: Water in a Changing World* (UNESCO and Earthscan, 2009), at 3.

9   Vandemoortele (note 7).

10 UN Secretary General, 'Keeping the Promise: A Forward-Looking Review to Promote an Agreed Action Agenda to Achieve the Millennium Development Goals by 2015' (February 12, 2010) UN Doc A/64/665.

11 See UN Conference on Sustainable Development (Rio de Janeiro, June 20–22, 2012), *The Future We Want* (June 19, 2012), UN Doc A/Conf.216/L.1, at paras 119, 245–51. Available online at www.uncsd2012.org/content/documents/727The%20Future%20We%20Want%2019%20June %201230pm.pdf (accessed March 27, 2013) (The Future We Want).

The role of the UNWC

The UNWC makes little explicit reference to development issues in its actual text. Yet, the Preamble takes note of the 'special situation and needs of developing countries'. This means that, when establishing the rights and duties of watercourse states, the UNWC implicitly takes into account the varying levels of technical and economic capacity among contracting parties. Drawing on these general links, the relationship between effective transboundary water management and development goals can be established.[12] This pertains ultimately to a more fundamental question, regarding the potential causal relationships between watershed management at the international level and developmental goals in a national or local context.

The UNWC has the explicit intention to contribute to overall international peace and security, as well as regional stability, through fostering regional coordination processes on water management. Collaboration on water resources is often perceived as highly conducive to overall regional cooperation and integration, which, in turn, support security and stability.[13] Security and stability are considered necessary preconditions for any type of economic progress and thus the fulfillment of any of the MDGs. Only in areas without open (military) conflict or protracted disputes related to the allocation of resources, such as water, will it be possible to concentrate on key development issues, such as the provision of education and health services.[14] Similarly, sustainable development and a functioning green economy are contingent on regional stability and peace.[15]

In addition to these benefits related to political stability, Articles 5–7, 20 and 24 of the UNWC include obligations pertaining to equitable and sustainable use and management of international watercourses and the protection and preservation of aquatic ecosystems. Healthy freshwater ecosystems are key foundations for the attainment of several MDGs, such as hunger alleviation. Health-related MDGs also crucially depend on ecosystem quality. Many infections and diseases are aggravated in poor environmental conditions. In this regard, the UNWC resonates with the MDGs related to environmental sustainability as well.

A specifically strong interrelation exists between the UNWC and the water-related MDG targets – i.e. access to water and sanitation. The UNWC does not explicitly recognize a right to water and sanitation.[16] Yet, Article 10(2) calls for the

---

12   Loures, F., Rieu-Clarke, A. and Vercambre, M. L. *Everything You Need to Know About the UN Watercourses Convention* (WWF International, 2010).

13   UN Environment Programme, *From Conflict to Peace Building: The Role of Natural Resources and the Environment* (UNEP, 2009).

14   Ashton, P. J. and Turton, A. 'Water and Security in Sub-Sahara Africa: Emerging Concepts and their Implications for Effective Water Resource Management in the Southern African Region' in Brauch, H. G., Spring, U. O., Grin, J., Mesjasz, C., Kameri-Mbote, P. Behera, N. C., Chourou, B., and Krummenacher, H. (eds), *Facing Global Environmental Change: Environmental, Human, Energy, Food, Health and Water Security Concepts* (Springer, 2009), at 661–74.

15   Brack, D. 'Introduction: Trade, Aid and Security: An Agenda for Peace and Development' in Brown, O., Halle, M., Moreno, S. P. and Winkler, S. *Trade, Aid and Security* (Earthscan, 2007), at 1–17.

16   See Chapter 24 for a thorough analysis of the extent to which the UNWC may advance the human right to water.

resolution of conflicts over international watercourses to give special regard to the requirements of vital human needs.

Furthermore, through fostering international cooperation on water resources management, including on water infrastructure development, the UNWC will support the practical implementation and realization of those rights to the benefit of under-served parts of the population. In other words, sustainable water resource management at the transboundary scale, as enabled under the UNWC, is expected to lead to coordinative mechanisms that contribute to ensuring a good basis for safe drinking water access. Two aspects are decisive in this regard: the protection of ecosystems and their functions and services, as a precondition for sustainably providing clean drinking water; and planning activities for infrastructure development, which can be conducted jointly or separately. In both cases, the UNWC provides guidance on coordination and collaboration on those issues, which would potentially benefit all riparians.

Cooperation on access to water and sanitation could also provide further benefits, such as mutual assistance with financing,[17] as well as knowledge transfer and thus potentially the proliferation of innovative technologies. The latter could be related to sanitation technologies, for example. More important, however, in the context of transboundary water management, is the issue of wastewater management and treatment. The Convention explicitly mandates that riparians must prevent, reduce and control pollution in international watercourses. Waterborne sewage, especially in the developing world, constitutes a key pollutant in international rivers.[18] An obligation to prevent this type of pollution could potentially trigger the installation and development of more advanced treatment technologies, to complement basic sanitation installations. The UNWC thus potentially introduces a systems perspective to wastewater management at a transboundary scale. Improvements in this regard are likely to have positive repercussions for environmental and human health.

These positive resonances between water and development goals are ultimately captured by the UNWC's implicit support for the concept of integrated water resources management (IWRM).[19] IWRM constitutes the internationally

17 Joyce, J., Granit, J. and Winpenny, J. *Challenges and Opportunities for Financing Transboundary Water Resources Management and Development in Africa* (SIWI, 2010), at 6–7.

18 Corcoran, E., Nelleman, C., Baker, E., Bos, R., Osborn, D. and Savelli, H. (eds), *Sick Water? The Central Role of Wastewater Management in Sustainable Development* (UNEP and UN-HABITAT, 2010), at 19. Available online at www.grida.no/publications/rr/sickwater/ebook.aspx (accessed March 27, 2013).

19 IWRM is a systematic process for the sustainable development, allocation and monitoring of water resources. The concept and principles of IWRM were articulated at the International Conference on Water and Environment, held in Dublin, in 1992, and in Chapter 18 of Agenda 21, a consensus document from the UN Conference on Environment and Development (UNCED), held in Rio, also in 1992. See International Conference on Water and the Environment (Dublin, January 26–31, 1992), 'Dublin Statement on Water and Sustainable Development' (January 31, 1992). Available online at www.wmo.int/pages/prog/hwrp/documents/english/icwedece.html (accessed March 27, 2013); UN Conference on Environment and Development (Rio de Janeiro, June 3–14, 1992), 'Agenda 21: A Programme for Action for Sustainable Development' (January 1, 1993) UN Doc A/CONF.151/26/Rev.1 (Vol. I), at 9.

recognized paradigm for sustainable water management and basically supports the integration of land, water and ecological resources, of water quality and quantity issues and of surface and underground waters. As outlined before, sound management of water resources, based on true and inclusive stakeholder involvement, is an input essential for achieving the MDGs, including those related to alleviation of poverty and hunger, gender equality, health, education and environmental degradation.

In recognition of these linkages, an important short-term target was agreed upon at the World Summit for Sustainable Development in Johannesburg in 2002 and included in the Johannesburg Plan of Implementation: 'To develop integrated water resources management and water efficiency plans by 2005, with support to developing countries' – or, in short, the 'IWRM Target'.[20] The IWRM paradigm promotes cross-sectoral cooperation at all levels to balance different interests and needs of various users in achieving sustainable water resources management. It also facilitates the mainstreaming of water issues in the political economy of a country and, as such, across all sectors. IWRM focuses on better allocation of water to different water user groups and, in so doing, stresses the importance of involving all stakeholders in the decision-making process. The application of IWRM as a guiding paradigm and implementation guideline can assist in addressing power disparities in terms of equitable access to and control over resources, benefits obtained and costs incurred, and thus lead to a better integration of water and other sectors, such as energy and land management.[21]

The IWRM target highlights the vital role of improving water management through IWRM for achieving the MDGs.[22] It is in this context that the Hashimoto Action Plans I and II – which aim to inform the UN Secretary-General of the key steps that, in the view of water experts, are needed for improving global access to water and sanitation – call on countries to ratify and implement the UNWC.[23]

Beyond the MDGs: The water, food and energy nexus

The process leading up to the Rio+20 Summit, including the 2011 Bonn Nexus conference, helped to firmly establish the water-food-energy nexus as a key topic in the context of the green economy discourse. 'A nexus perspective helps to identify

20   World Summit on Sustainable Development (Johannesburg, August 26–September 4, 2002), 'Johannesburg Plan of Implementation' UN Doc A/CONF.199/20, at 6.
21   ToolBox, Integrated Water Resource Management' (Global Water Partnership, 2008). Available online at www.gwptoolbox.org/index.php?option=com_content&view=article&id=8&Itemid=3 (accessed March 27, 2012).
22   UN-Water and Global Water Partnership (GWP), Roadmapping for Advancing Integrated Water Resource Management (IWRM) Processes (UN-Water and GWP 2007), at 1.
23   UN Secretary-General's Advisory Board on Water and Sanitation, 'Hashimoto Action Plan: Compendium of Actions' (UN DESA, March 2006), at 9. Available online at www.unsgab.org/content/documents/HAP_en.pdf (accessed March 27, 2013) (Hashimoto Action Plan); UNSGAB, 'Hashimoto Action Plan II: Strategy and Objectives Through 2012' (UN DESA, January 2010), at 5. Available online at www.preventionweb.net/files/12657_HAPIIen.pdf (accessed March 27, 2013) (Hashimoto Action Plan II).

mutually beneficial responses and provides an informed and transparent framework for determining trade-offs to meet demand without compromising sustainability and exceeding environmental tipping points'.[24] In advancing this perspective, the Rio+20 outcome document explicitly places water at the heart of sustainable development and all its economic, social and environmental dimensions.[25]

Through its support for sustainable water management practices, the UNWC provides an important impulse to 'nexus' thinking at a transboundary basin level by advocating a sustainable management perspective and promoting the optimal and rational utilization of water resources. The UNWC points to the interconnectedness of water management with other key development challenges, such as food and energy security.[26] This interconnectedness or nexus[27] is deemed one of the key prerequisites under the paradigm of a green economy in the context of sustainable development and poverty alleviation. Therefore, the UNWC essentially could support the operationalization of nexus thinking by highlighting the role of sound water management, including in a transboundary context, for the overall attainment of sustainable development, as per its fifth Preambular Recital.

In addition to this conceptual support, fostering transboundary cooperation, such as promoted by the UNWC, provides for an opportunity to map out interlinkages between different policy areas. Transboundary river basins offer specific insights into spatial and sectoral connections between water resources, land management and energy production. In this sense, transboundary basins epitomize the challenge of applying a nexus perspective in the pursuance of water, energy and food security. For example, when considering the water-energy nexus, Leaders of Academies of Sciences note that:

> Regional water cooperation is, in many cases, essential. Energy options are a complex mix of local resources (if any), global supply, and available/affordable technological options. The wide range of local circumstances means that the world needs a wide range of clean energy technology options, whose impacts on water need to be well understood and taken into account in the decision processes.[28]

The G-Science Academies Statements 2012 goes on to recommend that governments 'establish effective governance structures and clear policies to facilitate the

24 Bonn 2011 Conference: The Water, Energy and Food Security Nexus (Bonn, November 16–18, 2011) 'Policy Recommendations' (February 13, 2012), at 3. Available online at www.water-energy-food.org/documents/bonn2011_policyrecommendations.pdf (accessed March 27, 2013).
25 The Future We Want (note 11), at para. 119.
26 World Economic Forum Water Initiative, *Water Security: The Water-Food-Energy-Climate Nexus* (Island Press, 2011).
27 Hoff, H. 'Understanding the Nexus – Background Paper for the Bonn 2011 Conference: The Water, Energy and Food Security Nexus' (Stockholm Environment Institute, 2011).
28 G-Science Academies Statement 2012, 'Energy and Water Linkage: Challenges to a Sustainable Future' (Nexus May 12, 2012). Available online at www.water-energy-food.org/en/news/view__580/energy_and_water_linkage_challenge_to_a_sustainable_future.html (accessed March 27, 2013).

integrated management of energy, water, and agriculture systems'. Therefore, the relevance of the UNWC, as a 'governance structure', comes to bear when considering that the challenge of reaching an effective nexus between water, food and energy is compounded in cases where riparian countries follow diverging, often conflicting, interests and strategies in managing water resources for multiple uses.

The UNWC mandates that states coordinate on installations and any measure with potential (harmful) impacts on the water resource. This requirement practically obliges riparians to adopt a benefit sharing[29] approach. Under such an approach, coordinative mechanisms are employed to resolve potential conflicts deriving from issues emerging in relation to water, food and energy security, and to explore possible synergies, which might allow riparians to share and exchange values or benefits beyond water and thus alleviate potential conflicts.[30]

While these challenges are bound to increase with progressing climate change and expected impacts on water resources,[31] conflict resolution and benefit sharing mechanisms, such as promoted by the UNWC, could serve as a test case for aligning these issues at a broader scale in order to realize a green economy for sustainable development within the framework of the water-food-energy nexus.

Interest of the donor community

It becomes evident from this brief assessment that the UNWC has the potential to boost policies that are supportive of the attainment of the MDGs at the basin level. The Convention could go even beyond and pave the pathway to sustainable development by interlinking key development challenges, such as land and energy, with water challenges. These potential benefits constitute a strong motivator for bilateral and multilateral donors to become involved in activities related to transboundary water management. There is a general expectation that, through fostering institutions and coordinative mechanisms at the transboundary level, it will be possible to reap benefits towards the attainment of the MDGs in the respective national contexts as well. Donor support at the transboundary level is manifold, such as documented by the mapping of support given to set up cooperation in Southern African basins.[32]

---

29 The model is derived from welfare economics. Water is valuable only as a scarce resource with alternative uses, some more valuable than others. The transcendental objective of efficiency requires that the resource be allocated to the most valuable suite of uses. This means that some nations will have to forego the actual use of wet water but are entitled to monetary compensation for allowing other states to put the water to its most efficient use. This concept is usually associated with the 1961 Canada–United States Columbia River Treaty and has become a general principle of both international water law and environmental law. This idea has been applied in other basins where upstream states can store but divert water and downstream states consume the flow. In recent years, there have been increasing calls to shift the focus of international disputes from allocation to benefit sharing. For a thorough discussion of benefit sharing, see Chapter 26.

30 Sadoff and Grey (note 2), at 389–403.

31 Bates, B., Kundzewicz, Z. W., Wu, S. and Palutikof, J. (eds), *Climate Change and Water* (Intergovernmental Panel on Climate Change, 2008).

32 GTZ *Donor Activity in Transboundary Water Cooperation in Africa* (Deutsche Gesellschaft für Technische Zusammenarbeit, 2007).

During meetings on transboundary water management (e.g. African Ministers' Council on Water Technical Advisory Committee in South Africa, 2010), benefits of earlier support by the African Development Bank, European Union and UN Development Programme (UNDP) for transboundary IWRM mentioned by participants (mainly water directors and experts) included the harmonization of policies in and between countries in the same basin(s), conservation of lake waters, linking to regional economic cooperation, developing a common language and identity in a basin, as well as the prevention of transboundary conflicts and the strengthening of overall peace and security. This indicates to donors that promoting the principles put forward by the UNWC would benefit the harmonization of national policies and help prevent varying solutions for the same problem encountered by different states. It would also help to counter experiences, such as those in the Nile Basin, where development has not been linked sufficiently with issues of water resource management.[33]

At the same time, donor activities are subject to constant monitoring and control, such as in the case of German International Cooperation, Swedish International Development Agency and African Development Bank. This serves the purpose of continually adapting to new challenges and needs. While there is a strong focus on support to institutions and capacity-development, there is a growing understanding that more emphasis and attention must be devoted to the development of joint infrastructure, as well as the design of sustainable funding streams.[34] The UNWC provides a suitable reference and guidance for future donor programming in this regard. In this sense, the Convention resonates with the high-level commitments made by the environmental ministers of the European Union (EU), as proposed by the 2012 Hungarian EU Presidency, which explicitly highlight the connection of development cooperation and water management.[35]

Equity, gender and civil society participation and right to information

Sustainable water management has clear implications for poverty alleviation and sustainable development, but requires well-capacitated national and local authorities, as well as the involvement of all stakeholders. With all the positive implications

33 UNEP (note 13).
34 Joyce, Granit and Winpenny (note 17).
35 The Council of the European Union, 'Protection of Water Resources and Integrated Sustainable Water Management in the European Union and Beyond: Council Conclusions' (3103rd Environment Council meeting, Luxembourg, June 21, 2011). The EU is expected to decide in Nov 2012 on a new water strategy for Europe: the 'Blue Print'. Part of the strategy will cover global developments and EU policy. The EU is currently also restructuring its approach to development cooperation and water management. The 2nd Generation EU Water Initiative, to be developed by the EU Commission, will serve as the future guiding framework in this regard. '2nd Generation EU Water Initiative, Supporting the EU Policy and Implementation Framework on Water in 3rd Countries' (6th World Water Forum, Solutions for Water, March 9, 2012). Available online at www.solutionsforwater.org/solutions/2nd-generation-eu-water-initiative-supporting-the-eu-policy-and-implementation-framework-on-water-in-third-countries (accessed March 27, 2013).

of the UNWC, some constraints emerge here. The UNWC addresses states and governs their behavior. The question thus remains as to how to integrate non-state actors into the Convention's future implementation. Moreover, how can the Convention's international guidance be translated into practice at national and local levels, with local conditions varying greatly?

River basin authorities have a crucial role to play in this context: once established under the Convention's common framework, such organizations should be mandated and empowered to coordinate implementation and, in close collaboration with local authorities and stakeholders, give concrete meaning and effects to the Convention's rules, principles and procedures on the ground.

Against this background, this section examines how the role of non-state actors, including women, in transboundary water management is reflected in the UNWC. It is assumed that these non-state actors are critical in managing water resources sustainably, supplementing state negotiations and actions.[36]

Role of women and equity

In many definitions of good governance for sustainable development, involvement of all relevant groups is seen as one of the key criteria for success and often as a *conditio sine qua non*.[37] Particular awareness is growing of the importance of a gender approach: 'a world-wide consensus has been created around the idea that the empowerment of women is the most effective instrument for development, for reducing poverty'[38] and specifically for water supply, sanitation and sustainable water management.[39] Despite facts and policy statements over the years about the role of women in water management and commitments,[40] the result is meager: women are so far mainly seen as target groups and victims, linked to potable water and sanitation provision, but hardly ever as managers of water sources or as actors and decision makers.[41]

A lack of women's leadership and participation in decision-making related to natural resource management and sustainable development is another stumbling

36 Kranz, N. and Mostert, E. 'Governance in Transboundary Basins – the Roles of Stakeholders; Concepts and Approaches in International River Basins' in Earle A, Jägerskog, A. and Öjendal, J. (eds), *Transboundary Water Management: Principles and Practice* (Earthscan, 2010), at 91–105.

37 Strandeneas, J. G. *Sustainable Development Governance towards Rio+20: Framing the Debate* (Stakeholder Forum, 2010).

38 Rachel Mayanja, Secretary Ban Ki-moon's Special Adviser on Gender Issues and the Advancement of Women, 2010.

39 UNEP, *Women and the Environment* (UNEP and Women's Environment and Development Organization, 2004), at 60 (citing Francis, J. 'Gender and Water', paper prepared for Women and the Environment, Gender and Water Alliance, 2003).

40 For an overview, see Women's Environment and Development Organization and Women for Water Partnership, 'International Commitments on Gender, Poverty and Water: Overview of Formal International Commitments/Recommendations 1992–2004' (WEDO, 2003).

41 Van der Heide, E., Joki-Hubach, A. and Bouman-Dentener, A. 'Women: Agents of Change, Boosting Social Change by Kick-Starting Women Leadership' (Paper to the 54th Session of the UN Commission on the Status of Women, The Netherlands Council of Women, 2010).

block. Women make up only 18% of parliamentary representatives globally. And yet, Michelle Bachelet, the head of UN Women, points out that countries with higher female parliamentary representation are more likely to ratify international environmental treaties. Where women are in leadership, they are often concentrated in social ministries rather than in those responsible for the sciences and finance, both of which have significant bearing on environmental and sustainability decision making.[42]

Against this background, the principles of IWRM and their gender implications[43] potentially offer an opportunity to create a paradigm shift in water resources management. In the face of the global water crisis, involving both women and men in IWRM can help increase project effectiveness and efficiency and thus increase the likelihood of sustainable outcomes.[44]

The notions of active involvement of women, gender issues and equity, as well as the involvement of stakeholders in environmental and water issues, were, for the first time, explicitly introduced in 1992, in Dublin and Rio.[45]

The Dublin conference established several principles,[46] which are still valid today, and helped to inform subsequent international policy documents. Incorporated into Principle 2, the participatory approach involves raising awareness of the importance of water among policy-makers and the general public. It means that decisions are taken at the lowest appropriate level, with full public consultation and involvement of users in the planning and implementation. Principle 3 addresses the pivotal role of women as providers and users of water and guardians of the living environment. Such a role has seldom been reflected in institutional arrangements for managing water resources. Acceptance and implementation of this principle would require positive policies to address women's specific needs and to empower them to participate at all levels.

42 UN Development Programme Human Development Report, *Sustainability and Equity: A Better Future for All* (UNDP, 2011); Marston, A. 'Women and the Environment: The Forgotten Rio Principles?' *Outreach* (December 15, 2012). Available online at www.stakeholderforum.org/sf/outreach/index.php/outreach2012zerodraft2/73-cop17day12home/571-cop17day12item4 (accessed March 27, 2013). Evidence suggests that women's involvement in political institutions is associated with better local environmental management. Yet, women's mere presence in institutions is not enough to overcome entrenched disparities – additional changes and flexibility in institutional forms are needed to ensure that women can participate effectively in decision-making. See also 'World Poll' (Gallup). Available online at www.gallup.com/se/126848/worldview.aspx (accessed March 27, 2013).

43 Adapted from van Wijk-Sijbesma, C. *Participation and Education in Community Water Supply and Sanitation Programmes*, Technical Paper Series/IRC No. 12 (2nd edn, International Reference Centre for Community Water Supply and Sanitations, 1981, 1988).

44 GWA and UNDP, *Resource Guide: Mainstreaming Gender in Water Management* (Version 2.1, GWA and UNEP, 2006).

45 Shandra, J., Shandra, C. L. and London, B. 'Women, Non-Governmental Organizations, and Deforestation: A Cross-National Study,' *Population and Environment*, 2008; 38: 48–72.

46 International Conference on Water and the Environment (Dublin, January 26–31, 1992), 'Dublin Statement on Water and Sustainable Development' (January 31, 1992). Available online at www.wmo.int/pages/prog/hwrp/documents/english/icwedece.html (accessed March 27, 2013).

Building on the principles above, Principle 20 of the Rio Declaration on Environment and Development underlines that 'women have a vital role in environmental management and development and are therefore essential to achieve sustainable development'.[47] This was confirmed by the 1995 Beijing Declaration on the role of women in economic growth, social development and environmental protection,[48] the 2002 Johannesburg Plan of Implementation,[49] and the 2003 Proclamation of the 2005–2015 Water for Life Decade, calling for women participation in water-management and water-related development efforts.[50]

In the African regional context, increasing emphasis has been placed on the role of women in achieving sustainable water management. In order to implement the African Water Vision 2025,[51] the African Ministers' Council on Water adopted a gender-mainstreaming strategy for the region,[52] which is to be implemented at transboundary and national levels.[53] In particular, the strategy calls on women to take on roles as decision-makers and emphasizes stakeholder involvement, particularly women and youth, in order to: a) address specific gender issues in IWRM, transboundary waters and water and sanitation; b) determine gender differentiated access to, use of and control over water resources; and c) address perceptions on the effectiveness of existing institutional arrangements to enhance equitable access of water resources for women and men.

47 UN Conference on Environment and Development (Rio de Janeiro, June 3–14, 1992), 'Rio Declaration on Environment and Development' (January 1, 1993), UN Doc A/CONF.151/26/Rev.1 (Vol. I), at 3 (Rio Declaration).

48 4th World Conference on Women (Beijing, September 4–15, 1995), 'Beijing Declaration' (September 15, 1995), UN Doc A/CONF.177/20/Rev.1, at 2. Chapter K of the Beijing Platform for Action explicitly talks about the involvement of women in environmental issues: 'Eradication of poverty based on sustained economic growth, social development, environmental protection and social justice require the involvement of women in economic and social development, equal opportunities and the full and equal participation of women and men as agents and beneficiaries of people-centred sustainable development'. 4th World Conference on Women (Beijing, September 4–15, 1995), 'Beijing Platform for Action' (September 15, 1995), UN Doc A/CONF.177/20/Rev.1, at 103.

49 World Summit on Sustainable Development (Johannesburg, 26 Aug-4 Sep 2002), 'Johannesburg Plan of Implementation' UN Doc A/CONF.199/20, especially Paras 7(d), 40, 44(k) & 46 and 25, 54, 67, 102 & 164.

50 UNGA, Res 58/217 'International Decade for Action, "Water for Life", 2005–2015' (February 9, 2004) UN Doc A/Res/58/217, at 2.

51 UN-Water Africa and UN Economic Commission for Africa, Africa Water Vision for 2025: Equitable and Sustainable Use of Water for Socioeconomic Development (UN-Water Africa and others), at 2.

52 Economic and Social Council, 'Mainstreaming the Gender Perspective into All Policies and Programmes of the UN Systems' in UNGA, 'Report of the Economic and Social Council for 1997' (September 18, 1997) UN Doc A/52/3, at 27. According to the report, 'Gender mainstreaming is the process of assessing the implications for women and men of any planned action, including legislation, policies and programmes in all areas and at all levels. It is a strategy for making women's as well as men's concerns and experiences an integral dimension of the design, implementation, monitoring and evaluation of policies and programmes in all political, economic and societal spheres, so that women and men can benefit equally and inequality is not perpetuated. The ultimate goal is to achieve gender equality (by transforming the mainstream)'.

53 African Ministers' Council on Water, *Policy and Strategy for Mainstreaming Gender in the Water Sector in Africa: Strategic Objective 4* (AMCOW, 2010), at 24.

Beyond the declaratory level, increasing *climate change* highlights the relevance of women's involvement in water management. We must recall that women are the majority (up to 70 percent) of workers in agriculture and thus greatly influence irrigation practices and water use. At the same time, women tend to be affected most harshly by climate change due to their social role and prevailing lack of finances and information, in both OECD and non-OECD countries. Climate change can adversely affect crop yields and thus the livelihoods and food security of women who are largely responsible for food production and family nutrition. Women and children are also far more likely to die than men during extreme weather events linked to climate change owing to their (traditional) roles and thus greater vulnerability. In the case of Hurricane Katrina in the United States, those who were hardest hit and had the least ability to recover included women, who represent the majority of the poor, especially single mothers.[54]

On the other hand, women are most resourceful in adapting to climate change. As a result of their land-based work and knowledge of natural resources, women represent an untapped asset in coping with the effects of climate change on livelihoods.[55] In many cases, also local knowledge about (traditional) management of water sources rests with the women of the community. Coping with the damage of extreme weather events, such as storms, floods and cyclones, mostly falls under the responsibility of women as well, who hold together families and households.

Women face many challenges in taking on these important roles. One of those factors is limited and even diminishing right to land tenure, which is often the basis for access to water or granting water rights. In many countries, registration of ownership is non-existent and property and inheritance laws prevent women from owning land. Women often hold customary rights to land and no formal rights,[56] conflicting with their abovementioned role as main provider for the family. Increasingly, land resources are under pressure from foreign investors, seeking to gain access to land for export crops. This development is mainly affecting women farmers.

With women facing continued challenges with regard to accessing and managing water, the question emerges: Can their concerns be supported by the UNWC?

54 OECD, *Gender and Sustainable Development: Maximising the Economic, Social and Environmental Role of Women* (OECD, 2008), at 73.
55 UNEP (note 39).
56 Across Africa, water tends to be vested in, and managed by the state. In most places, local people have customary uses but do not hold formal rights. For example, fishermen do not hold a formal water right, nor do pastoralists who use floodplain pastures during the dry season. Even if local people have legally protected land use rights, they rarely have formal control over the water that they use, beyond recognition that supplying drinking water is a basic human requirement that cannot be refused. In most cases, traditional users of water simply accept water rights as a secure tradition and either see no need to formalize them, or are unable to access the process for doing so. Data on water rights are hard to quantify, although in one documented example on the Ruaha Basin, in Tanzania, some 40% of rights were held by government bodies, 28% by private land owners and only 10% by local water user associations. van Koppen, B., Sokile, C. S., Hatibu, N., Lankford, B. A., Mahoo, H. and Yanda, P. Z., *Formal Water Rights in Rural Tanzania: Deepening the Dichotomy?* Working Paper 71 (International Water Management Institute, 2004).

Additionally, can women's role be placed in the broader context of the role of civil society in managing water resources?

Civil society involvement in the management of water resources[57]

Public participation in water sources management is recognized as an important tool for identifying the measures most appropriate to achieve common objectives, through considering and carefully balancing interests of various groups. It is also expected that, with greater transparency in the establishment of objectives, the implementation of measures finds broader acceptance among those affected, thus leading to a better track record in implementation altogether. Depending on the specific case and purpose, civil society involvement can include the general public and interested parties, as well as non-governmental organizations.

In this sense, the 1992 Rio Declaration introduced several so-called major groups, i.e. those groups representing different societal interests, and established the principle of stakeholder involvement in the context of Agenda 21: '*Critical to the effective implementation of the objectives, policies and mechanisms agreed to by Governments in all programme areas of Agenda 21 will be the commitment and genuine involvement of all social groups*'.

In addition, the declaration underlines:

> One of the fundamental prerequisites for the achievement of sustainable development is broad public participation in decision-making. In the more specific context of environment and development, the need for new forms of participation has emerged. This includes the need of individuals, groups and organizations to participate in environmental impact assessment procedures and to know about and participate in decisions, particularly those which potentially affect the communities in which they live and work.[58]

A crucial factor when it comes to public participation is the right to information so as to enable prior informed choices and decisions. This idea is enshrined in Principle 10 of the Rio Declaration:

> At the national level, each individual shall have appropriate access to information concerning the environment that is held by public authorities, including information on hazardous materials and activities in their communities, and the opportunity to participate in decision-making processes. States shall facilitate and encourage public awareness and participation by making

---

57 'In getting our waters clean, the role of citizens and citizens' groups will be crucial'. See European Commission, Environment, 'Introduction to the New EU Water Framework Directive' (European Commission, February 23, 2012). Available online at http://ec.europa.eu/environment/water/water-framework/info/intro_en.htm (accessed March 27, 2013).

58 UN Conference on Environment and Development (Rio de Janeiro, June 3–14, 1992), 'Agenda 21: A Programme for Action for Sustainable Development' (January 1, 1993) UN Doc A/CONF.151/26/Rev.1 (Vol. I), at 373 (Agenda 21).

information widely available. Effective access to judicial and administrative proceedings, including redress and remedy, shall be provided.[59]

At the regional level, the above principles form the basis for the development of River Basin Management Plans under the EU Water Framework Directive. Such plans lay out the future management and quality goals for all rivers in Europe. The plans must be issued in draft and all the background documentation on which the decisions are based must be made accessible in order to allow for sufficient consultation when river basin management plans are established.[60]

Article 16 of the UNECE Water Convention makes references to the issue of public information as well: 'The Riparian Parties shall ensure that information on the conditions of transboundary waters, measures taken or planned to be taken to prevent, control and reduce transboundary impact, and the effectiveness of those measures is made available to the public'.[61] Article 9(2) of the UNECE Water Convention relates to the dissemination and exchange of information on water disasters, waste and environmental impacts, and establishes emergency and alert procedures. Finally, Article 11 refers to the right to information: 'The results of these assessments shall be made available to the public'.

The above provisions combined establish a clear obligation on states to supply adequate information in the UNECE region and give stakeholders access to judicial means.[62] Such provisions have been translated into practice, with the UNECE

59   UN Conference on Environment and Development (Rio de Janeiro, June 3–14, 1992), 'Rio Declaration on Environment and Development' (January 1, 1993), UN Doc A/CONF.151/26/Rev.1 (Vol. I), at 5 (Principle 10).

60   Directive 2000/60/EC of 23 October 2000 of the European Parliament and of the Council establishing a framework for community action in the field of water policy [2000] OJ L327/1 (EU Water Framework Directive). Article 14 reads: 'Member States shall encourage the active involvement of all interested parties in the implementation of this Directive, in particular in the production, review and updating of the river basin management plans. Member States shall ensure that, for each river basin district, they publish and make available for comments to the public, including users'.

61   UNECE Convention on the Protection and Use of Transboundary Watercourses and International Lakes (adopted March 17, 1992, entered into force October 6, 1996) 1936 UNTS 269, Articles 9(2) and 16 (UNECE Water Convention).

62   Decision 2005/370/EC of 17 February 2005 of the European Council on the conclusion, on behalf of the European Community, of the Convention on access to information, public participation in decision-making and access to justice in environmental matters [2005] OJ L124/1. The EC has been a party to the Aarhus Convention since May 2005. In 2003, two Directives concerning the first and second 'pillars' of the Aarhus Convention were adopted; they were to be implemented in the national law of the EU Member States by February 14 and June 25, 2005, respectively. Provisions for public participation in environmental decision-making are furthermore to be found in a number of other environmental directives, such as Directive 2001/42/EC of 27 June 2001 of the European Parliament and of the Council on the assessment of certain plans and programmes on the environment [2001] OJ L 197/30 (see also the 'Environmental Assessment' homepage) and EU Water Framework Directive. See also European Commission, 'Environment: Environmental Assessment' (European Commission). Available online at http://ec.europa.eu/environment/eia/home.htm (accessed March 27, 2013); European Commission, 'Environment: Water' (European Commission). Available online at http://ec.europa.eu/environment/water/index_en.htm (accessed March 27, 2013).

Water Convention Secretariat actively encouraging and enabling the participation of NGOs in working groups and as observers.

The role of the UNWC

Article 32 of the UNWC concerns the principle of non-discrimination, which pertains to the rights of persons in the context of transboundary threat or harm and access to judicial or other procedures:

> Unless the watercourse states concerned have agreed otherwise for the protection of the interests of persons, natural or juridical, who have suffered or are under a serious threat of suffering significant transboundary harm as a result of activities related to an international watercourse, a watercourse State shall not discriminate on the basis of nationality or residence or place where the injury occurred, in granting to such persons, in accordance with its legal system, access to judicial or other procedures, or a right to claim compensation or other relief in respect of significant harm caused by such activities carried on in its territory.

This provision introduces a broad principle of non-discrimination and thus can be interpreted as conducive to public consultation and equity concerns. Apart from this paragraph, there are no other concrete direct references to public participation, equity and gender in the text of the UNWC. How, therefore, would the UNWC or, more specifically, its ratification and entry into force, support women's actions and stakeholder involvement?

The Preamble of the UNWC recalls the principles and recommendations of the Rio Declaration and Agenda 21, which stipulate that 'any policies, definitions or rules affecting access to and participation by non-governmental organizations in the work of [UN] institutions or agencies associated with the implementation of Agenda 21 must apply equally to all major groups'.[63] It is thus justified to conclude that participatory approaches are implicitly a part of the UNWC.

Moreover, practical experience with the UNECE Water & Health Protocol,[64] the Espoo Convention[65] and the Aarhus Convention[66] in the Eastern Europe,

63 UNWC, at Preamble, Recital 8; UN Conference on Environment and Development (Rio de Janeiro, June 3–14, 1992), 'Agenda 21: A Programme for Action for Sustainable Development' (January 1, 1993) UN Doc A/CONF.151/26/Rev.1 (Vol. I), at para. 23.3.
64 Protocol on Water and Health to the 1992 Convention on the Protection and Use of Transboundary Watercourses and International Lakes (adopted June 17, 1999, entered into force August 4, 2005) 2331 UNTS 202.
65 UNECE Convention on Environmental Impact Assessment in a Transboundary Context (adopted February 25, 1991, entered into force September 10, 1997) 1989 UNTS 309 (1991 Espoo Convention).
66 UNECE Convention on Access to Information, Public Participation in Decision-Making and Access to Justice in Environmental Matters (adopted June 25, 1998, entered into force October 30, 2001) 2161 UNTS 447 (Aarhus Convention). See also UNECE, Meeting of the Parties to the Aarhus Convention (29 June 29–July 1, 2011), 'Chisinau Declaration' (July 1, 2011) UN Doc ECE/MP.PP/2011/CRP.4/rev.1.

Caucasus and Central Asia (EECCA) region has shown that this type of international agreement provides civil society and NGOs in the countries concerned with great leverage to start discussions with their national governments and regional authorities, hold them accountable, and thus, in effect, lead to (more) participatory approaches.

Also, in combination and through mutual enhancement with other UN treaties, including the Committee on the Elimination of Discrimination against Women Convention,[67] as well as selected principles of the Rio Declaration and recommendations of the UN Commission on Sustainable Development, the full implementation of the UNWC is likely to enhance the involvement and participation of non-state actors, and contribute to securing equitable access to and decision-making related to water resources by all major groups, which include women and youth.

Furthermore, it can be expected that, through the promotion of sustainable water management approaches (IWRM) and the overall beneficial effects of improved cooperation on transboundary river basins, women will be able to strengthen their position and increase their overall resilience to climate change impacts.

A requirement to engage stakeholders and ensure their rights is also implicit in the reference to Article 1 of the UN Charter, as well as within a number of obligations under the UNWC. For instance, the obligation in Article 7 to take *all appropriate* measures not to cause significant harm could be read to oblige states to ensure that stakeholders are involved in decisions on planned measures and notification. Similarly, the obligation in Article 12 to notify and provide 'available technical data and information, including the results of any environmental impact assessment' could imply engagement with potentially affected stakeholders. It could further be maintained that, if a particular state has ratified the UNWC and it has entered into force, it would be easier for marginal groups to hold that state accountable pursuant to the obligations contained therein.

Conclusion

In assessing the linkages between the UNWC and other policy initiatives in the developmental/environmental field, as well as with regard to equity and empowerment, we have identified interesting cross-linkages and resonances. In particular, the full implementation of the UNWC can lay the foundations to accelerate the process for attaining the MDGs and pave the way for the implementation of the future Sustainable Development Goals. The Convention can significantly contribute to those processes by fostering international cooperation on water resources management and thus, indirectly, lead to progress on issues dependent on sustainable water management, such as food and energy security, access to water and sanitation and poverty alleviation more generally.

---

67 Committee on the Elimination of Discrimination against Women (adopted December 18, 1979, entered into force September 3, 1981) 1249 UNTS 13, at Article 14 (CEDAW Convention).

Pertaining to international environmental goals and discussions leading up to and during Rio+20, the Convention requires states to cooperate in managing shared water and related resources for multiples uses and thus across sectors. These aspects are directly supportive of key elements of a Green Economy and the Nexus thinking based on the premises of the integrated management of water and cross-sectoral approaches for sustainable development.

Regarding stakeholder participation, we conclude that – in combination with other UN treaties (e.g. the Committee on the Elimination of Discrimination against Women Convention), selected principles adopted at UNCED, and recommendations of the Commission on Sustainable Development – the implementation of the UNWC will:

- provide stakeholders with an entry point for their involvement, dialogue with regional authorities and national governments and participation of non-state actors;
- as a minimum, provide stakeholders with (timely) information and access to 'justice', especially in the case of transboundary harm and disasters, through the principle of non-discrimination;
- enhance information exchange and dialogue between states and thus improve data availability to their citizens; and
- further strengthen equity and women's voice in water management and water policy.

At the same time, improved public participation in transboundary water management processes would actively contribute to the advancement of negotiations and coordination among riparians and thus support the implementation of the UNWC itself.

Part 5

Beyond entry into force

Strengthening the role and relevance of the
UN Watercourses Convention

22 An institutional structure to support the implementation process

Alistair Rieu-Clarke and Flavia Rocha Loures

In 1990, Stephen McCaffrey, Special Rapporteur to the International Law Commission (ILC) on the Law of the Non-Navigational Uses of International Watercourses, proposed that the ILC Draft Articles that preceded the UNWC include an Annex related to *Implementation of the Articles*.[1] A key provision of this proposed Annex was Article 7 – Conference of the Parties. Pursuant to Article 7, states would be obliged, 'not later than two years after the entry into force of the present articles', to convene a meeting of the Conference of the Parties, and to hold regular meetings at least once every two years, to: (i) consider and adopt amendments to the articles; (ii) receive and consider any reports presented by any party or by any panel, commission or other body established by the articles; and, where appropriate, (iii) make recommendations for improving the effectiveness of the articles.[2] The inclusion of such a provision was justified by the Special Rapporteur on the grounds that, 'several recent conventions relating to the environment or transboundary harm contain provisions for regular meetings of a "conference of the parties"'.[3]

In his commentary, the Special Rapporteur highlights the role that secretariats play in other conventions, namely 'convening and servicing meetings of the conference of the parties and conducting studies and research at the request of the parties'.[4] However, the Special Rapporteur was hesitant to propose such a mechanism in the context of the Draft Articles, 'in connection with what is envisaged as a framework agreement'.[5] He went on to state that, 'if a convention is eventually concluded on the basis of the present draft articles, the parties may certainly establish such an institution if they so desire'.[6]

---

1 McCaffrey, S. C. (Special Rapporteur), 'Sixth Report on the Law of the Non-Navigational Uses of International Watercourses' (February 23 and June 7, 1990), (1990) II(1) Extract from the Yearbook of the ILC 41, at 64, UN Doc A/CN.4/427. Available online at http://untreaty.un.org/ilc/documentation/english/a_cn4_427.pdf (accessed March 27, 2013).
2 Ibid.
3 Ibid.
4 Ibid.
5 Ibid.
6 Ibid.

Draft Article 7 was debated in the meetings of the ILC during its 42nd Session in 1990.[7] There was a general feeling that the Article was 'very ambitious' for the ILC Draft Articles, and 'it was not for the Commission to deal with arrangements subsequent to entry into force of the draft articles'.[8] Ultimately, a provision for a Conference of the Parties was not included within the text of the ILC Draft Articles, nor within the UNWC itself. A question thus remains as to whether the UNWC would benefit from such a mechanism and, if so, what such a mechanism would look like.

The value of institutional mechanisms in terms of reporting, information exchange, participation and compliance strategies is considered in Chapter 8 of this book. That chapter also examines the extent to which the text of the UNWC provides for global institutional design features. The conclusion derived from Chapter 8 is that such features are limited. Here, therefore, we take the analysis one step further to ask whether an institutional structure could be introduced at the global level to support the implementation of the UNWC, and what that institutional structure might look like.

The potential benefits of institutional mechanisms at the global level

Regular meetings would allow the parties to the UNWC to better coordinate the management of water and related natural resources; promote interaction among joint water bodies; enable networking and exchange of information, expertise and technologies; assess funding needs and create a forum for coordination among donors and recipient countries so as to facilitate access to financial resources; explore and develop common guidelines and approaches to sustainable and cooperative water management; and maintain water issues as a priority item on the global policy agenda.

As noted in previous chapters, the UNWC has a key role to play in the management and sustainable use of transboundary freshwater resources. The Convention incorporates the concepts of sustainable and equitable water use and adopts an integrated approach to river basin management. Additionally, Chapter 8 suggests that the role of the UNWC could be significantly enhanced by the existence of institutional mechanisms that, as consensus grew, would support the Convention's progressive development, coherent application and effective implementation.

Specifically with regard to the progressive development of international law, there are certain areas that the UNWC does not address well enough. For example, the UNWC does not contain detailed provisions on the *joint* management of transboundary water resources (e.g. functions of joint water bodies, requisites of

---

7   ILC, 'Summary Records of the Meetings of the 42nd Session' (May 1–July 20, 1990) Yearbook of the ILC Vol. 1 (1990), UN Doc A/CN.4/SER.A/1990 (1990), at 84–115. Available online at http://untreaty.un.org/ilc//publications/yearbooks/Ybkvolumes(e)/ILC_1990_v1_e.pdf (accessed March 27, 2013).
8   Ibid., at 111. See also comments at 105–6, 114 and 121.

joint monitoring and assessment programmes, joint research and development, common databases, etc.). Other than incorporating the non-discrimination principle, the UNWC lacks mechanisms for public participation in interstate discussions and does not clarify how states are to apply the polluter-pays principle in a transboundary context.

In order to address these gaps, an amendment to the UNWC could create procedures for the adoption of additional amendments, protocols and annexes. The two protocols adopted under the UNECE Water Convention illustrate the importance of the issue. A protocol on water and health aims to ensure access to drinking water and adequate sanitation throughout the UNECE region for a high level of protection against water-related diseases.[9] Another protocol creates a comprehensive civil liability regime, under the polluter-pays principle, applicable to transboundary effects on international waters resulting from industrial accidents.[10]

Furthermore, institutional mechanisms have great value for the interpretation and implementation of multilateral environmental agreements (MEAs). The various activities carried out by parties through these mechanisms demonstrate the benefits from building upon the original text of a convention as consensus increases.

For example, at their 7th Meeting, the parties to the Ramsar Convention[11] adopted guidelines regulating Article 5, which requires interstate consultations on transboundary wetlands.[12] Such guidelines illustrate how parties to international environmental treaties in force may interpret and regulate their obligations to facilitate implementation.

The same is true with strategic or work plans, which define programme areas and may create working groups to coordinate certain activities. These implementation plans have been adopted by the conference of the parties of a variety of international environmental conventions.[13] Successive strategic plans are useful to guide implementation and focus activities on identified priority areas. As targets are

---

9 Protocol on Water and Health to the 1992 Convention on the Protection and Use of Transboundary Watercourses and International Lakes (adopted June 17, 1999, entered into force August 4, 2005) 2331 UNTS 202 (Protocol on Water and Health).

10 Protocol on Civil Liability and Compensation for Damage Caused by the Transboundary Effects of Industrial Accidents on Transboundary Waters to the 1992 Convention on the Protection and Use of Transboundary Watercourses and International Lakes and to the 1992 Convention on the Transboundary Effects of Industrial Accidents (adopted May 21, 2003). Available online at www.unece.org/env/civil-liability/documents/protocol_e.pdf (accessed March 27, 2013).

11 Convention on Wetlands of International Importance Especially as Waterfowl Habitat (adopted February 2, 1971, entered into force December 21, 1975) 996 UNTS 245 (Ramsar Convention).

12 Ramsar Convention CoP-7 (San José, Costa Rica, May 10–18, 1999) 'Guidelines for International Cooperation under the Ramsar Convention', Resolution VII.19. Available online at www.ramsar.org/pdf/res/key_res_vii.19e.pdf (accessed March 27, 2013).

13 See, for example, Ramsar Convention CoP-8 (Valencia, Spain, May18–26, 2002) 'The Ramsar Strategic Plan 2003–2008', Resolution VIII.25. Available online at www.ramsar.org/cda/en/ramsar-documents-resol-resolution-viii-25-the/main/ramsar/1-31-107%5E21439_4000_0__ (accessed March 27, 2013).

achieved, new ones are established,[14] facilitating the assessment of progress towards a treaty's major goals.[15]

A conference of the parties is also the forum under which subsidiary bodies may be created. These bodies may meet periodically to offer scientific or legal advice, monitor a convention's implementation or otherwise assist countries in complying with their obligations.[16] Under the UN Economic Commission for Europe (UNECE) Water Convention,[17] for instance, a number of bodies have been established in addition to the conference of the parties and the secretariat, such as the working groups on integrated water resources management and monitoring and assessment, respectively; a legal board; a task force on water and climate; a joint ad hoc expert group on water and industrial accidents; and an international water assessment center.[18]

In addition, conferences of the parties explore common goals among different international environmental treaties.[19] As experience shows, once those opportunities are identified, memoranda of cooperation can be adopted to regulate and frame collaboration between executive bodies[20] and with other relevant actors.[21] These

14 See, for example, the four work plans adopted by the Meeting of the Parties to the UN Economic Commission for Europe (UNECE) Convention on the Protection and Use of Transboundary Watercourses and International Lakes, for the periods 1997–2000, 2000–2003, 2004–2006, 2007–2009 and 2010–2012. Available online at www.unece.org/env/water/meetings/convention_meeting.html (accessed March 27, 2013).

15 See, for example, Convention on Biological Diversity, CoP-7 (Kuala Lumpur, Malaysia, February 9–20, 2004), 'VII/30: Strategic Plan: Future Evaluation of Progress' (April 13, 2004) UNEP/CBD/COP/DEC/VII/30. Available online at www.cbd.int/doc/decisions/cop-07/cop-07-dec-30-en.pdf (accessed March 27, 2013), adopting a framework for the evaluation of progress in the implementation of the Strategic Plan.

16 See UN Framework Convention on Climate Change (adopted May 9, 1992, entered into force March 24, 1994) 1771 UNTS 107, Articles 7(2)(i), 9 and 10.

17 UNECE Convention on the Protection and Use of Transboundary Watercourses and International Lakes (adopted March 17, 1992, entered into force October 6, 1996) 1936 UNTS 269 (UNECE Water Convention).

18 See UNECE Water Convention, 'About the Convention: Convention Bodies'. Available online at www.unece.org/env/water (accessed March 27, 2013).

19 CBD CoP-8 (Curitiba, March 20–31, 2006), 'VIII/16: Cooperation with Other Conventions and International Organizations and Initiatives' (June 15, 2006) UNEP/CBD/COP/DEC/VIII/16. Available online at www.cbd.int/doc/decisions/cop-08/cop-08-dec-16-en.pdf (accessed March 27, 2013).

20 See, for example, 'Memorandum of Cooperation between the Bureau of the Convention on Wetlands of International Importance, especially as Waterfowl Habitat (Ramsar, Iran, 1971) and the Secretariat of the Convention on Biological Diversity (Nairobi, 1992)' (January 19, 1996). Available online at www.ramsar.org/cda/en/ramsar-documents-mous-memorandum-of/main/ramsar/1-31-115^16060_4000_0 (accessed March 27, 2013). See also subsequent documents, including four joint work plans covering the period 1998-2010, and a 2nd Memorandum of Cooperation. Available online at www.ramsar.org/cda/en/ramsar-documents-mous/main/ramsar/1-31-115_4000_0__ (accessed March 27, 2013).

21 See, for example, 'Memorandum of Cooperation between the Bureau of the Convention on Wetlands (Ramsar, 1971) and the Lake Chad Basin Commission (LCBC)' (November 23, 2002). Available online at www.ramsar.org/cda/en/ramsar-documents-mous-memorandum-of-21474/main/ramsar/1-31-115^21474_4000_0__ (accessed March 27, 2013), promoting joint initiatives and solutions to transboundary issues at the basin scale.

memoranda allow for the development of *joint* programmes of work that enhance synergies and avoid duplication of effort through adequate coordination of activities. In this sense, an executive body under the UNWC would consult with other MEA secretariats on possible areas for collaboration, coordination and mutual reinforcement between different and yet intimately connected environmental goals. After all, effective transboundary cooperation on the sustainable management of water and related resources would support a variety of environmental regimes.

For example, a Ramsar decision makes a special reference to the importance of the UNWC when discussing possible partnerships with relevant conventions, organizations and initiatives for the achievement of common goals.[22] Likewise, a recent Conference of the Parties decision under the Convention on Biological Diversity (CBD) urges parties and other governments to ratify and implement the UNWC to support activities on the control and prevention of invasive species.[23] Article 22 of the UNWC requires preventive action from states regarding the introduction of species, alien or new, into international rivers, and could play a crucial role in addressing invasive species at the basin level. Currently, the Ramsar Convention and the CBD deal jointly with the topic.[24] Expanding cooperation on invasive species to include activities under a global water treaty could optimize and accelerate results. The UNWC would refocus the attention of *basin* states on the problem and promote cooperative action to address it through the applicable preventive and mitigation measures.

Invasive species represent just one example of a subject matter linking the UNWC to other environmental regimes. In the absence of governing bodies to promote the implementation of the UNWC, there would be no room for exploring issues that this Treaty may have in common with other environmental agreements.

Conferences of the parties may also approve soft-law instruments such as guidelines and codes of conduct. Under the UNECE Water Convention, countries have relied on the guidelines adopted by the Meeting of the Parties to establish and implement monitoring and assessment programmes of transboundary rivers, lakes and aquifers, and to exchange knowledge and experiences across basins.[25] These

---

22   Ramsar Convention CoP-7 (San José, Costa Rica, May 10–18, 1999) 'Resolution VII.18: Guidelines for Integrating Wetland Conservation and Wise Use into River Basin Management', at 21. Available online at www.ramsar.org/cda/en/ramsar-documents-resol-resolution-vii-18/main/ramsar/1-31-107^20586_4000_0__ (accessed March 27, 2013).

23   CBD CoP-8 (Curitiba, March 20–31, 2006), 'VIII/27: Alien Species that Threaten Ecosystems, Habitats or Species (Article 8 (h)): Further Consideration of Gaps and Inconsistencies in the International Regulatory Framework' (June 15, 2006) UNEP/CBD/COP/DEC/VIII/27, at 3. Available online at www.cbd.int/doc/decisions/cop-08/cop-08-dec-27-en.pdf (accessed March 27, 2013).

24   See 'Joint Work Plan between Ramsar and the Convention on Biological Diversity: Historical Note' (May 6, 2002), at paras 49–52. Available online at www.ramsar.org/cda/en/ramsar-documents-mous-joint-work-plan-between-15850/main/ramsar/1-31-115^15850_4000_0__ (accessed March 27, 2013).

25   See UNECE Water Convention, 'List of Publications'. Available online at www.unece.org/env/water/publications/pub.html (accessed March 27, 2013).

guidelines are not legally binding, but are useful to promote management activities consistent with an ecosystem approach and enable the harmonization of monitoring strategies and reporting practices.

Some conferences of the parties manage trust funds for voluntary or compulsory contributions. Trust funds support the promotion and effective implementation of international agreements, but can only exist if there is a body under the treaty to function as a manager.[26] In the case of the UN Convention to Combat Desertification, a financial mechanism operates under the authority and guidance of the conference of the parties. This mechanism supports the identification of available financial mechanisms and cooperation programmes, assists and advises parties on financial issues and facilitates activities of resource allocation.[27] The possibility to amend the UNWC would allow parties to consider creating a funding mechanism to support, for example, the implementation of watercourse agreements.

Finally, a key feature of institutional mechanisms of MEAs relates to implementation and compliance.[28] Numerous MEAs provide basic obligations whereby contracting parties must report periodically on their activities towards implementation.[29] MEA institutions can then use the resultant reports to analyze patterns of behavior and identify specific and general implementation challenges.[30] Additionally, MEA institutions may adopt non-compliance procedures. For instance, under the Protocol on Water and Health to the UNECE Water Convention, a compliance committee has been established.[31]

Conclusion

The above analysis demonstrates that there is considerable precedence and experience related to the establishment of institutional mechanisms for the implementation of MEAs. There are numerous pathways by which such mechanisms have been adopted – such as formally through their inclusion within the original treaty text, or subsequent development following the entry into force of a particular agreement. Ultimately, such mechanisms provide an important means by

---

26 See UNECE Water Convention MoP-3 (Madrid, November 26–28, 2003), 'Establishment of a Trust Fund under the Convention' (April 8, 2004) ECE/MP.WAT/15/Add.1, at 6. Available online at   www.unece.org/fileadmin/DAM/env/documents/2004/wat/ece.mp.wat.15.e.add1.pdf (accessed March 27, 2013).

27 UN Convention to Combat Desertification in Those Countries Experiencing Serious Drought and/or Desertification, Particularly in Africa (adopted June 17, 1994, entered into force December 26, 1996) 1954 UNTS 3, Art 21(4)–(5) (UNCCD).

28 Raustiala, K. *Reporting and Review Institutions in 10 Multilateral Environmental Agreements* (UN Environment Programme, 2001), at 9.

29 Ibid.

30 Ibid., at 10.

31 Protocol on Water and Health MoP-1 (Geneva, January 17–19, 2007), 'Review of Compliance' (July 3, 2007) UN Doc ECE/MP.WH/2/Add.3. Available online at www.unece.org/fileadmin/DAM/env/documents/2007/wat/wh/ece.mp.wh.2_add_3.e.pdf (accessed March 27, 2013).

which to strengthen the implementation of an agreement, coordinate activities between institutions and further develop law and policy in a particular field. While many of the mechanisms described above were in their infancy during the negotiation of the text of the UNWC in the 1970s to 1990s, a significant opportunity now exists to draw from the experiences of MEAs to support the progressive development and effective implementation of the UNWC.

23 Filling gaps

A protocol to govern groundwater resources of relevance to international law

Joseph W. Dellapenna and Flavia Rocha Loures

This chapter illustrates the potential for further strengthening and developing the content and scope of the UN Watercourses Convention (UNWC) upon its entry into force. We do so by looking at one of the Convention's gaps that has been widely recognized: the need to address the special vulnerabilities and characteristics of groundwater systems in a transboundary context. If the parties to the Convention eventually agreed upon the necessary procedures, such a development process could take place through the adoption of protocols, amendments and annexes.

As per Article 2(a), the UNWC applies to 'a system of surface waters and ground waters constituting by virtue of their physical relationship a unitary whole and normally flowing into a common terminus'. In this sense, the UNWC highlights the need for an integrated approach to water management: it applies to watercourses that cross international boundaries, including major rivers, their tributaries and connected lakes and aquifers,[1] even when these components are entirely located within a single state. The Convention's systemic approach is scientifically sound, given the interconnections among the mainstream and its tributaries and between surface and underground waters.

Yet, the UNWC lacks rules dealing specifically with groundwater. In addition, the Convention does not apply to groundwater that, albeit flowing under the territories of two or more countries, is not related to surface water systems. These types of groundwater, which include but are not limited to fossil groundwater, were intentionally excluded from the Convention's scope during the drafting process.[2]

Recognizing the gaps in the UNWC and the importance of groundwater resources, the UN General Assembly (UNGA) asked the International Law Commission (ILC), in 2001, to continue the study of the law of transboundary aquifers in the context of its work on shared natural resources.[3] After all, 'the

---

1 Components of freshwater systems that may fall under the UNWC's scope, when connected to one another, include rivers, lakes, aquifers, glaciers, reservoirs and canals. International Law Commission (ILC), 'Report on the Work of Its Forty-sixth Session' (May 2–July 22, 1994) UN Doc A/49/10 (1994), at 90. Available online at http://untreaty.un.org/ilc/documentation/english/A_49_10.pdf (accessed March 27, 2013) (1994 ILC Report).
2 Ibid.
3 UNGA, Resolution 55/152 'Report of the ILC on the Work of Its Fifty-second Session' (January 19, 2001) UN Doc A/RES/55/152, at 2.

vulnerability of groundwater, especially fossil groundwater, to depletion and pollution calls for the development of norms of international law that contain stricter standards . . . than those applied to surface waters'.[4] Under its mandate to contribute to the codification and progressive development of international law, the ILC then engaged in a process that culminated, in 2008, with the adoption of the Draft Articles on the Law of Transboundary Aquifers (ILC Draft Articles).[5] The Draft Articles are a set of 19 provisions pertaining to the use, protection and management of water flowing underneath the territories of two or more countries.

Upon receiving the Draft Articles from the ILC, also in 2008, the UNGA (i) took note of the Draft Articles; (ii) commended them to the attention of governments, encouraging that they be considered in the negotiation of specific aquifer agreements; and (iii) decided provisionally to examine the question of the final form that might be given to those articles at its 66th Session, in 2011–12.[6] As per its 6th Committee's recommendation upon considering the topic,[7] the UNGA has once again postponed a decision on the Draft Articles' final form and adoption, which is now to take place at its 68th Session.[8]

Against this background, this chapter puts forward and develops the idea of a future groundwater protocol to the UNWC, with a view to supplementing and further developing the regulatory framework under the mother Convention. The protocol would merely apply and adjust the Convention's provisions to the extent needed to address the special characteristics of groundwater resources. As the most authoritative framework for the codification and development of international water law, the UNWC serves as subsidiary guidance for the law governing groundwater. For this reason, the protocol would include a provision clarifying that the norms of the UNWC apply to all groundwater of relevance to international law. Where necessary, the protocol would refer back to relevant provisions of the Convention. This approach cements the notion *that there is only one legal regime for internationally shared freshwater systems*, which includes a *subset of specific rules governing groundwater only to the extent necessary to support sustainable groundwater management*.

We start by discussing the importance of having a single legal regime governing all freshwaters that may be of relevance to international law. The chapter then goes on to consider the potential scope and content of such a protocol, focusing on the instances in which such a protocol should depart from the ILC Draft Articles.

---

4 Yamada, C. (Special Rapporteur), 'Shared Natural Resources, 1st Report on Outlines' (April 30, 2003) UN Doc A/CN.4/533, at 3 (Yamada 1st Report).
5 'Draft Articles on the Law of Transboundary Aquifers, with Commentaries' (ILC Draft Articles) in ILC, 'Report on the Work of Its 63rd Session' (May 5–June 6 and July 7–August 8, 2008) UN Doc A/63/10 (2008), at 13–79 (2008 ILC Report).
6 UNGA, Resolution 63/124 'The Law of Transboundary Aquifers' (January 15, 2009) UN Doc A/RES/63/124, at 1–2.
7 UNGA 6th Committee, 'The Law of Transboundary Aquifers' (November 9, 2011) UN Doc A/66/477, at 3.
8 UNGA, Resolution 66/104 'The Law of Transboundary Aquifers' (January 12, 2012) UN Doc A/RES/66/104, at 1–2.

Why a single water treaty regime at the global level

This section considers the value of a future protocol to the UNWC to govern groundwater. In so doing, we must consider the possible scenarios for the final shape of the ILC Draft Articles: non-binding guidelines, a separate binding convention or, with the necessary adjustments, a protocol to the UNWC.[9] As per our analysis, scientific, legal, administrative and policy factors call for a careful consideration of the 'protocol' scenario, as a means to further strengthen the role and relevance of the UNWC beyond entry into force, as well as to ensure the coherent and integrated codification and development of international water law. The UNWC and its groundwater protocol would create a broader regulatory framework governing all freshwater systems of relevance to international law, without overlooking the specifics of groundwater.

In the case of transboundary groundwaters connected to international watercourses, the UNWC and the ILC Draft Articles apply to the same natural resource, only at different stages in the hydrologic cycle. In such cases, states overlying an aquifer may cause harm to countries located beyond the area of the geological formation, but still within the larger river basin, and vice-versa. For example, overuse of an aquifer may compromise the water levels in a lake, to the detriment of the states bordering the lake in question including those riparian states not overlying the aquifer itself. Sound water management must take these physical interconnections into account. Thus, '[s]overeignty over groundwater must be restricted in the same way as it is over surface water... [T]he hydrologic, economic, and engineering variables involved are essentially the same'.[10]

The international community recognizes the importance of integrated river basin planning and management[11] – that is, conjunctive use and management of interconnected surface and underground waters.[12] Even aquifers that receive negligible amounts of contemporary recharge – and thus can be considered a

---

9   During the elaboration of the ILC Draft Articles, it was argued that it would not be appropriate to link the latter to the UNWC, since the Convention may lack support from the international community, as evidenced by its non-entry into force more than ten years after its adoption. ILC, 'Report on the Work of Its Fifty-sixth Session' (May 3–June 4 and July5–August 6, 2004), Chapter VI: Shared Natural Resources, UN Doc A/59/10 (2004), at 135. Available online at http://untreaty.un.org/ilc/reports/2004/english/chp6.pdf (accessed March 27, 2013) (2004 ILC Report); ILC, 'Report on the Work of Its Fifty-seventh Session' (May 2–June 3 and July 11–August 5, 2005), Chapter IV: Shared Natural Resources, UN Doc A/60/10 (2005), at 41. Available online at http://untreaty.un.org/ilc/reports/2005/english/chp4.pdf (accessed March 27, 2013) (2005 ILC Report). Such a discussion seems less relevant in the context of this chapter, which focuses on the UNWC and how to further strengthen its content and scope beyond entry into force, rather than on debates around the final shape of the ILC Draft Articles.

10   Dellapenna, J. W. 'The Customary International Law of Transboundary Freshwaters', *International Journal of Global Environmental Issues*, 2001; 1 (3/4): 264–305, at 274.

11   See UN Conference on Environment and Development (Rio de Janeiro, June 3–14, 1992), 'Agenda 21: A Programme for Action for Sustainable Development' UN Doc A/CONF.151/26/Rev.1 (Vol. I), at 9–479 (Agenda 21).

12   See ILA, 'Berlin Rules on Water Resources' (Berlin Rules) in Report of the 71st Conference (Berlin 2004), at 3, 13–14 (2004 ILA Report).

non-renewable (or 'fossil') resource[13] – usually have some level of interaction with the hydrological cycle.[14] Moreover, such virtually non-recharging aquifers are still a *current* source of water.[15] Comprehensive, integrated planning must consider all available water sources within a border area so that all such resources can be developed and sustained in an integrated, optimal fashion.[16]

From a policy standpoint, creating a single international water law regime would facilitate the elaboration and implementation of integrated river basin and water management plans in a transboundary context. Governing all internationally relevant water resources together would also represent an economy in costs: a single regime would require just one secretariat and one set of meetings for parties to attend, if such mechanisms were to be incorporated into the UNWC.

The unified legal treatment of surface and underground waters also offers benefits to the broader international legal system as an enabling factor for transboundary water cooperation: it would facilitate implementation of international water law and avoid the unnecessary proliferation of global treaties, all the while accounting for the need to manage freshwater systems as hydrological units and, even in the case of fossil aquifers, in an integrated manner. Moreover, the possibility of dual applicability of two independent international conventions – the UNWC and a groundwater convention, in the case of transboundary aquifers connected to international watercourses – would raise an unnecessary risk of confusion as to which treaty to apply. In fact, it could increase the potential for conflict, when international law should really serve as a dispute prevention mechanism. Furthermore, while science and basic notions of rational use call for integrated water resources management, international law could be forcing countries to move in the opposite direction if it adopts two independent and separately evolving legal regimes. Instead, the role of international law is to offer legal instruments and norms that support and guide states in applying proper water management strategies in a transboundary context.

In this sense, when it submitted what became the UNWC to the UNGA, the ILC presented a resolution encouraging countries to apply the principles of the Convention to all transboundary groundwaters, including to those not connected to surface waters.[17] In other words, the ILC then recognized that shared groundwaters should be regulated fundamentally in the same manner as other transboundary freshwater resources. Accordingly, at the beginning of the development process of the

---

13 To illustrate the point, the ILC Draft Articles define a 'recharging aquifer' as 'an aquifer that receives a non-negligible amount of contemporary water recharge'. ILC Draft Articles (note 5), Article 2(e).

14 As experts have noted, 'absolutely zero recharge is extremely rare'. Foster, S. and Loucks, D. P. (eds), *Non-Renewable Groundwater Resources: A Guidebook on Socially-Sustainable Management for Water-Policy Makers* (UNESCO, 2006), at 13.

15 Ibid., at 14.

16 In this sense, the ILC Draft Articles require states to 'establish . . . a comprehensive utilization plan, taking into account present and future needs of, and alternative water sources for, the aquifer States'. ILC Draft Articles (note 5), Article 4(c).

17 ILC, 'Resolution on Confined Transboundary Groundwater' in 1994 ILC Report (note 1), at 135.

Draft Articles, the ILC aimed merely to adapt the general principles of the UNWC to the special needs of groundwater. The Convention was then described as the 'the basis upon which to build a regime for groundwater'.[18]

Nonetheless, as ultimately adopted by the ILC, the Draft Articles focus excessively on transboundary aquifers per se. In so doing, the ILC Draft Articles 'almost completely fail to provide for the situations where surface and groundwaters form a single unit and should be managed as such – situations in which aquifer and watercourse states have correlated obligations and rights'.[19] This is further discussed below, with respect to the proposed scope of a future protocol to the UNWC. For now, it suffices to note that, as a result of being drafted to become a stand-alone document, the ILC Draft Articles often:

- Overlook the relations between aquifer states and non-aquifer states that are nonetheless hydrologically interconnected within a system of surface and underground waters. For example, the draft articles do not extend to those other states the obligations and rights regarding data generation, which are far more detailed than those in the UNWC.
- Unnecessarily reiterate certain provisions of the Convention [that] would apply equally to surface and underground waters. This is the case with Draft Article 18, on the protection of transboundary aquifers in time of armed conflict. In such instances, the draft articles take attention away from their very own provisions that aim to address the specifics of groundwater resources . . . , [making] implementation more difficult.
- Mix together provisions of the UNWC sometimes failing to include key aspects contained in that instrument. For example, Draft Article 5(2) attempts to condense into one provision Articles 6(3) and 10(2) of the Convention, but fails to clarify the relationship between groundwater uses other than those needed to address vital human needs. The Convention makes clear that no water use enjoys inherent priority over others, including existing activities. ILC Draft Article 5(2) also fails to require aquifer states to consult with each other in applying the principle of reasonable and equitable use, when needed, in the absence of joint management arrangements.

---

18   See Yamada, C. (Special Rapporteur), 'Second Report on Shared Natural Resources: Transboundary Groundwaters' (March 9, 2004) UN Doc A/CN.4/539, at 3 (Yamada 2nd Report); ILC, 'Report on the Work of Its Fifty-eighth Session' (May 1–June 9 and July 3–August 11, 2006) UN Doc A/61/10 (2006), at 194. Available online at http://untreaty.un.org/ilc/reports/2006/english/chp6.pdf (accessed March 27, 2013) (2006 ILC Report). When considering the first version of the draft articles, many delegations 'noted with approval that [they] had been largely modeled on the [UNWC] and reiterated the value of that instrument'. UNGA, 'Topical Summary of the Discussion Held in the 6th Committee' (January 19, 2007) UN Doc A/CN.4/577, at 5–6 (2007 Topical Summary).

19   Dellapenna, J. W. and Loures, F. 'Transboundary Aquifers: Towards Substantive and Process Reform in Treaty-Making' in Benidickson, J., Boer, B., Benjamin, A. H. and Morrow, K. (eds), *Environmental Law and Sustainability after Rio* (Edward Elgar, 2011), 212–234, at 222.

- Instead of advancing the law by taking the principles applicable to water-courses as minimum starting points, create less strict provisions for aquifers.[20]

Therefore, negotiating a protocol to the UNWC that would govern groundwaters of relevance to international law is a logical solution for ensuring the sustainable and integrated use, management and protection of these precious resources. Such a protocol would address the hydrological unity of freshwater systems, as well as the related need for integrated water resources management, while acknowledging the special characteristics of groundwater. The protocol would also be less likely to create the problems stated above, enabling coherence in the progressive development of international water law.

In sum, 'implemented in an integrated fashion, the convention and its protocol would better support transboundary freshwater cooperation while avoiding the unnecessary overlapping provisions and the danger of subtle inconsistencies and gaps that have arisen between [the UNWC and the ILC Draft Articles] as stand-alone documents'.[21] Counting now 30 contracting states, the Convention is only five shy of the number required for entry into force.[22] This makes the argument for considering the adoption of a future protocol on groundwater even more compelling. The Convention and its protocol would create a single regime to govern all non-navigational uses of transboundary freshwater resources, as interrelated, mutually supportive stepping-stones in law development. The current political context is ideal for an extra push to bring the UNWC into force and consolidate it as the universally agreed framework for the continuous evolution of international water law.

Proposed scope of a future protocol to the UNWC[23]

The ILC Draft Articles determine their scope when defining the term 'aquifer', limiting it to a geological formation and the waters contained within it. ILC Draft

---

20 Ibid., at 223. The authors note further that, 'as currently drafted, the ILC Draft Articles miss the point of creating special rules for groundwater without overlooking their connections with surface waters and the legal relations emerging from hydrological connections between watercourse states and aquifer states. In other words, the draft articles inadvertently create a gap between the law of international watercourses and that of transboundary aquifers, making it harder for states to promote their integrated management. The adoption of a carefully crafted protocol to the UNWC would aid in addressing those problems.' Ibid., at 224.

21 Ibid., at 224.

22 UN Treaty Collection, 'UN Convention on the Law of the Non-Navigational Uses of International Watercourses: Ratification History' (UN, 2012). Available online at http://treaties.un.org/pages/ViewDetails.aspx?src=TREATY&mtdsg_no=XXVII-12&chapter=27&lang=en (accessed March 27, 2013).

23 In this chapter, we focus on the scenario of a domestic aquifer with recharge zones located in another state's territory or connected to international watercourses. The discussion does not apply to discharge zones, as the chances of identifying a domestic aquifer with discharge zones across the border are negligible. As Eckstein explains, this 'is more of an academic issue. Based on the current definitions of aquifer and discharge zone, the discharge zone would merely be the interface between the aquifer and the outlet (e.g. watercourse, lake, ocean). If that interface existed in a second country, the aquifer would have to be transboundary since the aquifer is on the other side of the interface. The only scenario of a domestic aquifer with a discharge zone in another country that I can invent (and I know of no real life example) is where the boundary lay precisely along the edge of a vertical cliff and aquifer water discharged from the side of that cliff into the second country'. Email from Gabriel Eckstein to authors (August 16, 2007).

Article 2(a) defines an aquifer as 'a permeable water-bearing geological formation underlain by a less permeable layer and the water contained in the saturated zone of the formation'. This definition excludes an aquifer's recharge and discharge zones, which are addressed separately (and less inclusively) in later provisions. Such a definition is deficient: 'protection of the recharge and discharge zones is crucial to the protection of the aquifer because of the prominent causal relationship between what occurs in the two zones and the health of the aquifer'.[24] In addition, as per ILC Draft Article 2(c), the articles apply to *transboundary* aquifers; i.e. aquifers, 'parts of which are situated in different States'. Therefore, domestic aquifers with recharge zones located in a neighbouring state do not fall within the scope of the ILC Draft Articles.

In this context, the proposed protocol would create an opportunity to shift the focus from the aquifer (a geologic formation within which exploitable groundwater is found) to the groundwater itself – consistent with the approach of the UNWC.[25] The problem posed by the ILC Draft Articles is that managing geologic formations requires different rules than managing water – rules that, to a significant extent, are incompatible with the approach of the UNWC.

While managing groundwater will require managing aquifers, the focus on the geological formation per se in the ILC Draft Articles underlies the unsatisfactory nature of the Draft Articles in fully addressing the circumstances of integrated and conjunctive water management. The ILC Draft Articles' approach ignores the hydrological reality and puts aquifers at risk of problems originating in recharge areas located beyond the aquifer states' territories. Therefore, a definition of groundwater in the proposed protocol would recognize that all components of a groundwater system are interconnected and that, through them, impacts on the larger hydrological system may occur. From a legal standpoint, the proposed definition would acknowledge the role of states not overlying an aquifer's geological formation, but with recharge zones within their territories. For example, a recharge zone could be located within state A for an aquifer situated wholly within state B. Runoff from agricultural activities developed in state A could contaminate the surface recharge area and eventually the aquifer in state B, or developments in state A could block groundwater recharge to the detriment of state B.

A related issue arises from ILC Draft Article 11(1), which requires *aquifer* states (i.e. states overlying the aquifer, but not necessarily states in whose territory a recharge zone lies) to 'take appropriate measures to prevent and minimize detrimental impacts on the recharge and discharge processes'. The same provision, in Paragraph 2, requires *non-aquifer* states with recharge zones in their territories to cooperate 'to protect the aquifer... and related ecosystems'. Hence, the Draft Articles impose obligations on states in which a recharge zone is located, without

24 Eckstein, G. E. 'Protecting a Hidden Treasure: The UN International Law Commission and the International Law of Transboundary Ground Water Resources', *Sustainable Development Law and Policy*, 2005; 5 (1): 5–12, at 7.
25 McCaffrey, S. C. 'The International Law Commission Adopts Draft Articles on Transboundary Aquifers', *American Journal of International Law*, 2009; 103: 272–93, at 282.

conferring any rights on such states. In the above example, ILC Draft Article 11(2) would require state A to 'cooperate with the aquifer States to protect the aquifer or aquifer system'. State B, however, would not have to consider the equitable and reasonable rights of state A in demanding that state A refrained from its agricultural activities capable of interfering with the recharge process. This is so because, under ILC Draft Article 4, such rights extend only to aquifer states. So what would the incentive be for state A to comply with the requirement in question?

With the focus on the aquifer and with the recharge zone not overlying the aquifer, state A would not be entitled to share in the benefits of development benefits or changing its behavior – in this case, stopping or changing its agricultural practices, so as not to put the aquifer at risk.[26] Shift the focus to the groundwater itself and these conceptual problems largely go away. State A and state B are part of the groundwater system ... even if state A does not overlay the aquifer itself. As such, state A would have rights as well as duties.

In line with the comments above, states have endorsed the inclusion of recharge and discharge zones in the definition of an aquifer in the ILC Draft Articles. As Lebanon has pointed out, this would 'give States in those zones a role in management, thereby ensuring that water management would be sound and comprehensive'.[27] States have also noted that, as currently drafted, the ILC Draft Articles make it difficult to give effect to other provisions therein, such as Draft Article 11, on the identification of recharge and discharge zones.[28] Expanding the scope of the Draft Articles in that regard is a necessary response to the role of recharge zones in replenishing these water sources.[29] The most direct approach to doing so would be to redefine the focus of concern from aquifers to the groundwater as such – which at least obviates any possible debate about whether a recharge zone is technically part of the aquifer it recharges.

As stated above, this could best be accomplished in the context of a protocol to the UNWC by defining the object of regulation and management as the groundwater system rather than the geological formation (aquifers). Activities that significantly affect the management of groundwater must be considered for the effective management of the resource, even if such activities do not relate directly to the aquifer itself, but only to a recharge zone. With this expanded scope, the proposed future protocol would cover all transboundary groundwaters, including groundwater in domestic aquifers with recharge zones located beyond the borders of the state overlying the geological formation.

Another consideration pertaining to scope involves domestic aquifers linked to international watercourses. As Loures and Dellapenna explain, 'an aquifer located entirely within one country's territory, but connected to an international river basin, although not transboundary per se, is part of a transboundary hydrological

---

26 In this sense, Colombia proposed including economic incentives for states with recharge zones to protect them. UN Secretary-General, 'Report: The Law of Transboundary Aquifers' (June 29, 2011) UN Doc A/66/116, at 6 (2011 SG Report).

27 Ibid., at 12 (comments of Lebanon).

28 Ibid., at 11 (comments of Lebanon).

29 Ibid., at 6, 9, 11, 14 (comments of Colombia, El Salvador, Lebanon, and The Philippines).

unit'.[30] The proposed protocol to the UNWC would thus apply to a domestic aquifer connected to an international watercourse.

For example, an international watercourse may discharge into a domestic aquifer, subjecting that aquifer to pollution from activities along the watercourse. Or, in dry periods, when aquifers can become the main recharge source of a river, groundwater overexploitation may reduce the volumes of discharge into surface waters and have an impact on dependent ecosystems and water uses downstream, with effects on non-aquifer basin states. Where surface waters feed into underground flows, states upstream of recharge zones may divert water from rivers and disrupt the aquifer's recharge process and hydrological balance, potentially affecting populations across the border dependent upon that groundwater.

Although these situations would come within the sweep of the UNWC, the Convention alone does not explicitly address the specific characteristics of groundwater, such as the need to protect an aquifer's outlets and recharge zones. Under ILC Draft Article 2, the international border must cross the aquifer itself for the Draft Articles to apply. Thus, the Draft Articles exclude purely domestic aquifers linked to international watercourses. In this instance, therefore, neither the ILC Draft Articles nor the UNWC adequately protect these domestic aquifers. The proposed protocol would address the gap. As one state summarized in recent comments on the Draft Articles,

> aquifer systems extend beyond States' political boundaries, and the approach taken in managing aquifers should be based on the catchment basin, as the behaviour of water [is] closely linked within river basin and catchment basin hydrogeology and river basin topographic boundaries.[31]

With this broad scope, the protocol would cover all groundwater through which transboundary harm might be caused, including certain domestic formations. This expansion in scope should be accompanied by a more careful consideration of the rights and duties of states without direct access to an aquifer, but in whose territory connected surface waters or recharge zones are located. In this sense, the ILC itself has noted that, 'to be effective, some draft articles would have to impose obligations on states which do not share the transboundary aquifer in question and in certain cases give rights to the latter states towards the states of that aquifer'.[32] These countries are not aquifer states if no portion of such an aquifer is situated within their territories, but they share the larger freshwater system. Such states can affect an aquifer's recharging sources or be affected through its discharging outlets.

After all, any agreement on transboundary groundwater must be attractive to all states within the larger hydrological system, not only to those with direct ability to

---

30   Dellapenna and Loures (note 19), at 227.
31   2011 SG Report (note 26), at 14 (comments of The Philippines). See also Colombia's proposal to change the definition of aquifer system to mean 'a series of two or more aquifers that are hydraulically connected and their hydraulic connection with surface water'. Ibid., at 5.
32   2006 ILC Report (note 18), at 194.

extract the resource. For all the relevant states, such an agreement needs to recognize and frame their rights over and duties toward freshwater and related resources located within their territories. In this sense, the proposed protocol would create a legal framework for benefit-sharing and regional cooperation to prevent transboundary harm, and enable integrated water and land management. In other words, the future protocol would make provisions for all states with a significant relationship to the groundwater to the extent necessary to protect and sustainably manage and develop the resource. In so doing, the protocol would extend to relevant aquifer and basin states not only the obligations thereunder, but also the corresponding rights.

For example, when detailing equitable and reasonable use, the proposed protocol would establish that: a) all relevant states, including those not overlying the aquifer but with recharge zones within their territories, are entitled to utilize and develop their portions of the groundwater system of relevance to international law; b) states have the right to develop and utilize, in a reasonable and equitable manner, areas in and around the recharge zones located within their own territories, taking into account the rights of other states to the equitable and reasonable use of the groundwater associated with such zones; c) all such states are under a duty to participate in the protection, preservation and management of the resource, in an equitable and reasonable manner.

With regard to the watercourse states concerned – i.e. the basin states in an international watercourse in relation to a connected aquifer – the proposed protocol would better clarify their rights and duties, as follows: a) the watercourse states concerned have the right to protect, develop and utilize such international watercourses in a reasonable and equitable manner; b) in so doing, such states must take into account the right of the states concerned to the equitable and reasonable use of the groundwater associated with such watercourses; c) the determination of whether the use of groundwater by a state is equitable and reasonable would have to consider, among other factors, the right of the watercourse states concerned to protect and make equitable and reasonable use of a connected international watercourse; d) the watercourse states concerned must diligently prevent the causing of significant transboundary harm to groundwater states when undertaking activities that have or are likely to have an impact on their watercourses, and vice-versa; e) all such states must cooperate with each other, in an equitable and reasonable manner, where necessary to protect and preserve ecosystems within, or dependent upon, international watercourses and connected groundwaters of relevance to international law.

Sustainability of aquifers[33]

Under Article 5(1) of the UNWC, 'watercourse States shall in their respective territories utilize an international watercourse in an equitable and reasonable

---

33   For a detailed discussion on the ILC Draft Articles and the sustainable use of recharging aquifers, see Allan, A., Loures, F. and Tignino, M. 'The Role and Relevance of the Draft Articles on the Law of Transboundary Aquifers in the European Context', *Journal for European Environmental and Planning Law*, 2011; 8: 231–51, at 235.

manner... with a view to attaining optimal and *sustainable* utilization thereof and benefits therefrom'. Hence, the Convention places the notion of sustainability at the core of equitable and reasonable use. The same approach is merited in the case of a recharging aquifer, defined in ILC Draft Article 2(f) as 'an aquifer that receives a non-negligible amount of contemporary water recharge'.

For recharging aquifers, the average annual extractions and discharge rates should not exceed annual recharge, natural or artificial. Countries should only exceed actual average recharge rates in emergency situations, compensating for the overexploitation to the extent feasible. For instance, recharge during wet seasons or wet years, when groundwater requirements are commonly lower, could make up for excessive extractions during a dry season or dry years when recharge is at its minimum and water needs tend to be higher. Of course, there must be a threshold of how much overexploitation can sustainably occur in dry seasons without any irreversible detrimental impacts on the resource. Such thresholds differ for each aquifer and are related to the hydro-geo-climatological parameters of the landscape. In any such case, information is needed to determine what is *sustainable* under all the relevant considerations. This is a vital requirement as 'rapidly falling water tables might not appear until some years after a serious overdraft begins, by which time it might be too late to do much about it'.[34]

In the case of non-recharging aquifers, planning for their long-term use or development is difficult, but no more difficult than defining extraction rates for oil, natural gas or hard-rock minerals. The question in such cases is how quickly, if at all, the resource should be depleted. Resolving this question requires consideration of available substitutes for the depleting resource and how quickly and affordably the transition can be accomplished.

ILC Draft Article 4 applies the idea of long-term benefits maximization to recharging and non-recharging aquifers alike:

> Article 4. Equitable and reasonable utilization
> Aquifer States shall utilize transboundary aquifers or aquifer systems according to the principle of equitable and reasonable utilization, as follows:
> (a) They shall utilize transboundary aquifers or aquifer systems in a manner that is consistent with the equitable and reasonable accrual of benefits therefrom to the aquifer States concerned;
> (b) *They shall aim at maximizing the long-term benefits derived from the use of water contained therein*;
> (c) They shall establish individually or jointly a comprehensive utilization plan, taking into account present and future needs of, and alternative water sources for, the aquifer States; and
> (d) They shall not utilize a recharging transboundary aquifer or aquifer system at a level that would *prevent continuance of its effective functioning*.

---

34   Ibid., at 237.

In its commentaries to this provision, the ILC clarifies that, while 'it is desirable to plan a much longer period of utilization' for a recharging aquifer, 'it is not necessary to limit the level of utilization to the level of recharge'.[35] In our view, this approach threatens the sustainability of recharging aquifers and allows great latitude for discretion and potential abuses. For example, states with direct access to an aquifer could decide among themselves that such an aquifer will provide groundwater for irrigation purposes for a period of 20 years, admitting the resource's exhaustion beyond that time. Such a decision could have serious impacts on dependent ecosystems, but neighboring countries would have little to say in light of Draft Article 4 and the limited scope of the ILC Draft Articles. In such case, states overlying the aquifer would be in compliance with treaty law, especially if harm to another state was not considered significant or inequitable under the circumstances. The governments involved in exploiting or approving the exploitation of the aquifer would decide on their own which uses were inequitable – governments that might not attend to the needs of vulnerable human communities or aquatic ecosystems.

Recharging aquifers are renewable resources, subject, therefore, to the principle of sustainable use. 'For [such] aquifers, extractions that consider only the formation's storage capacity over the years, i.e. which do not reflect current recharge... rates, disregard the aquifer's capacity for natural renewal, leading to its gradual exhaustion'.[36] In this sense, the Berlin Rules require states to 'give full effect to the principle of sustainability in managing aquifers, taking into account natural and artificial recharge'.[37]

This idea has found support among states, too. For example, at the 61st Session of the UNGA, 'it was suggested that the term *reasonable* be replaced with the term *sustainable*...in conformity with recent practice in international environmental law. A preference was also expressed for a specific reference to sustainable *utilization* because utilization, as opposed to exploitation, could be sustainable in the case of transboundary aquifers'.[38] Another suggestion has been to replace *effective* with *sustainable* in Draft Article 4(d), so that a recharging aquifer is not used in a way that would 'prevent continuance of its *sustainable* functioning'.[39] Accordingly, the proposed protocol would require that overall extractions, combined with discharge rates, not *normally* exceed *overall* recharge for recharging aquifers.

Another issue is that Article 4 does not link sustainability to a requirement on groundwater users to prevent or control pollution or other forms of ecological impairment. For recharging aquifers, that may be a larger threat to sustainability than abstraction. While there are provisions in the ILC Draft Articles that address the protection of aquifers from pollution and environmental degradation, these do not speak in terms of sustainability. The proposed future protocol would provide

35 2008 ILC Report (note 5), at 42.
36 Allan, Loures and Tignino (note 33), at 239.
37 Berlin Rules (note 12), Article 40(1).
38 2007 Topical Summary (note 18), at 7.
39 2011 SG Report (note 31), at 11 (comments of Lebanon).

an opportunity to craft an article that would consider all aspects of sustainability, in the context of equitable and reasonable use.

Data exchange and monitoring

Monitoring and the regular exchange of information are essential to the management of shared freshwater resources. The need for effective monitoring and data exchange is even more pressing in relation to groundwater, given its special vulnerability to irreversible harm, as well as due to groundwater being a *hidden* treasure.[40] Monitoring requires states to gather, on an ongoing basis, the necessary information to manage their groundwater. Data exchange ensures that all states sharing groundwater have access to that information. Hence, the proposed future protocol would require states to gather and exchange data on planned groundwater uses and measures of protection, preservation and management, as well as any other activities that may impact their condition. It would also encourage states to consider forming a joint body for data gathering, processing and exchange.

In this sense, the proposed protocol would mostly match the ILC Draft Articles' provision on data exchange, with one crucial difference: ILC Draft Article 8(1) requires only the exchange of data relating to the *condition* of transboundary aquifers. Data related to uses of, and other activities that may have an impact on, these resources are 'dealt with in later draft articles'.[41] However, those later articles (on monitoring and planned uses) do not require aquifer states to exchange information on an aquifer's *present uses*. Excluding present uses from a data exchange requirement is a serious omission. Many of the requirements of transboundary groundwater management rely on a basic level of information sharing, from determining what is equitable and reasonable to assessing whether a particular activity can harm groundwater resources. For example, ILC Draft Article 5(1)(f) includes 'the existing and potential utilization of the aquifer or aquifer system' as a factor to be considered when determining equitable and reasonable utilization. However, with no correlating requirement to exchange data on existing uses, states would not be able to give that provision full effect.

The same gap in the ILC Draft Articles prevents the requirements on planned activities from fulfilling their precautionary intent. ILC Draft Article 15 only obliges an aquifer state to assess the possible effects of *planned* activities, if and when it has *reasonable grounds* for believing that such an activity may result in *significant* transboundary harm. This provision does not cover information exchange on existing, potentially harmful activities, including groundwater uses. Neither does it apply to planned utilizations that may cause *insignificant* transboundary harm individually, but which could represent a significant impact on groundwater resources when cumulative effects are considered.

---

40   Dellapenna, J. W. 'The Physical and Social Bases of Quantitative Groundwater', in Beck, R. E. (ed.), *Waters and Water Rights* (3rd edn, Michie, 2010), at Ch. 18.
41   2008 ILC Report (note 5), at 52.

Hence, neighboring countries would lack knowledge of existing sources of environmental degradation when evaluating planned measures. Neither would they be aware of minor planned activities. In such a scenario, a state would be unable to assess the cumulative impacts on the groundwater deriving from various sources, uses and activities, and thus to realistically evaluate the risk of significant transboundary harm. A similar rationale applies to internal measures of management, conservation and protection addressing existing activities that may have an impact on such groundwaters.

Information exchange should be as wide and frequent as possible, and not limited to major new measures or to the natural conditions of groundwaters. The proposed protocol's expansive data-sharing obligation would be consistent with its broad applicability to groundwater utilization, activities likely to impact groundwater and measures for its protection, preservation and management, in line with the broad scope of application determined by ILC Draft Article 1.

Similarly, the proposed future protocol would require extensive monitoring of groundwaters. This obligation includes monitoring the current condition of the groundwater, its ongoing uses (major or minor), any activities likely to have an impact upon groundwaters, as well as the effectiveness of protection and management measures, such as artificial recharge. Comparatively, ILC Draft Article 13(2) is less specific, allowing states to set their own parameters for monitoring. The provision goes only so far as to recommend that such parameters include the condition of the groundwater and its uses. In this sense, the Draft Articles fail to address the monitoring of activities that could have a serious impact on groundwater resources. For example, farming in the recharge zone of an aquifer would not have to be monitored with respect to potential groundwater pollution. If the pollution continued for years, it could then be so pervasive as to render the groundwater unusable for decades. Under the proposed protocol, therefore, the use of fertilizers in or around recharge areas would be monitored as a potentially harmful activity.

Of course, just as for data-sharing, any monitoring requirement has to take into account the financial and technical capabilities of the states involved, as well as the political situation.[42] This ensures that all states have access to enough information to make good decisions about the use of their groundwaters of relevance to international law. Finally, the proposed protocol would extend the rights and duties under its provisions on monitoring and data exchange to the watercourse states concerned, as appropriate, to ensure the availability of data on the relation of groundwaters to connected international watercourses.

Planned activities

Data-sharing, consultations and negotiations are particularly relevant with respect to planned activities. The proposed protocol would follow the standards and procedural requirements in Part III of the UNWC, recognizing that those provisions

---

42 See 2008 ILC Report (note 5), at 60.

apply to surface and underground waters alike. Articles 14(b) and 17(3) of the UNWC require six-month waiting periods after notification and through consultations, during which implementing activities are to remain suspended. These provisions demonstrate a cautious approach regarding the development of transboundary watersheds. Such caution should extend to groundwaters of relevance to international law. Underground freshwater systems are in fact more vulnerable to irreversible harm than surface bodies of water.

ILC Draft Article 15 governs planned measures through more modest provisions. This is one of the instances in which the Draft Articles afford less protection for aquifers than that under the UNWC. Draft Article 15 draws from the UNWC, but is less strict and detailed. That provision contains requirements on notification, consultations and negotiations, without, however, requiring the suspension of implementation activities pending that process. This omission hampers the efficacy of all provisions regulating groundwater protection, preservation and management. In this sense, the Dutch Government has underscored the special vulnerability of groundwater systems, requesting repeatedly the insertion of a requirement to refrain from implementing planned measures during the course of consultations.[43]

The proposed protocol would address this gap by simply referring back to the provisions of the UNWC, better fulfilling the intent of protecting groundwater resources. The scarcity of state practice with respect to groundwaters is no excuse for weakening the approach under the UNWC. 'Planned measures' provisions in existing watercourse agreements, many of which include groundwater within their scope, should have been considered. The framework character of the Convention and its role as a basis for further law development call for its application to all internationally shared waters, except to the extent necessary to adequately address the specific characteristics of groundwaters of relevance to international law.

Conclusion

Adopting a protocol on groundwater would fill an important gap in the UNWC, while demonstrating a continued commitment by the international community to develop international water law. Water insecurity is a growing threat – a progressive international legal regime, including the UNWC and a protocol on groundwater, would help mitigate such a threat. The UNWC and its protocol, serving and evolving as a single international regime for both surface and underground waters, would best reflect the interconnections within the hydrological cycle. The protocol, in particular, would ensure due consideration of the particular vulnerability of groundwater to pollution and overexploitation, applying and adjusting the UNWC's provisions only to the extent necessary to address such vulnerabilities.

---

43 See, e.g. 2011 SG Report (note 31), at 7; ILC, 'Shared Natural Resources: Comments and Observations by Governments on the Draft Articles on the Law of Transboundary Aquifers' (March 26, 2008) UN Doc A /CN.4/595, at 42 (comments of The Netherlands).

As for the ILC Draft Articles, they represent an important step towards strengthening the international legal protection of groundwater. However, their content and scope would need to be improved before they can serve as the basis for a protocol to the UNWC or even for a separate binding convention. Continued debate on their shape, form and content before final adoption would allow for the ILC Draft Articles to be further improved and refined and for some of the issued raised above to be addressed. As discussed in this chapter, such a debate should consider a future protocol to the UNWC that would codify, clarify and progressively develop international groundwater law.

24 Reconciling the UN Watercourses Convention with recent developments in customary international law

Owen McIntyre and Mara Tignino

The text of the UN Watercourses Convention (UNWC) is close to 16 years old. The origins of most of its provisions lie even further in the past, resulting from over 20 years of prior work by the International Law Commission (ILC) that culminated in the adoption of the 1994 ILC Draft Articles on the Law of the Non-Navigational Uses of International Watercourses (1994 ILC Draft Articles)[1] shortly before the adoption of the UNWC itself.[2]

Given the substantial body of work which preceded the adoption of the UNWC, it is hardly surprising that so broad a range of developments in international water law should have taken place in the years following. These developments include judgments by the International Court of Justice (ICJ) in cases involving interstate disputes over international watercourses,[3] many new watercourse agreements,[4] a new restatement on water resources law by the

---

1 ILC, 'Draft Articles on the Law of the Non-Navigational Uses of International Watercourses and Commentaries thereto and Resolution on Transboundary Confined Groundwater' (1994 Draft Articles) in ILC, 'Report on the Work of its Forty-sixth Session' (May 2–July 22, 1994) UN Doc A/49/10(Supp) (1994). Available online at http://untreaty.un.org/ilc/documentation/english/A_49_10.pdf (accessed March 28, 2013) (1994 ILC Report).

2 For more information on the development process of the UNWC, see 1994 ILC Report (note 1); ILC, 'The Law of the Non-Navigational Uses of International Watercourses – Extracts from the Report of the International Law Commission', *Environmental Policy and Law*, 1994; 24: 335–68.

3 *Case Concerning the Gabčíkovo-Nagymaros Project (Hungary/Slovakia)* (Judgment) [1997] ICJ Rep 7 (Gabčíkovo); *Case Concerning Pulp Mills on the River Uruguay (Argentina v Uruguay)* (Judgment) [2010] ICJ Rep 14 (*Pulp Mills*). See also McIntyre, O. 'Environmental Protection of International Rivers: Case Concerning the Gabčíkovo-Nagymaros Project (Hungary/Slovakia)', *Journal of Environmental Law*, 1998; 10: 79–91; McIntyre, O. 'The Proceduralisation and Growing Maturity of International Water Law: Case Concerning Pulp Mills on the River Uruguay (Argentina v. Uruguay), ICJ, 20 Apr 2010', *Journal of Environmental Law*, 2010; 22: 475–97.

4 See, for example, Charte des Eaux du Fleuve Sénégal [Senegal River Waters Charter] (Mali-Mauritania-Senegal), Organisation pour la Mise en Valeur Du Fleuve Sénégal (OMVS) Resolution 005, adopted by the Conference of Heads of State and Government (May 18, 2002), at Article 4 [French]. Available online at http://lafrique.free.fr/traites/omvs_200205.pdf (accessed March 28, 2013); La Charte de l'Eau du Bassin du Niger [Niger Basin Water Charter], adopted by the 8th Summit of the Niger Basin Authority Heads of State and Government (April 30, 2008) [French]. Available online at www.abn.ne/attachments/article/39/Charte%20du%20Bassin%20du%20Niger%20version%20finale%20francais_30-04-2008.pdf (accessed March 28, 2013).

International Law Association (ILA)[5] and the adoption of successive instruments endorsing the existence of a human right to water and sanitation by various UN organs.[6]

Notwithstanding the above legal developments and that entry into force of the UNWC is still pending, much of what the Convention contains is widely regarded as a seminal codification of customary international water law.[7] Indeed, the ICJ cited the UNWC as evidence of 'modern development of international law' in the *Gabčíkovo case* case,[8] despite the fact that it had only been concluded four months earlier and signed by a mere four states.[9] The Court did so in a manner reminiscent of its earlier reliance on a draft provision of the UN Convention on the Law of the Sea.[10] In addition, in the more recent *Pulp Mills* case, both parties to the

---

5 ILA, 'Berlin Rules on Water Resources' (Berlin Rules) in Report of the 71st Conference (Berlin 2004) (2004 ILA Report).

6 See Committee on Economic, Social and Cultural Rights, 'Substantive Issues Arising in the Implementation of the International Covenant on Economic, Social and Cultural Rights: General Comment No 15, The Right to Water (Articles 11 12, International Covenant on Economic, Social and Cultural Rights)' (November 26, 2002) E/C.12/2002/11. Available online at www.unhchr.ch/tbs/doc.nsf/0/a5458d1d1bbd713fc1256cc400389e94/$FILE/G0340229.pdf (accessed March 28, 2013) (General Comment 15); UN Economic and Social Council (ECOSOC), Sub-Commission on the Promotion and Protection of Human Rights, 'Draft Guidelines for the Realization of the Right to Drinking Water and Sanitation' (July 11, 2005) UN Doc E/CN.4/Sub.2/2005/25. Available online at http://www2.ohchr.org/english/issues/water/docs/Sub_Com_Guisse_guidelines.pdf (accessed March 28, 2013); UN Human Rights Council (UNHRC), Decision 2/104, 'Human Rights and Access to Water' (November 27, 2006). Available online at http://ap.ohchr.org/documents/E/HRC/decisions/A-HRC-DEC-2-104.doc (accessed March 28, 2013); UNGA, Res 64/292 'The Human Right to Water and Sanitation' (July 28, 2010) UN Doc A/RES/64/292; UNHRC, Res 15/9 'Human Rights and Access to Safe Drinking Water and Sanitation' (September 30, 2010) UN Doc A/HRC/RES/15/9.

7 Because the UNWC is the product of over 20 years of exhaustive research, deliberation and consultation with states by the ILC, it is likely to be considered highly persuasive in aiding in the identification and interpretation of relevant rules of customary international law. See McIntyre, 'Environmental Protection of International Watercourses under International Law' (note 3), at 2.

8 *Gabčíkovo* (note 3), at para. 85.

9 See McCaffrey, S. C. *The Law of International Watercourses* (2nd edn, Oxford University Press, 2007), at 193–4, who suggests, on the question of why the Court accorded such authority to the Convention, that 'the answer might lie to some extent in the process that produced the Convention – twenty years' work by the ILC, culminating in a diplomatic negotiation which produced an agreement that closely tracks the ILC's draft articles. Whatever the case may be, the Court's invocation of the Convention constitutes a strong endorsement of the treaty as an authoritative instrument in the field, and seems likely to lead states to refer to it in support of their positions concerning internationally shared water resources'.

10 UN Convention on the Law of the Sea (adopted December 10, 1982, entered into force November 16, 1994) 1833 UNTS 3 (UNCLOS). In the *Case concerning the Continental Shelf (Tunisia v. Libyan Arab Jamahiriya)* (Judgment) [1982] ICJ Rep 38, the Court, basing its decision on a draft provision in the Convention before it had been adopted and well before it came into force, stated, at paragraph 24, that it 'could not ignore any provision of the draft convention if it came to the conclusion that the content of such a provision is binding upon all members of the international community because it embodies or crystallizes a pre-existing or emergent rule of customary law'. See Mensah, T. A. 'Soft Law: A Fresh Look at an Old Mechanism', *Environmental Policy and Law*, 2008; 38: 50–6, at 51.

dispute invoked UNWC provisions in their written pleadings. Yet, the Court's decision does not even mention the Convention, focusing instead on the water-course agreement in question – the Statute on the River Uruguay.[11]

The developments of the last 16 years may, therefore, put into question the status of the UNWC as an up-to-date codification of current international water law in certain areas, including, for the purposes of this chapter: a) procedural requirements – among other developments pertaining to procedural rules, it now seems clear that some form of transboundary environmental impact assessment (EIA) is required in respect of any planned use of international watercourses, or other projects which might impact significantly upon such waters, as a matter of generally applicable customary international law; b) human right to water – the vibrant and developing discourse regarding the emerging human right to water raises the question of the normative implications of such a right for states which share water resources.

The ongoing development of the law does not automatically mean that the UNWC is obsolete. Recent ICJ judgments exemplify how existing treaties can nonetheless be responsive to new state practice.[12] As the ILC explains, treaties are intended to 'provid[e] stability to their parties and . . . fulfil the purposes which they embody';[13] treaties 'can therefore change over time, must adapt to new situations, evolve according to the social needs of the international community and can, sometimes, fall into obsolescence'.[14]

In this context, this chapter investigates whether or not the text of the UNWC is consistent with current law and practice. In the two sections that follow, major developments related to procedural responsibilities, including the obligation to conduct a transboundary EIA, and the emerging recognition of the human right to water, are both elaborated in an effort to understand their compatibility with and contribution to a model which treats the UNWC as a living global legal framework for international water law. With that purpose, we first review the *Pulp Mills* case, highlighting the ICJ's recognition of an international customary law duty with respect to a transboundary EIA. We then consider the implications of the recognition of the human right to water in relation to the UNWC.

Transboundary environmental impact assessment

The dispute between Argentina and Uruguay in the *Pulp Mills* case arose out of authorizations granted by Uruguay for the construction of two pulp mills on the banks of the River Uruguay. Argentina claimed that 'the authorization, construction

---

11   Statute of the River Uruguay (adopted February 26, 1975, entered into force September 18, 1976) 1295 UNTS 331 (1975 River Uruguay Statute).

12   *Gabčíkovo* (note 3) at para. 112; *Pulp Mills* (note 3), at para 204.

13   Nolte, G. 'Treaties over Time: In Particular Subsequent Agreement and Practice' in ILC, 'Report on its 60th Session' (May 5–June 6, and July 7–August 8, 2008) UN Doc A/63/10(Supp), at 365. Available online at http://untreaty.un.org/ilc/reports/2008/english/annexA.pdf (accessed March 28, 2013).

14   Ibid.

and future commissioning of two pulp mills on the River Uruguay' breached the 1975 River Uruguay Statute with reference in particular to 'the effects of such activities on the quality of the waters of the River Uruguay and on the areas affected by the river'.[15] Uruguay argued that Argentina had demonstrated neither harm nor risk of harm to the river or its ecosystem resulting from alleged violations of the 1975 River Uruguay Statute sufficient to warrant the dismantling of the plants.

While the Court concluded that Uruguay had breached its procedural obligations to inform, notify and negotiate, it also found

> no conclusive evidence to show that Uruguay [had] not acted with the requisite degree of due diligence or that the discharges of effluent from the [operative] Orion (Botnia) mill [had] had deleterious effects or caused harm to living resources or to the quality of the water or the ecological balance of the river since it started its operations in November 2007.[16]

Beyond its impact on relations between Argentina and Uruguay, the *Pulp Mills* decision also provided useful judicial clarification of the role and significance of the key procedural rules of international water law, as set out under the UNWC. However, since Uruguay and Argentina have not joined the UNWC, which, in turn, is not yet in force, the 1975 River Uruguay Statute constitutes the specific source of law governing this case. Consequently, the UNWC cannot of its own force create binding obligations relevant to this case.

Nevertheless, Uruguay's position was that reference to principles of general international law, as embodied in the UNWC, can aid in the interpretation of the specific provisions of the 1975 Statute, in line with Article 31(3)(c) of the Vienna Convention on the Law of Treaties.[17] That the UNWC bears manifold similarities to the Statute, was adopted by the UN General Assembly (UNGA) by a vote of 106 in favor and only three against and had its importance recognized by the ICJ itself in the Gabčíkovo case[18] were used by Uruguay to reinforce its claim.

Articles 11–19 of the UNWC establish a framework of procedural obligations applicable to watercourse states planning projects likely to impact on shared watercourses or liable to cause transboundary harm. These provisions assist watercourse states in maintaining an equitable balance between the respective uses of an international watercourse.[19] The very general nature of both the principle of equitable

---

15 *Case Concerning Pulp Mills on the River Uruguay (Argentina v. Uruguay)* (Application Instituting Proceedings) General List No 135 [2006] ICJ 1, at 5.

16 *Pulp Mills* (note 3), at para. 265.

17 Vienna Convention on the Law of Treaties (adopted May 23, 1969, entered into force January 27, 1980) 1155 UNTS 331, at Article 31(3)(c).

18 *Gabčíkovo* (note 3), at para. 85.

19 McCaffrey, S. C. (Special Rapporteur), '2nd Report on the Law of the Non-Navigational Uses of International Watercourses' (March 19, May 12 and 21, 1986), (1986) 1987 II(1) Yearbook of the International Law Commission 88, at134, UN Doc A/CN.4/399 and Add 1 and 2. Available online at http://untreaty.un.org/ilc/documentation/english/a_cn4_399.pdf (accessed March 28, 2013) (McCaffrey 2nd Report).

utilization and the obligation to prevent transboundary harm makes procedural rules, such as those articulated in the Convention's aforementioned provisions, a crucial aspect of reconciling conflicting uses between watercourse states, as well as of managing the risks of damage to the environment.[20] As early as 1987, the ILC special rapporteur on non-navigational uses of international watercourses had indicated that procedural rules are an essential part of the overall structure of international water law and form an 'integrated whole' with its substantive principles.[21]

In similar reasoning, in the *Pulp Mills* case, the ICJ accepted Argentina's contention that the procedural and substantive obligations contained in the 1975 River Uruguay Statute were 'intrinsically linked'.[22] Despite this 'functional link', the Court was not prepared to accept their 'indivisibility',[23] concluding that a breach of procedural obligations does not automatically entail a breach of substantive duties. Rather, the ICJ found that the states may be 'required to answer for those obligations separately'.[24]

In acknowledging the 'functional link' between procedural and substantive obligations intended to ensure the equitable and sustainable management of a shared natural resource, the ICJ helped to clarify the relationship between the substantive and procedural rules found in the UNWC. The generality of the principle of equitable and reasonable utilization and the due diligence nature of the obligations contained in the duty of prevention of transboundary harm require that they be made normatively operational by means of a number of procedural requirements, including the duties to 'notify, consult and negotiate' and to exchange information – all more detailed obligations that flow from the general duty to cooperate.

In another part of its ratio relevant to UNWC interpretation, the ICJ held that Argentina and the Administrative Commission on the River Uruguay (CARU)[25] should have been informed at a very early stage, prior to the authorization or implementation of the project. The Court explicitly found that 'the duty to inform . . . will become applicable at the stage when the relevant authority has had the project referred to it with the aim of obtaining initial environmental authorization and before the granting of that authorization'.[26] On the facts of the case, therefore, the Court had little difficulty in concluding that Uruguay, by issuing the initial

---

20   Ibid., at 134 and 139.
21   McCaffrey, S. C. (Special Rapporteur), '3rd Report on the Law of the Non-Navigational Uses of International Watercourses' (March 30, April 6 and 8, 1987), 1987 II(1) Yearbook of the International Law Commission 16, at 23, UN Doc A/CN.4/406 and Add 1 and 2. Available online at http://untreaty.un.org/ilc/documentation/english/a_cn4_406.pdf (accessed March 28, 2013) (McCaffrey 3rd Report).
22   *Pulp Mills* (note 3), at para. 68. The Court further stated that, 'it is by co-operating that the States concerned can jointly manage the risks of damage to the environment that might be created by the plans initiated by one or other of them, so as to prevent the damage in question, through the performance of both the procedural and substantive obligations laid down in the 1975 Statute'. Ibid., at para. 77.
23   Ibid., at para. 72.
24   Ibid., at paras 78–79.
25   CARU is the joint body created under Articles 49–57 of the 1975 Statute.
26   *Pulp Mills* (note 3) at para. 99.

environmental authorizations and the authorizations for construction before noti-fying Argentina, had wrongly given priority to its own legislation over its procedural obligations under the 1975 River Uruguay Statute.[27] According to the ICJ, there-fore, Uruguay had failed to comply with its conventional obligation to notify.[28]

In this respect, both Argentina and Uruguay referred to two relevant articles of the UNWC. Article 11 creates a duty 'to exchange information and consult each other and, if necessary, negotiate on the possible effects of planned measures on the condition of an international watercourse'. Accordingly, under Article 12, 'before a watercourse State implements or permits the implementation of planned measures which may have a significant adverse effect upon other watercourse States, it shall provide those States with *timely* notification thereof'. Commenting on the corre-sponding provision in the 1994 ILC Draft Articles, the ILC specifies that '[t]he term "timely" is intended to require notification sufficiently early in the planning stages to permit meaningful consultations and negotiations under subsequent articles'.[29] The ICJ's judgment, therefore, greatly clarifies the meaning of the intended time-lines connected to the duty to cooperate and, more specifically, the duty to notify.

Moreover, the ICJ's decision in the *Pulp Mills* case recognized the central signifi-cance of cooperative machinery, such as CARU, for both the principle of equitable and reasonable utilization,[30] which it once again linked to the concept of sustainable development,[31] and the duty to prevent transboundary damage.[32] Consistent with the enhanced significance attributed to procedural rules and Argentina's contentions, the ICJ emphasizes the role of institutional arrangements, without which it is difficult to ensure effective cooperation[33] – an understanding of the significance of joint institu-tions for cooperation that corresponds with that of many leading commentators.[34] In sum, the Court stressed the interrelated nature of the procedural obligations arising under international water law, as well as the central importance of institutional machinery in facilitating compliance with such obligations.

27   Ibid., at para. 121.
28   Ibid at paras 121–22.
29   1994 ILC Draft Articles (note 1), at 111.
30   According to the ICJ, 'the object and purpose of the 1975 Statute, set forth in Article 1, is for the Parties to achieve 'the optimum and rational utilization of the River Uruguay' by means of the 'joint machinery' for co-operation, which consists of both CARU and the procedural provisions contained in Articles 7 to 12 of the Statute'. *Pulp Mills* (note 3) at para. 75.
31   See *Gabčíkovo* (n 3) at Para 140; ibid., at 88–111 (Separate Opinion of Vice-President Weeramantry); *Pulp Mills* (note 3), at paras 75–6.
32   *Pulp Mills* (note 3), at para. 77.
33   Ibid., at para. 89. The Court stressed, in particular, the authority of formally instituted international organizations, such as river basin organizations (RBOs) like CARU, which it described as 'governed by the "principle of speciality", that is to say, they are invested by the States which create them with powers, the limits of which are a function of the common interests whose promotion those States entrust to them'.
34   See, e.g. Thomas Franck, noting that 'sophist principles', such as that of equitable and reasonable utilization, 'usually require an effective, credible, institutionalized, and legitimate interpreter of the rule's meaning in various instances'. Franck, T. M. *Fairness in International Law and Institutions* (Oxford University Press, 1997), at 81–2; McIntyre, O. *Environmental Protection of International Watercourses under International Law* (Ashgate, 2007), at 377.

The UNWC also provides guidance on the role institutional mechanisms play in facilitating cooperation on measures and procedures jointly adopted by watercourse states.[35] While Articles 11–19 do not contain any reference to joint commissions, such institutions may be indispensable for the implementation and compliance with procedural requirements, such as the obligations of notification and consultation. In this regard, Article 8 of the Convention contains a paragraph specifically suggesting that states 'consider the establishment of' joint commissions to realize and facilitate cooperation on relevant measures and procedures. While not even an updated interpretation of the UNWC can force states to create joint commissions, the ICJ's judgment not only contributes to an understanding of best practice in the area, but may also inform future decisions on duties and rights of such commissions when they are created.

The most significant aspect of the judgment in the *Pulp Mills* case for the development of international water law, and of international environmental law generally, is its finding that EIA is absolutely essential for effective notification of neighboring states in respect of planned activities or projects which might cause transboundary harm.[36] The Court recognized the central importance of the EIA in the context of Uruguay's obligation to notify Argentina under the 1975 River Uruguay Statute.[37] The ICJ also underscored that this notification, based on an EIA process that takes into account transboundary impacts, must take place prior to issuing any initial environmental authorizations.[38]

Consistent with its understanding of the interrelated role of procedural and substantive obligations in this area, the Court also found that states *must* conduct an EIA, to satisfy the due diligence requirements of the duty of prevention.[39] Most importantly, the Court described the obligation to conduct an EIA as a 'requirement under general international law'.[40] The Court considered that, for the parties 'to comply with their obligations for the purposes of protecting and preserving the aquatic environment with respect to activities which may be liable to cause transboundary harm', they 'must carry out an environmental impact assessment'.[41]

On its own, the UNWC appears somewhat more restrained. Article 12 provides only that notification 'shall be accompanied by available technical data and information, *including* the results of any EIA, in order to enable the notified States to

35   UNWC, at Articles 8 and 24.
36   *Pulp Mills* (note 3) at paras 119–21.
37   According to the ICJ, 'the environmental impact assessments which are necessary to reach a decision on any plan that is liable to cause significant transboundary harm to another State must be notified by the party concerned to the other party, through CARU, . . . to enable the notified party to participate in the process of ensuring that the assessment is complete, so that it can then consider the plan and its effects with a full knowledge of the facts'. Ibid., at para. 119.
38   Ibid., at para. 121. The Court notes that 'this notification must take place before the State concerned decides on the environmental viability of the plan, taking due account of the environmental impact assessment submitted to it'. Ibid., at para. 120.
39   Ibid., at paras 121 and 204.
40   Ibid., at para. 204.
41   Ibid.

evaluate the possible effects of the planned measures' [emphasis added]. The language of the Article, therefore, recognizes the utility of an EIA process for the purpose of discharging the obligation to notify, but stops short of imposing a mandatory requirement on states. The Convention text was a compromise between states who wanted to incorporate a separate EIA requirement in the Convention and those which opposed its inclusion. Still, the UNWC did go further than the 1994 ILC Draft Articles, which contain no reference to the EIA at all.[42]

What the Court's opinion in the *Pulp Mills* case suggests is that the consideration of transboundary dimensions during an EIA is absolutely pivotal in facilitating realization of many of the rights and obligations elaborated in the UNWC.

On the procedural side, this is eminently sensible: while some states may have preferred that performance of an EIA remain discretionary, it is hard to argue that meaningful consultation and negotiation (as mandated under Articles 12–19 of the UNWC) can be conducted absent relevant information about the likely environmental impacts of a proposed project.

Moreover, the UNWC not only endorses the adoption and application of a corpus of procedural rules, but places procedural obligations 'front and center', requiring their effective integration with substantive rules; otherwise, the desired outcomes under the substantive rules cannot be ensured. Thus, under the UNWC, an EIA is necessary for the meaningful implementation of its procedural rules regarding planned measures. An EIA should also be viewed as a criterion for assessing compliance with the due diligence requirement entailed by the duty to prevent significant transboundary harm under Article 7. Finally, the EIA is a means for ensuring accountability in meeting the requirements of equitable and reasonable use and participation under Articles 5–6 of the Convention.

Accordingly, the text of the UNWC can be regarded as having anticipated or implied the existence of a binding requirement to conduct transboundary EIA. Okowa, for example, underscores that, 'even in those instances where no specific provision is made, EIA may be taken to be implicit in other procedural duties, in particular the duty to notify other States of proposed activities that may entail transboundary harm'.[43] Therefore, the UNWC is to be interpreted in accordance with current customary law, including in the sense that states must carry out a transboundary EIA when there is a risk that an activity may have a significant adverse impact on a shared resource – a practice that, in recent years, has gained widespread acceptance, culminating with the *Pulp Mills* decision.

The human right to water

Doubts persist about the precise legal status of the human right to water under international law. Yet, a considerable degree of international consensus has now

42 McCaffrey (note 9), at 475.
43 Okowa, P. 'Procedural Obligations in International Environmental Agreements', *British Yearbook of International Law*, 1996; 67: 275–336, at 279.

been achieved with the recent adoption of resolutions by the UNGA and the Human Rights Council, respectively.[44]

As regards the origins of the recognition of a human right of access to water in international law and its broad normative content, emphasis should be given to the adoption of General Comment 15[45] to the International Covenant on Economic, Social and Cultural Rights (ICESCR)[46] by the Committee on Economic, Social and Cultural Rights (CESCR) of the UN Economic and Social Council (ECOSOC). General Comment 15 represents the CESCR's definitive position on the subject and 'is the first recognition by a UN human rights body of an independent and generally applicable human right to water'.[47]

Though CESCR general comments do not formally impose legal obligations on ICESCR states parties, let alone other states, General Comment 15 constitutes a 'highly authoritative interpretation of the Covenant' and of the legal implications which flow from its key relevant provisions.[48] As a non-binding interpretation, General Comment 15 may be used to determine whether states have met their treaty obligations.[49] McCaffrey characterizes General Comment 15 as being 'more in the nature of a statement *de lege ferenda* rather than *lex lata*'. He then cautions that the interpretation of Articles 11–12 contained in General Comment 15 'must be accepted by the States parties to the Covenant in order to be binding upon them'.[50]

Though a human rights-based approach may be conceptualized 'in terms of society's obligations to respond to the inalienable rights of individuals',[51] fundamental questions persist about the normative status, substantive or procedural

44 UNGA, Resolution 64/292 'The Human Right to Water and Sanitation' (July 28, 2010) UN Doc A/RES/64/292; UNHRC, Resolution 15/9 'Human Rights and Access to Safe Drinking Water and Sanitation' (September 30, 2010) UN Doc A/HRC/RES/15/9; UNHRC, Resolution 16/2 'The Human Right to Safe Drinking Water and Sanitation' (March 24, 2011) UN Doc A/HRC/RES/16/2.

45 Committee on Economic, Social and Cultural Rights, 'Substantive Issues Arising in the Implementation of the International Covenant on Economic, Social and Cultural Rights: General Comment No 15, The Right to Water (Articles 11–12, International Covenant on Economic, Social and Cultural Rights)' (November 26, 2002) UN Doc E/C.12/2002/11. Available online at http://daccess-ods.un.org/access.nsf/Get?Open&DS=E/C.12/2002/11&Lang=E   (accessed March 28, 2013) (General Comment 15).

46 International Covenant on Economic, Social and Cultural Rights (adopted December 16, 1966, entered into force January 3, 1976) 993 UNTS 3. Available online at http://treaties.un.org/doc/publication/UNTS/Volume%20993/v993.pdf (accessed March 28, 2013) (ICESCR).

47 McCaffrey, S. C. 'The Human Right to Water' in Brown Weiss, E., Boisson de Chazournes, L. and Bernasconi-Osterwalder, N., *Fresh Water and International Economic Law* (Oxford University Press, 2005), at 101.

48 Ibid., at 94.

49 Williams, M. 'Privatization and the Human Right to Water: Challenges for the New Century', *Michigan Journal of International Law*, 2007; 28: 465–504, at 475. See also Bluemel, E. B. 'The Implications of Formulating a Human Right to Water', *Ecology Law Quarterly*, 2004; 31: 957–1006, at 972.

50 McCaffrey (note 47), at 103. He further points out: 'This is also true, a fortiori, of States that are not parties to the Covenant'.

51 UN Development Programme, Integrating Human Rights with Sustainable Human Development (UNDP, 1998), at 173–74, cited in Filmer-Wilson, E. 'The Human-Rights-Based Approach to Development: The Right to Water', *Netherlands Quarterly of Human Rights*, 2005; 23: 213–41, at 213.

requirements and justiciability of the rapidly emerging rights-based approach to water entitlements. However, even at this early stage in its legal development, a human right to water may empower individuals and communities by means of enhanced rights of participation in water-related decision making.

Such a right may also impact on key normative requirements under international water resources law and international environmental law. For example, the recognition of a human right to water contributes to the implementation of the concept of vital human needs, included in Article 10(2) of the UNWC, by formalizing the rights of individuals in the management of shared water resources. Such a recognition presents a unique opportunity to enhance and supplement the position, already well-established in the body of international water law, whereby vital human needs enjoy effective priority over other actual or potential water uses. This provides states with a strong and clear incentive to systematically develop and quantify credible plans to provide for the vital human needs of their citizens, in the context of interstate negotiations or dispute settlement over allocations and/or benefit-sharing in an international watercourse.

While the human right to water may not, at least as yet, impose a formal and enforceable obligation on a state to ensure that the vital human needs of its own citizens are adequately catered for, a water rights discourse goes some way towards internationalizing the relationship between the citizen and the state by bringing the needs of each state's citizens within the scope of international water law. This might be viewed as touching issues once thought to be part of a state *domaine réservé*, by providing individuals and communities with entitlements over transboundary water resources. Such a trend would be consistent with developments in international water law, whereby conventions anticipate and provide for a significant role for private recourse by adversely affected private individuals to domestic courts and remedies in the avoidance and resolution of disputes over international watercourses.

In this sense, Article 32 of the UNWC is intended to encourage watercourse states to provide equality of access to national judicial or other procedures to victims of transboundary harm regardless of whether they are citizens or non-citizens/residents or non-residents. Private domestic remedies are preferred on policy grounds. Domestic remedies generally bring relief more expeditiously and cost-effectively than interstate dispute settlement processes, and prevent an interstate dispute from becoming unnecessarily politicized.

Similarly, the discourse on the human right to water is likely to inform and enhance existing obligations under international and national environmental law to protect freshwater resources, as almost any elaboration of its content includes express requirements on the part of states 'to ensure that natural water resources are protected from contamination' and 'to prevent threats to health from unsafe and toxic water conditions'.[52] Indeed, '[e]nvironmental law, in contrast to international development law, has proved particularly suitable for the use of a "rights and duties"

---

52 See General Comment 15 (note 45), at para. 8.

language', which provides environmental law, and the values inherent therein, with 'autonomy' or something of an absolute character so that 'policy considerations are generally excluded from the interpretation and application of the law'.[53]

In addition to the very broad range of global and regional human rights instruments commonly cited in support of the emergence of a human right to water,[54] a number of international instruments concerning the utilization of freshwater resources shared by two or more states would also appear to support the existence of a human right to water. At the global level, Article 10(1) of the UNWC codifies the general position that 'no use of an international watercourse enjoys inherent priority over other uses', but, per Article 10(2), states must give 'special regard ... to the requirements of vital human needs'. As McCaffrey points out, 'this language falls short of an explicit recognition of a human right to water, [but] expresses a fundamental idea behind such a right, namely that in making allocation decisions governments must not forget basic needs of humans for water'.[55]

Moreover, the UNWC aims to address the obligations of watercourse states rather than the rights of individuals. Nonetheless, one would expect the 'requirements of vital human needs' to correspond closely with the obligations of states and the entitlements of individuals under the human right to water. In this sense, the Statement of Understanding on Article 10(2) of the UNWC provides that, 'in determining "vital human needs", special attention is to be paid to providing sufficient water to sustain human life, including both drinking water and water required

53 Fuentes, X. 'International Law-Making in the Field of Sustainable Development: The Unequal Competition between Development and the Environment', *International Environmental Agreements: Politics, Law and Economics*, 2002; 2: 109–33, at 118.

54 See Universal Declaration of Human Rights (adopted December 10, 1948) UNGA Resolution 217 A (III) (UDHR) Articles 22, 25; Geneva Convention (III) on the Treatment of Prisoners of War (adopted August 12, 1949, entered into force October 21, 1950) 75 UNTS 135, Articles 20, 26, 29 and 46; Geneva Convention (IV) on the Treatment of Civilian Persons in Time of War (adopted August 12, 1949, entered into force October 21, 1950) 75 UNTS 287 Articles 23, 55, 59, 85, 89 and 127; International Covenant on Civil and Political Rights (adopted December 16, 1966, entered into force March 23, 1976) 999 UNTS 171 (ICCPR) Article 6; ICESCR, at Articles 11–12; Convention on the Elimination of All Forms of Discrimination Against Women (adopted December 18, 1979, entered into force September 3, 1981) 1249 UNTS 13, Article 14(2); ILO Convention 169 concerning Indigenous and Tribal Peoples in Independent Countries (adopted June 27, 1989, entered into force September 5, 1991) (1989) 28 ILM 1382, Article 15(1); Convention on the Rights of the Child (adopted November 20, 1989, entered into force September 2, 1990) 1577 UNTS 3, Article 24(2)(c); Additional Protocol to the American Convention of Human Rights in the Area of Economic, Social and Cultural Rights (adopted November 17, 1988, entered into force November 16, 1999), Article 11, reprinted in *Basic Documents Pertaining to Human Rights in the Inter-American System* OEA/SER./L.V/II.82 doc. 6 rev. 1, at 67; African Charter on the Rights and Welfare of the Child (adopted July 1990, entered into force November 29, 1999) OAU Doc CAB/LEG/24.9/49 (1990), Article 14; Convention on the Rights of Persons with Disabilities (adopted December 13, 2006, entered into force May 3, 2008) 2515 UNTS 3, Article 28(2).

55 McCaffrey (note 47), at 100–1.

for production of food in order to prevent starvation'.[56] The indirect endorsement of the concept of vital human needs under the Convention is particularly significant given that much of what is contained therein, notwithstanding its lack of entry into force, is widely regarded as a codification of customary international water law.[57]

Further aiding in the interpretation of the UNWC in this regard, the Protocol on Water and Health[58] to the UN Economic Commission for Europe (UNECE) Water Convention expressly requires the parties, under Article 4(2)(a), to take 'all appropriate measures for the purpose of ensuring…adequate supplies of wholesome drinking water'. The protocol further provides, in Article 6(1)(a), that the parties 'shall pursue the aims of…access to drinking water for everyone'.

At the level of river basin agreements, the 2002 Senegal Water Charter[59] and the 2008 Niger Basin Water Charter[60] include explicit reference to the right to water. The latter agreement recognizes, in Article 4, that one of the criteria of equitable utilization is 'the right to water of local riparian populations'. In this sense, these and perhaps other basin agreements are ahead of the Convention in expressly codifying the right to water and its applicability in a transboundary context.

Beyond official legal sources, the International Law Association's 2004 Berlin Rules on Water Resources Law, which revise and update the ILA's seminal 1966 Helsinki Rules,[61] give clear and formal priority to vital human needs.[62] The Berlin

---

56 UNGA, 'Convention on the Non-Navigational Uses of International Watercourses: Report of the 6th Committee Convening as the Working Group of the Whole' (April 11, 1997) UN Doc A/51/869, at 5. See the oral report of the coordinator of the informal consultations on Article 10(2), UN Doc A/C.6/51/SR.57 (1997), at 3. See further, Tanzi, A. and Arcari, M. *The UN Convention on the Law of International Watercourses* (Kluwer Law International, 2001), at 139.

57 The UNWC is the product of over 20 years of exhaustive research, deliberation and consultation with States by the ILC. It is thus likely to be highly persuasive in identifying and interpreting relevant rules of customary international law. See McIntyre (note 34), at 2.

58 Protocol on Water and Health to the 1992 Convention on the Protection and Use of Transboundary Watercourses and International Lakes (adopted June 17, 1999, entered into force August 4, 2005) 2331 UNTS 202.

59 Charte des Eaux du Fleuve Sénégal [Senegal River Waters Charter] (Mali-Mauritania-Senegal), Organisation pour la Mise en Valeur Du Fleuve Sénégal (OMVS) Resolution 005, adopted by the Conference of Heads of State and Government (May 18, 2002), at Article 4. Available online at http://lafrique.free.fr/traites/omvs_200205.pdf (accessed March 28, 2013).

60 Niger Basin Water Charter, adopted by the 8th Summit of the Niger Basin Authority Heads of State and Government (April 30, 2008) [French]. Available online at www.abn.ne/attachments/article/39/Charte%20du%20Bassin%20du%20Niger%20version%20finale%20francais_30-04-2008.pdf (accessed March 28, 2013).

61 ILA, 'Berlin Rules on Water Resources' (Berlin Rules) in Report of the 71st Conference (Berlin 2004) (2004 ILA Report). According to the Special Rapporteur's commentary, at 20, the Berlin Rules represent 'a comprehensive revision of the Helsinki Rules and related rules approved from time to time by the Association…[and] set about to provide a clear, cogent, and coherent statement of the customary international law that applies to waters of international drainage basins…[and] also undertake the progressive development of the law needed to cope with emerging problems of international or global water management for the twenty-first century'.

62 Article 14(1) provides: 'In determining an equitable and reasonable use, States shall first allocate waters to satisfy vital human needs'.

Rules define that expression in Article 3(20) as the 'waters used for immediate human survival, including drinking, cooking, and sanitary needs, as well as water needed for the immediate sustenance of a household'. In conjunction with this formal priority for vital human needs in interstate allocation of waters, the Berlin Rules include a dedicated Article 17 on 'The Right of Access to Water', which provides, inter alia, that 'every individual has a right of access to sufficient, safe, acceptable, physically accessible, and affordable water to meet that individual's vital human needs'. Thus, a learned body such as the ILA expressly links the human right to water to the position, widely acknowledged in international codifications and accepted in state and judicial practice, that uses required for the satisfaction of vital human needs take priority over other, less urgent uses.

General Comment 15 is also relevant for assessing the significance of the emergence of a human right to water for international water law – and thus for interstate practice in relation to the utilization and environmental protection of international watercourses. For example, Paragraph 31 provides:

> [t]o comply with their international obligations in relation to the right to water, States parties have to respect the enjoyment of the right in other countries. International cooperation requires States parties to refrain from actions that interfere, directly or indirectly, with the enjoyment of the right to water in other countries.[63]

This statement emphasizes states' obligations arising by virtue of the human right to water. It might easily be read as well as an attempt to inform the requirement to consider vital human needs within the framework of the principle of equitable and reasonable utilization[64] – the cardinal principle of international water law.[65] Also, by referring to the requirements of 'international cooperation', the CESCR could be alluding to the firmly established customary obligation on states to cooperate in

---

63 General Comment 15 (note 45), at para. 31.
64 A footnote to Para 31 of General Comment 15 expressly creates such linkage: 'The Committee notes that the [UNWC] requires that social and human needs be taken into account in determining the equitable utilization of watercourses, that States parties take measures to prevent significant harm being caused, and, in the event of conflict, special regard must be given to the requirements of vital human needs: see arts 5, 7 and 10 of the Convention'. Ibid., at note 25.
65 According to the ILC commentary to Article 5 of its 1994 ILC Draft Articles, 'a survey of all available evidence of the general practice of States, accepted as law, in respect of the non-navigational uses of international watercourses – including treaty provisions, positions taken by States in specific disputes, decisions of international courts and tribunals, statements of law prepared by intergovernmental and non-governmental bodies, the views of learned commentators and decisions of municipal courts in cognate cases – reveals that there is overwhelming support for the doctrine of equitable utilization as a general rule of law for the determination of the rights and obligations of States in this field'. See 1994 Draft Articles (note 1), at 222. See further, McIntyre (note 34), at 53–86.

the utilization and environmental protection of shared natural resources, as established under Article 8 of the UNWC.[66]

The CESCR envisages a particularly important role for international water resources law in the realization of the human right to water, through the ongoing development of the concept of vital human needs in a manner consistent with said human right. This view is commendable when one considers the significance for international water law of the apparent priority accorded to vital human needs in relevant state and conventional practice. For example, in outlining the importance of Article 10(2) of the UNWC for the interstate allocation of shared water resources pursuant to its Articles 5–6, leading commentators explain that:

> the protection of vital human needs entails a 'presumptive' priority over all the other factors listed in Article 6, such presumption being rebuttable only on the basis of the specific circumstances of the individual case. That is to say that watercourse States, in discussing the equitable allocation of shared watercourses, cannot avoid starting negotiations, taking the water supplies needed to support vital human needs as a fixed parameter.[67]

Couching the protection of vital human needs in human rights terms might do much to make this presumption irrefutable in any circumstances and the protection of vital human needs even more of a *sine qua non* in international water resources law. Accordingly, when clarifying the significance of Article 10(2) in the context of equitable and reasonable use, the commentary to the 1994 ILC Draft Articles confirms that a use of the water resources in question that is at variance with vital human needs is 'inherently inequitable and unreasonable'.[68]

In addition, Article 21(2) of the UNWC expressly prohibits 'the pollution of an international watercourse that may cause significant harm to other watercourse states or their environment, including harm to human health or safety'.

Against this background, experts also underscore the potentially dramatic impact that the emergence of a human right to water might have on interstate water resources law. For example, though focusing more generally on the policy consequences of a rights-based approach to environmental protection for international law relating to climate change and international watercourses, Shelton clearly articulates the potential significance of such an approach for legal principles based on equitable allocation of rights and responsibilities:

---

66 See, e.g. Dupuy, P.-M. 'Overview of the Existing Customary Legal *Regime Regarding International Pollution'*, in *Magraw, D. B. (ed.), International* Law and Pollution (University of Pennsylvania Press, 1991), at 61, who notes, at 70, that 'co-operation is the general means by which States will implement the substantive rights and duties regarding the use of transboundary natural resources'. See also McIntyre (note 34), at 221–9.

67 Tanzi and Arcari (note 56), at 141.

68 1994 Draft Articles (note 1), at 242. See also Tanzi and Arcari (note 56); McIntyre (note 34), at 163–4.

In climate change negotiations and in issues relating to international water-courses, debate often centres on equitable allocation of rights and responsibilities. Traditional international law is state-based and would divide rights and responsibilities according to sovereign equality, using historic responsibility for harm, wealth, capacity or other factors to determine the 'common but differentiated responsibilities' of each party. A rights-based approach on the other hand might suggest a *per capita* allocation based on the equal rights and responsibilities of each individual, giving certain countries considerably more permissible greenhouse gas emissions and reducing the allowances of others.[69]

The implications for the practical application of the principle of equitable and reasonable utilization, and for the priority accorded to vital human needs, are multiple. Linkage to the international rules on the utilization of shared water resources[70] might operate to augment significantly the scope of application of obligations relating to the human right to water. Shelton suggests that this might lead to a 'potentially vast expansion of the territorial scope of state obligations' beyond the traditional position whereby, 'presently, human rights instruments require each state to respect and ensure guaranteed rights "to all individuals within its territory and subject to its jurisdiction"'.[71]

General Comment 15 would appear to recognize the possibility of extra-territorial application of the human right to water in Paragraphs 31 and 34–35. Indeed, in the context of a more general discussion of a human rights-based approach to environmental protection, Shelton states that 'fear of potentially vast liability makes many states reluctant to embrace environmental protection as a human right'.[72] This may be equally true in respect of the human right to water.

In any case, for the purposes of this chapter, we conclude that Article 10(2) of the UNWC is consistent with the aim of ensuring to all peoples access to safe and sufficient water supplies, as set out in Chapter 18 of Agenda 21.[73]

Once the Convention becomes effective, that provision is expected to have an important role in reflecting and supporting the ongoing discourse in general international law on a human right to access to water, including in a transboundary context and as a precursor to discussions around a right to sanitation and its applicability within transboundary water systems.

Within the framework of the UNWC, therefore, international water law and human rights law can complement each other to a significant degree. On the one

---

69   Shelton, D. 'Human Rights and the Environment: Problems and Possibilities', *Environmental Policy and Law*, 2008; 38: 41–51, at 46.

70   See General Comment 15 (note 45), at note 25.

71   Shelton (note 69), at 46.

72   Ibid.

73   UN Conference on Environment and Development (Rio de Janeiro, June 3–14, 1992), 'Agenda 21: A Programme for Action for Sustainable Development' (January 1, 1993) UN Doc A/CONF.151/26/Rev.1 (Vol. I), at 9 (Agenda 21).

hand, the concept of 'vital human needs', now firmly established under international water resources law, provides a solid basis for the effective implementation of a human rights-based approach to the management of shared waters. On the other, the ongoing intense discourse on the human right to water can do much to elaborate upon this concept and to ensure that the priority already afforded to vital human needs is given further normative endorsement. Also, the requirements put forward under the rubric of the human right to water would generally function to enhance the continually developing obligations for environmental protection of shared waters set out under the UNWC.

Conclusion

Since the adoption of the UNWC in 1997, international water law has seen many developments, among them, the recognition that EIA is required by general international law and the ongoing elaboration of a human right to water. In the *Pulp Mills* case, the reference by both Argentina and Uruguay to the UNWC was entirely logical as the Convention constitutes a source of general international law for the management and protection of transboundary water resources. Therefore, general principles of the Convention may be helpful in resolving ambiguities or filling lacunae in the text of specific water agreements.

The UNWC appears to be consistent with a requirement under general international law to conduct transboundary EIA where potential harm to transboundary water resources may occur. It is reasonable to characterize this general requirement as an inherent entailment of the obligations of consultation, notification and negotiation as set out under the UNWC. As states have an obligation to provide prior notification concerning planned activities that could have adverse transboundary impacts, they must first determine whether there is a possibility that the activities will have such effects – a determination best made through an EIA process. The position adopted by the ICJ in regard to transboundary EIAs enhances the effectiveness of the customary obligation of prior notification, as included among the customary principles elaborated in the UNWC.

In much the same way, the emergence of a human right to water does not conflict with, but rather complements and confirms, the international water law principle granting formal priority to 'vital human needs' in interstate water allocation processes, in line with the UNWC. The human right to water can assist in the normative elaboration of this concept and help promote its legal implementation, by formalizing the rights of individuals and communities in the management of the water resources they share. A human right to water may also contribute to empowering individuals and communities, enhancing their rights of participation in water-related decision-making. Such rights impact on norms of international water law and international environmental law, including Article 32 of the UNWC codifying the principle of non-discrimination.

To use the language of the ICJ in the *Gabčíkovo* case, a 'treaty is not static, and is open to adapt to emerging norms of international law', when it contains 'evolving provisions'.[74] As a framework agreement, the UNWC will be able to ensure the objective of offering a stable and predictable legal framework, which is also adaptable as the law progressively develops.

---

74   *Gabčíkovo* (note 3), at para. 112.

Part 6

Emerging challenges and future trends

25 Governing international watercourses in an era of climate change

Jamie Pittock and Flavia Rocha Loures

The climate and hydrological cycles are intimately linked, and pay little heed to human institutions. Increased anthropogenic emissions of greenhouse gases are warming the atmosphere, changing the climate and, as a consequence, beginning to change the hydrological cycle. Managing fresh water, along with the impacts of climate change, involves complex, multidisciplinary challenges. Even before climate-induced changes are considered, freshwater resources are limited and often overexploited and badly managed.[1] Water management – like responding to climate change – is a 'wicked problem', with particularly incomplete, contradictory and changing requirements; complex interdependencies; and difficulties in defining the problems themselves, and in identifying and reaching consensus on solutions.[2]

In international watercourses, the challenge of managing freshwater systems in a changing climate is multiplied by the many jurisdictions concerned and potentially conflicting interests at stake. A vast number of water-sharing treaties and arrangements have been agreed to between states, within nations (between provincial and local jurisdictions) and between border communities. These mechanisms, however, have largely been crafted under the assumption that the planet's hydrology was stable. With climate change, stationarity is dead.[3]

Considerable research has looked at the likely impacts of climate change and subsequent need for enhanced governance of international watercourses, and numerous studies call for mechanisms to deal more effectively with the growing severity of climate change-related water problems and natural disasters.[4] Such studies, however, often overlook the potential for existing framework legal instruments,

1 Millennium Ecosystem Assessment, *Ecosystems and Human Well-being: Wetlands and Water Synthesis* (World Resources Institute, 2005); World Water Assessment Programme, *Water in a Changing World* (UNESCO and Earthscan, 2009).

2 Turnpenny, J., Lorenzoni, I. and Jones, M. 'Noisy and Definitely Not Normal: Responding to Wicked Issues in the Environment, Energy and Health', *Environmental Science and Policy*, 2009; 12 (3): 347–58.

3 Milly PCD and others, 'Stationarity Is Dead: Whither Water Management?', *Science*, 2008; 319: 573–4, at 574.

4 Tarlock, A. D. 'How Well Can International Water Allocation Regimes Adapt to Global Climate Change?', *Journal of Land Use and Environmental Law*, 2000; 15: 423–49; Lakshmi, V. 'Dealing with the Increased Problems Associated with Trans-boundary Water', *Eos*, 2003; 84: 505–6.

such as the UN Watercourses Convention (UNWC), to help states in better co-managing such change.[5]

Against this background, this chapter investigates the role of international water law in a changing climate and how that field of law would need to evolve with climate-induced hydrological changes in order to support the peaceful, equitable and sustainable use, management and protection of transboundary waters. We first outline water-related climate change impacts, and then consider the likelihood of conflict over shared waters in the context of a changing hydrology. The chapter then reviews options for our societies to adapt water management to climate change, in order to illustrate the range of desirable actions that watercourse agreements should enable. Next, we examine some water treaties before analyzing the strengths and weaknesses of the UNWC. We conclude with an assessment of the ability of the Convention to aid our societies in achieving the sound management of international watercourses in an era of climate change.

Direct and indirect climate change impacts on water resources

In relation to water, climate change is likely to: a) change the seasonality and decrease precipitation in many – usually, heavily populated – regions, and increase rainfall in other places, often where few people live; b) diminish runoff from glaciers and snow-packs, reducing the base flow of rivers and increasing seasonal variation; c) increase evapotranspiration and diminish runoff in many regions; d) increase the frequency of extreme storms, floods and droughts, often beyond the engineering thresholds for which current infrastructure was designed; and e) contribute to the degradation of water quality through greater erosion, diffuse pollution and eutrophication.[6]

Among the areas at particular risk from such impacts, many are transboundary watersheds. In Africa, for example, a 2007 study identifies specific basins with likely substantial decreases and increases in runoff and changes in the frequency and magnitude of floods and droughts.[7] A 2008 biophysical assessment creates global maps illustrating potential changes in discharge and water stress.[8] The assessment shows that rivers impacted by dams or extensive development will require more management interventions compared to free-flowing rivers. Furthermore, the areas 'almost certain' to require action include the Lake Chad, Nile, Indus, Ganges-Brahmaputra, Tigris-Euphrates, Colorado and Columbia Basins.

As discussed in greater detail below, in those and other basins, especially where the existing governance systems are weak or inadequate, international water law in

---

5   Lakshmi (note 4).
6   Bates, B., Kundzewicz, Z. W., Wu, S. and Palutikof, J. (eds), *Climate Change and Water*, Technical Paper of the Intergovernmental Panel on Climate Change VI (IPCC, 2008), at 21.
7   Goulden, M., Conway, D. and Persechino, A. 'Adaptation to Climate Change in International River Basins in Africa: A Review', *Hydrological Sciences Journal*, 2009; 54: 805–28.
8   Palmer, M. A., Reidy Liermann, C. A., Nilsson, C., Flörke, M., Alcamo, J., Lake, P. S. and Bond, N. 'Climate Change and the World's River Basins: Anticipating Management Options', *Frontiers in Ecology and the Environment*, 2008; 6: 81–9.

general can mediate interstate disputes and ensure continued dialogue as a basis for sustainable water management.[9] The UNWC, in particular, as a codification of customary law, can have an impact even among non-contracting states.

Just an example, the fragile governance of the Nile Basin, combined with anticipated climate change impacts, is likely to exacerbate vulnerability of freshwater ecosystems and dependent communities, and increase the risk of conflicts between riparian states.[10] Although climate change models vary in their forecasts for precipitation, higher evaporation rates may exacerbate water scarcity, while the rapidly expanding population will increase demand for water.[11] On the governance side, an agreement between Egypt and Sudan divides the entire river flow between those two downstream states.[12] In addition to being inequitable to the upper riparians, that agreement fails to account for the possibility of changing flows. Despite the establishment of the Nile Basin Initiative in 1999, an effective water management agreement endorsed by all basin states is still lacking. Egypt has even threatened military action in the event of upstream water diversions.[13]

In addition to the direct water-related impacts of climate change discussed above, we should not overlook the indirect water demands of mitigation and adaptation measures: a) climate change mitigation policies often favor energy technologies with high water impact, such as biofuels and hydropower;[14] b) most carbon sequestration proposals involve more water consumption, including afforestation;[15] c) many water adaptation measures are energy-intensive, such as desalination, or have significant environmental impacts, such as inter-basin water transfer schemes.[16]

9   Examining the adequacy of the legal regime in place in each of those watersheds falls outside the scope of this chapter. See Chapter 18, for an example of weak governance, the relations between the Ganges Basin states and implications for river sustainability in the context of climate change.

10   See Chapter 12 for a detailed discussion on the role and relevance of the UNWC in East Africa, including the Nile Basin.

11   Tarlock (note 4); Conway, D. 'From Headwater Tributaries to International River: Observing and Adapting to Climate Variability and Change in the Nile Basin', *Global Environmental Change*, 2005; 15: 99–114.

12   Agreement between the Republic of the Sudan and the United Arab Republic of Egypt for the Full Utilization of the Nile Waters (adopted November 8, 1959, entered into force December 12, 1959) 453 UNTS 66 (1959 Nile Agreement). See Tarlock (note 4); Conway (note 11).

13   Conway (note 11).

14   de Fraiture, C., Giordano, M. and Liao, Y. 'Biofuels and Implications for Agricultural Water Use: Blue Impacts of Green Energy', *Water Policy*, 2008; 10: 67–81; Pittock, J. 'A Pale Reflection of Political Reality: Integration of Global Climate, Wetland, and Biodiversity Agreements', *Climate Law*, 2010; 1: 343–73.

15   Herron, N., Davis, R. and Jones, R. 'The Effects of Large-scale Afforestation and Climate Change on Water Allocation in the Macquarie River Catchment, NSW, Australia', *Journal of Environmental Management*, 2002; 65: 369–81; van Dijk, A. I. J. M. and Keenan, R. J. 'Planted Forests and Water in Perspective', *Forest Ecology and Management*, 2007; 251 (1/2): 1–9; Pittock (note 14); Pittock, J. 'National Climate Change Policies and Sustainable Water Management: Conflicts and Synergies', *Ecology and Society*, 2011; 16: 25.

16   Prime Minister's Science, Engineering and Innovation Council, *Challenges at Energy-Water-Carbon Intersections* (PMSEIC, 2010); Pittock (note 15).

Conflicts, international watercourses and climate change adaptation

Could climate change lead to more conflicts between states in international watercourses? The underlying reasons for water-related controversies have been attributed to issues of quantity, quality and timing of water supplies.[17] However, at the international scale, Wolf *et al.* argue:

> institutional capacity within a basin, whether defined as water management bodies or treaties, or generally positive international relations are as important, if not more so, than the physical aspects of a system.... [V]ery rapid changes, either on the institutional side or in the physical system, which outpace the institutional capacity to absorb that change, are at the root of most water conflicts.[18]

Therefore, climate-resilient watercourse agreements in general, supplemented by a framework convention like the UNWC, are needed to reduce the risk for conflict, by codifying and contributing to the progressive development of international law in response to evolving circumstances, including climate change.

Directly or indirectly, water treaties can help tackle key barriers to freshwater climate adaptation.[19] Measures undertaken through treaties can influence physical and economic conditions, e.g. by providing for the construction of physical structures like dams or creating payments for the use of transboundary waters. Watercourse agreements can also have direct impacts on sociopolitical and institutional capacity to adapt to climate change. In this sense, cooperative management mechanisms are required to provide a forum for joint negotiations, consideration of different perspectives, trust- and confidence-building and the adoption of informed and participatory decisions. Accordingly, Wolf *et al.* propose, as indicators of stress, recently internationalized basins and those with unilateral development projects in the absence of cooperative regimes, in particular the Ganges-Brahmaputra, Han, Incomati, Kunene, Kura-Araks, Lake Chad, La Plata, Lempa, Limpopo, Mekong, Ob (Ertis), Okavango, Orange, Salween, Senegal, Tumen and Zambezi Basins.[20]

17 Wolf, A. T., Kramer, A., Carius, A. and Dabelko, G. D. 'Managing Water Conflict and Cooperation', in Worldwatch Institute, *State of the World 2005: A Worldwatch Institute Report on Progress Toward A Sustainable Society* (W. W. Norton, 2005), at 80.
18 Wolf, A. T., Yoffe, S. B. and Giordano, M. 'International Waters: Identifying Basins at Risk', *Water Policy*, 2003; 5: 29–60.
19 Kundzewicz, Z. W. and Mata, L. J. (lead authors) 'Freshwater Resources and Their Management', in Parry, M. L., Canziani, O. F., Palutikof, J. P., van der Linden, P. J. and Hanson, C. E. (eds), *Climate Change 2007: Impacts, Adaptation and Vulnerability. Contribution of Working Group II to the Fourth Assessment Report of the Intergovernmental Panel on Climate Change,* (Cambridge University Press, 2007), at 173; Adger, W., Dessai, S., Goulden, M., Hulme, M., Lorenzoni, I., Nelson, D. R., Naess, L. O., Wolf, J. and Wreford, A. 'Are There Social Limits to Adaptation to Climate Change?', *Climatic Change*, 2009; 93: 335–54.
20 Wolf *et al.* (note 17), at 80.

In closing their analysis, Wolf *et al*. underscore: '[t]he historical record proves that international water disputes get resolved even among enemies, and even as conflicts erupt over other issues'.[21] Yet, the study in question does not explicitly consider climate change, and thus that the loss of hydrological stationarity will challenge water treaties adopted under the assumption of unchanging average flows, where these agreements lack the flexibility to adapt to a new era. This raises the question: with climate-induced changes expected to push river systems beyond conditions previously experienced, exacerbating historical water scarcity in many parts of the world, and with most transboundary watersheds still poorly governed, might these factors combined tip the balance away from cooperation towards more conflict? In illustrating this question's pertinence, let us consider a US State Department cable released by Wikileaks:

> Australia's top intelligence agency believes south-east Asia will be the region worst affected by climate change by 2030, with decreased water flows from the Himalayan glaciers triggering a 'cascade of economic, social and political consequences'. The dire outlook was provided by the Office of National Assessments... which... predicted increased conflict in the Kashmir region because of a decrease in flows into the Indus River. 'Internal migrations in multi-ethnic countries may cause more problems than cross-border migra tion', the cable said.... It assessed China as potentially the biggest loser because of decreased river flows, an event that could lead to international confrontations with states sharing the Mekong system.[22]

In the region, the Indus Waters Treaty between India and Pakistan[23] is often cited as an example of cooperation on water even amidst broader conflict. Climate change, however, threatens the treaty's durability and thus the relations between those two countries and, more broadly, political stability in South Asia. In the words of a former World Bank advisor, that treaty 'will come crashing into conflict sooner rather than later' in view of the added pressure posed by climate change.[24]

Adaptation options to be considered in the drafting of watercourse agreements

Climate change adaptation is a relatively new and contested field.[25] Emerging from climate science, many adaptation proponents begin by considering a system's

21   Ibid., at 95.
22   Dorling, P. and Baker, R. 'Climate Change Warning over South-east Asia' (*The Age*, December 16, 2010). Available online at www.theage.com.au/national/climate-change-warning-over-southeast-asia-20101215-18y6b.html (accessed March 28, 2013).
23   The Indus Waters Treaty (adopted September 19, 1960, entered into force January 12, 1961) 419 UNTS 125.
24   Wheeler, W. 'The Water's Edge' (*Good Magazine*, August 1, 2009). Available online at http://awards.earthjournalism.org/finalist/the-waters-edge-indus (accessed March 28, 2013).
25   Dovers, S. R. and Hezri, A. A. 'Institutions and Policy Processes: The Means to the Ends of Adaptation', *Climate Change*, 2010; 1: 212–31.

sensitivity to particular climate impacts by the extent of exposure; specific interventions should then be identified to reduce the resulting vulnerability. This is often suggested as a separate and centralized climate change adaptation strategy. The problem with this approach is the limited precision of climate change forecasts for any one water basin; in fact, different global climate models may suggest different precipitation and runoff outcomes for the same river basin. In the case of Australia's Murray-Darling Basin, for example, a 2008 well-funded, government-led modeling project concludes that annual average runoff could increase by seven percent or decline by 37 percent by 2030.[26] Those forecasts were exceeded in the short-term during the 2002–2010 drought, when runoff declined by 63 percent and the river ceased flowing to the sea.[27]

Consequently, river basin management strategies to prepare for climate change must consider significant uncertainty and thus include 'no- and low-regret measures' – interventions that would be desirable without climate change at no or low cost, such as using water more efficiently.[28]

In addition, climate change adaptation to different impacts in different places in different sectors has to be embedded in existing institutions at all scales, and should be decentralized.[29] In this regard, a question for assessing the adequacy of watercourse agreements is whether they enable subsidiarity – that is, facilitate management at the smallest relevant scale within an overarching framework.

With respect to institutional resilience, one must assess whether existing treaties foster resistance and maintenance of the status quo, allow for change at the margins or enable openness and adaptability.[30] For example, treaties that are overly prescriptive, such as those that specify volumetric water allocations, are likely to fail with climate-induced changes; those that empower a technical body to perform periodic reviews in water allocation and provide advice on necessary adjustments should be more successful.

Further illustrating the need for institutional resilience, Dovers and Hezri classify the levels of potential climate change impacts and how institutions may respond: a) climate variability similar to existing variability, and capable of being managed within existing institutional capacities; b) significantly exacerbated climate variability, resulting in a difficult adjustment, but capable of being addressed with boosted application of existing capacities; and c) climate change and variability beyond current experience and memory, representing a step change that threatens ecosystems and society.[31] As a minimum, therefore, watercourse

26   CSIRO, *Water Availability in the Murray-Darling Basin: A Report from CSIRO to the Australian Government* (CSIRO, 2008).
27   Pittock, J. and Connell, D. 'Australia Demonstrates the Planet's Future: Water and Climate in the Murray-Darling Basin', *International Journal of Water Resources Development*, 2010; 26: 561–78.
28   Pittock, J. 'Lessons for Climate Change Adaptation from Better Management of Rivers', *Climate and Development*, 2009; 1: 194–211.
29   Dovers and Hezri (note 25), at 212.
30   Ibid.
31   Ibid.

agreements must be capable of managing significantly exacerbated climate variability, such as large and more frequent floods and droughts.

Moreover, climate change adaptation measures can be categorized as 'hard path' and 'soft path'.[32] Hard path measures involve engineering interventions, such as the construction of flood levees, desalination plants, water storage dams and inter-basin transfer schemes. These are evident in countries like Australia and China, which have invested in desalination plants and inter-basin transfer schemes.[33] Soft path measures involve managing the demand for water and maintaining sustainable river flows for protecting ecosystems, their functions and services. Examples of these measures are the restoration of floodplains to manage flood risks along the Danube and Yangtze rivers,[34] and better watershed management in Brazil and Mexico.[35] Other options to manage greater hydrological variability resulting from climate change include conjunctive management of surface and groundwater resources,[36] as well as more cautious, better planned development of water infrastructure and periodic reoperation or removal of existing infrastructure.[37]

Other basic requirements for sound freshwater climate adaptation that experts have identified are as follows: a) generate multiple, including immediate, benefits; b) communicate the ability for all stakeholders to start adaptation; c) local ownership of measures; d) institute iterative, adaptive management cycles, on the basis of plans and targets; f) draw support from measures on local to national to global scales; g) seize policy reform windows, such as political changes or natural disasters; and h) ensure that adaptation institutions are partly self-funded, e.g. through local taxes.[38]

Against this background, the next section examines how water agreements and arrangements, at various levels, particularly the UNWC, may enable the adaptation

32 Gleick, P. H. 'Water Management: Soft Water Paths', *Nature*, 2002; 418 (6896): 373; Pittock (note 28), at 194.

33 Barnett, J. and O'Neill, S. 'Maladaptation', *Global Environmental Change*, 2010; 20: 211–13, at 212; Pittock, J. and Xu, M. *Controlling Yangtze River Floods: A New Approach* (World Resources Institute, 2011), at 4.

34 Ebert, S., Hulea, O. and Strobel, D. 'Floodplain Restoration Along the Lower Danube: A Climate Change Adaptation Case Study', *Climate and Development*, 2009; 1: 212–19; Yu, X. and others, 'Freshwater Management and Climate Change Adaptation: Experiences from the Central Yangtze in China', *Climate and Development*, 2009; 1: 241–8; Pittock and Xu (note 33), at 2.

35 Barrios, J. E., Rodriguez-Pineda, J. A. and de la Maza Benignos, M. 'Integrated River Basin Management in the Conchos River Basin, Mexico: A Case Study of Freshwater Climate Change Adaptation', *Climate and Development*, 2009; 1 (3): 249–60; Pereira, L. F. M., Barreto, S. and Pittock, J. 'Participatory River Basin Management in the Sao Joao River, Brazil: A Basis for Climate Change Adaptation?', *Climate and Development*, 2009; 1: 261–8.

36 Kabat, P. and van Schaik, H. *Climate Changes the Water Rules: How Water Managers Can Cope with Today's Climate Variability and Tomorrow's Climate Change* (Dialogue on Water and Climate, 2003), at 75.

37 Pittock, J. and Hartmann, J. 'Taking a Second Look: Climate Change, Periodic Re-licensing and Better Management of Old Dams', *Marine and Freshwater Research*, 2011; 62: 312–20; Hallegatte, S. 'Strategies to Adapt to an Uncertain Climate Change', *Global Environmental Change*, 2009; 19: 240–7.

38 Pittock (note 28), at 194.

and water management practices described above. We structure our analysis according to institutional attributes, taking into account the numerous studies that call for: a) a treaty and a river basin organization; b) information exchange; c) equitable water allocation and reallocation mechanisms; c) broader benefit transfers; d) provisions on extreme events; e) policy-making; f) dispute prevention and settlement; and g) governance mechanisms for implementation, including procedures for treaty amendment and financing tools.[39]

The UN Watercourses Convention and freshwater adaptation to climate change

The UNWC does not make specific reference to climate change. The Convention was adopted by the UN General Assembly in 1997, well before the current awareness of the extent of challenges climate change poses for water management. Yet, the Convention addresses some vital aspects of sustainable water management and thus, to some extent, directly or indirectly, tackles the concerns identified above. The Convention does so both through specific individual provisions[40] and as a global framework instrument to inform climate change adaptation and mitigation in international watercourses.

Watercourse agreements

Since AD805, some 3,600 water treaties have been adopted.[41] Even if imperfect, these agreements provide riparian states with principles, rules and procedures that can help reduce tensions and facilitate cooperation in addressing the impacts of climate change, including extreme events. Yet, such agreements cover only 40 percent of international watercourses, and often involve only two states, even when more riparians share the basin.[42] Moreover,

---

39   Goldenman, G. 'Adapting to Climate Change: A Study of International Rivers and Their Legal Arrangements', *Ecology Law Quarterly*, 1990; 17: 741–802; Ansink, E. and Ruijs, A. 'Climate Change and the Stability of Water Allocation Agreements', *Environmental and Resource Economics*, 2008; 41: 249–66; Drieschova, A., Giordano, M. and Fischhendler, I. 'Governance Mechanisms to Address Flow Variability in Water Treaties', *Global Environmental Change*, 2008; 18: 285–95; Raadgever, G. T., Mostert, E., Kranz, N., Interwies, E. and Timmerman, J. G. 'Assessing Management Regimes in Transboundary River Basins: Do They Support Adaptive Management?', *Ecology and Society*, 2008; 13 (1): 14; Cooley, H., Christian-Smith, J., Gleick, P. H., Allen, L. and Cohen, M. *Understanding and Reducing the Risks of Climate Change for Transboundary Waters* (Pacific Institute, 2009); De Stefano, L., Duncen, J., Dinar, S., Stahl, K., Strzepek, K. and Wolf, A. T. *Mapping the Resilience of International River Basins to Future Climate Change-induced Water Variability*, Water Sector Board Discussion Paper Series, No. 15. (World Bank, 2010).
40   See Chapter 18 for a detailed analysis of the UNWC provisions that may be relevant for tackling the effects of climate change on water resources in a transboundary context.
41   Wolf *et al.* (note 17), at 80.
42   Loures, F., Rieu-Clarke, A. and Vercambre, M. L. *Everything You Need to Know About the UN Watercourses Convention* (WWF International, 2010), at 5.

[e]xisting agreements are sometimes not sufficiently effective to promote integrated water resources management due to problems at the national and local levels such as inadequate water management structures and weak capacity in countries to implement the agreements as well as shortcomings in the agreement themselves (for example, inadequate integration of aspects such as the environment, the lack of enforcement mechanisms, limited – sectoral – scope and non-inclusion of important riparian States).[43]

In a scenario where most of the world's transboundary waters lack adequate legal protection, the UNWC can provide overarching, global-scale governance standards. In addition, Article 3 of the Convention encourages the development of basin-specific agreements. An effective UNWC would serve as a more compelling, flexible and broadly accepted basis for interstate negotiations on new or revised watercourse treaties.

River basin organization

Basin-specific institutions are important for strengthening the information basis, developing common visions and promoting cooperative management. Treaties increasingly include such organizations, although the power granted to them can vary widely.

For example, the US–Mexico International Boundary and Water Commission (IBWC) is responsible for distributing water from the Colorado River and the Rio Grande/Bravo, protecting riparian lands from floods through levee and floodway projects and addressing sanitation and water quality problems. The IBWC makes proposals relating to those basins and the border region, and submits them to each Government for approval. Once approved, those proposals become binding obligations.[44] In contrast, the joint committee created under the 1996 Treaty on the Sharing of the Ganges between India and Bangladesh is charged with observing and recording daily flows, but is not empowered to make binding decisions. Under the treaty, the two governments are to meet 'at appropriate levels' to decide upon further action, based on the committee's yearly reports.[45]

In the UNWC, Article 8(2) *encourages* the establishment of river basin organizations. In turn, Article 9(2) of the UNECE Convention on the Protection and Use of Transboundary Watercourses and International Lakes[46] explicitly requires parties to adopt watercourse agreements, which must provide for the establishment of 'joint bodies'.

---

43 UN-Water, *Transboundary Waters: Sharing Benefits, Sharing Responsibilities* (Thematic Paper, UN-Water, 2008), at 6.

44 See Treaty Relating to the Utilization of Waters of the Colorado and Tijuana Rivers and of the Rio Grande (United States–Mexico) (February 3, 1944) 3 UNTS 313, Articles 6–8, 13, 16.

45 Treaty Between the Government of the People's Republic of Bangladesh and the Government of the Republic of India on the Sharing of the Ganga/Ganges Waters at Farrakka (December 12, 1996) (1996) 36 ILM 519, Article 6 (1996 Ganges Treaty).

Given the importance of these interstate organizations for the sustainable and equitable management of international watercourses, especially in a changing climate, let us consider briefly the status of the duty included in Article 9(2) of the UNECE Water Convention under general customary law. According to Tanzi, 'a general customary obligation for co-riparians of international watercourses to establish joint bodies cannot be assessed due to the lack of a wide *opinio juris* in that direction. The consistent practice to that effect can be ascertained to be based at least on a consistent *opinio necessitatis*'.[47] A related question, then, is whether the recent entry into force of the amendments to the UNECE Water Convention, in February 2013, open it for accession by all UN member states, may have an impact on the status of such a duty. These issues, however, fall outside the scope of the present chapter.

Information exchange

Changing hydrology with climate change increases the importance of timely data-sharing to facilitate water management. In fact, greater access to information across an entire basin is vital for climate change adaptation in a transboundary context. Decisions on water use, basin-wide management strategies and climate adaptation plans require information on the entire watercourse – information which depends on continued cooperation and dialoguie among riparians to be produced, processed and shared in a timely manner.

While many agreements provide for some sort of data-sharing, often they lack the details needed to make the provisions operational.[48] The UNWC can fill in these gaps in watercourse agreements, as well as offer a starting point for cooperation among states on basins without specific treaties. Article 9 of the UNWC codifies and details the duty of states to regularly exchange readily available information on the condition of the watercourse, including meteorological data and related forecasts. The only exception to that duty refers to information vital to national defense or security. The Convention also requires states to consult over planned measures, as discussed further below.

Water allocation, flow variability and permanent changes in river flows

Treaties governing international watercourses often have a limited scope, such as the definition of borders, implementation of joint projects or water allocation. Allocation provisions, in particular, are popular because they provide a degree of certainty, which is important for states negotiating over a shared natural resource. Whether provisions on the apportionment of water rights can help states adapt to

46   UNECE Convention on the Protection and Use of Transboundary Watercourses and International Lakes (adopted March 17, 1992, entered into force October 6, 1996) 1936 UNTS 269.

47   Tanzi, A. and Arcari, M. *The UN Convention on the Law of International Watercourses* (Kluwer Law International, 2001), at 191.

48   See Part III for examples of gaps and failings in existing watercourse agreements that the UNWC could help address, in particular with regard to provisions on information exchange and collection.

climate change depends on how they are structured. Those that allocate fixed volumes of water regardless of the actual flow are less likely to withstand the test of climate change; in turn, water agreements that allocate water based on a percentage of the total flow offer greater flexibility – a key quality in adaptive water management regimes, as water availability becomes increasingly unpredictable.

The 1996 Ganges Treaty is a hybrid of those two approaches to water allocation. That treaty allocates water in fixed amounts, but those amounts change depending on the availability of water. The volume is split fairly evenly during average and wet years, with India getting the surplus in the former and Bangladesh receiving the surplus in the latter. During dry years, India and Bangladesh receive 50% of the total flow, rather than a fixed amount. In extraordinarily dry years, the treaty requires the parties to enter into immediate consultations. From a climate adaptation perspective, this agreement can be considered well-designed: fixed allocations provide certainty and contribute to enforceability; at the same time, the three-volume levels and the required consultations for extremely low flows provide the flexibility needed to adapt to variations in river flows. The key problem, however, is that the treaty does not include Nepal, located upstream in the basin.[49]

Under the Senegal River Charter water is allocated according to the principle of benefit-sharing, rather than by quantity. If climatic or other changes make current uses less beneficial, riparian states can reallocate water rights accordingly, as per advice of the basin institution. This same structure allows for reallocation based on changing knowledge – for example, if irrigation practices improve, the surplus water will be reallocated to the most beneficial uses, not necessarily to expand agriculture.

For those basins without agreements or with treaties lacking adequate allocation provisions, the UNWC offers some guidance. The Convention's central principle, equitable and reasonable utilization, embodies flexibility, requiring an ongoing assessment of all relevant circumstances. As per Article 6, such circumstances include climatic factors and related forecasts. On the one hand, this flexibility does not offer the same certainty as fixed allocations would; on the other, given the differences among river basins, a global treaty could not have gone into much greater detail or been more prescriptive than what is contained in the UNWC.

In addition, Article 6(2) of the UNWC requires states to enter into consultations when the need arises to determine equitable and reasonable utilization. This principle is crucial in the context of climate change and must be duly considered when states negotiatie treaty provisions on water allocation. The requirement in question ensures that states are prepared to meet and agree on periodical revisions and the implementation of the necessary adjustments to respond to climate variability and change. In other words, that provision implicitly acknowledges flow variability and calls for sustained cooperation, which allows riparians to monitor and identify temporary and permanent changes in river flows.

---

49 1996 Ganges Treaty, at Annexure I.

Broader benefit sharing on an equitable and sustainable basis

Watercourse agreements may gain in effectiveness and resilience if their scope is expanded to encompass transfers of a broader range of benefits, beyond immediate freshwater management issues.[50] This may include, for example, regional integration and trade towards basin-wide water, food and energy security.

The UNWC does not refer specifically to the term 'benefit-sharing', but does require and create the legal basis for states to cooperate towards broader benefit transfers. In Article 5(1), the UNWC requires states to use an international watercourse 'with a view to attaining the optimal and sustainable utilization thereof and benefits therefrom'. In the second paragraph of Article 5, the Convention determines that states 'shall participate in the use, development and protection of an international watercourse in an equitable and reasonable manner... [This] includes both the right to utilize the watercourse and the duty to cooperate in the protection and development thereof'. In determining the manner of such participation, states are to consider the factors and conditions included in Articles 6 and 10, just as they would for implementing equitable and reasonable use. Finally, when significant transboundary harm occurs, despite the adoption of all appropriate prevention measures, Article 7(2) of the Convention calls on the harming state to act diligently towards eliminating or mitigating such harm and, where appropriate, 'to discuss the question of compensation'.

Harmful conditions and emergencies

In addition to water allocation and benefit-sharing provisions, the increased likelihood of floods, droughts and other emergencies with climate change has raised the question of the extent to which existing treaties deal with water-related disaster prevention and mitigation.[51]

At the basin level, an assessment of agreements found that 34 out of 50 treaties explicitly mention flow variability – an indicator of openness and adaptability.[52]

In its turn, the UNWC addresses harmful conditions and emergencies and related management issues in some detail. Article 25 promotes cooperation in the design and construction of regulation works, such as dams, which are often deployed to manage risks from floods, particularly if managed jointly. In Article 27, the Convention requires states to 'take all appropriate measures', individually and, where necessary, jointly, to prevent and mitigate harmful conditions, including floods and droughts. The UNWC also lays out procedures to respond to an emergency, which Article 28 defines as 'a situation that causes, or poses an imminent

---

50   Sadoff, C.W. and Grey, D. 'Beyond the River: The Benefits of Cooperation on International Rivers', *Water Policy*, 2002; 4: 389–403.

51   See Chapter 26 for a detailed discussion around benefit sharing in the UNWC and general international water law.

52   Drieschova, Giordano and Fischhendler (note 39).

threat of causing, serious harm to watercourse States and that results suddenly from natural causes, such as floods'. In this situation, a state must notify potentially affected neighbours and immediately take all appropriate measures to prevent, mitigate and eliminate the emergency's harmful effects. Where necessary, states must develop joint contingency plans.

Furthermore, Articles 8(2) and 24(1) of the UNWC promote the establishment of joint mechanisms to facilitate cooperation and the joint management of the watercourse, as mentioned above. However, joint commissions are not always adequately set up or sufficient, on their own, to enable adequate responses to extreme events. For example, Article 4(1) of the 1959 Nile Agreement charges the Egypt-Sudan Permanent Joint Technical Commission with drawing up plans and submitting them to both governments in times of drought. While this provision suggests some flexibility for extreme flow variability, the lack of a specific procedure is problematic. Furthermore, during a prolonged drought in the 1980s, the joint commission failed to prepare such plans.[53] This example illustrates the relevance of the UNWC provisions on water allocation, reallocation, harmful conditions and emergencies examined above.

Policy

The loss of stationarity with climate change suggests that national governments may need new, flexible, longer term policies for the management and use of fresh water. In other words, states need to consider the long-term viability and collateral effects of projects like dams and reservoirs. Aligning and harmonizing national policies with broader river basin strategies, agreed among all relevant riparians, would help in this regard. In this sense, for example, the International Commission for the Protection of the Rhine makes policy decisions, which members must then implement.[54]

The UNWC does not contain an explicit general requirement for co-riparian states with regard to national policies. Still, climate change responses will often demand coordinated and harmonized policies across international borders, in order to ensure compliance with the basic principles of the Convention. In this sense, a policy-related obligation could be seen as implicit in the general duty to cooperate in Article 8, if required by circumstances of the specific case. In addition, Article 21, dealing with pollution prevention, reduction and control, expressly mandates states to 'take steps to harmonize their policies in this connection', and to 'consult with a view to arriving at mutually agreeable measures and methods' to address pollution in an international watercourse.

---

53   Goldenman (note 39), at 755.
54   Raadgever and others (note 39), at 14.

Conflict prevention and resolution

As mentioned above, climate change adaptation measures include those involving engineering interventions, such as the incorporation of climate change considerations into the planning, siting, design and operation of water infrastructure. Soft path measures, in turn, entail managing demand for water services and restoring the natural capacity of ecosystems to store, transport, regulate and purify fresh water for people and nature.

On international watercourses, infrastructure development to store water or control floods, in the context of climate adaptation, can affect river flows and result in significant transboundary impact. As the environment changes and states increasingly resort to unilateral measures, more disputes over international watercourses are thus likely to occur. To aggravate things, most existing agreements lack dispute resolution mechanisms, and few have the procedures needed to pre-empt conflict and bring states together during the early stages of planning infrastructure development.[55]

For example, the 1996 Ganges Treaty refers disputes first to the joint committee. If the joint committee cannot resolve it, it is escalated to the river commission. The final say rests with the two governments, as per Article 7 – the same parties that could not agree in the first place and, because of political concerns, are probably least able to address it.

Within the framework of the UNWC, planned measures are subject to well-developed rules and procedures on notification, information exchange, consultations and negotiations. These provisions have a special role to play in ensuring that climate adaptation measures comply with the basic principles of international water law, pertaining to reasonable and equitable use and participation, as well as the prevention and mitigation of significant transboundary harm.

Furthermore, in Article 33, the Convention lays out detailed procedures and timeframes for resolving disputes, which can include seeking the good offices of a third party or a joint body, or submitting the case to the International Court of Justice. By involving a neutral third-party and designing a clear procedure, the UNWC can supplement watercourses agreements and thus contribute to dispute prevention and settlement.

Governance mechanisms

As hydrology changes and knowledge improves, there will be an increasing need to adjust previous basin-specific agreements. In Article 3(2), the UNWC invites contracting parties to consider harmonizing pre-existing treaties with the Convention and thus may generate a policy reform window for national governments to amend and enhance existing treaties to manage climate change impacts.[56]

---

55   Wolf *et al.* (note 17), at 80; Drieschova, Giordano and Fischhendler (note 39), at 285; Ansink and Ruijs (note 39), at 249; Raadgever *et al.* (note 39), at 14.
56   Tarlock (note 4), at 423.

The Convention itself, however, does not include procedures for the adoption of amendments, protocols and annexes, which would allow its scope and content to evolve progressively. Given that the UNWC is a framework instrument, there are issues that the Convention does not contemplate or covers only to a limited manner. An example is groundwater: Article 2(a) of the Convention includes groundwater in the definition of an international watercourse, thus encouraging the integrated management of the watershed. However, the Convention does not contain separate provisions reflecting the greater vulnerability and special characteristics of groundwater resources. For this and other aspects, a provision on the possibility for parties to agree on how and when to develop the Convention's text would have been useful.[57]

Furthermore, adequate resources are an obvious prerequisite for effective water governance, including for the operation of basin institutions and the implementation of treaty obligations. As a general rule, the duty to participate in the development and protection of an international watercourse under Article 5(2) would require states to share the relevant costs and benefits related to implementation and compliance in an equitable manner. Article 25 contains a specific application of that general rule to the case of 'hydraulic works or any other continuing measure to alter, vary or otherwise control the flow of the waters of an international watercourse'. Otherwise, the Convention lacks a dedicated funding mechanism to support its future implementation, including for the initial set up of river basin organizations.

With only five additional ratifications required for entry into force, it may thus be time for the contracting states to start considering the potential costs and benefits associated with the creation of governance mechanisms to aid states in their efforts to comply with and implement the UNWC.[58]

Conclusion

The world's transboundary waters are on the whole poorly managed and degrading. Climate change will affect hydrology and exacerbate the existing threats from a growing human population that is consuming more water-intensive products. These precious freshwaters support rich ecosystems that generate critical services for humanity, and must be better managed.

A considerable number of intergovernmental institutions exist for the management of international watercourses at the regional, basin and bilateral levels, but they are often inadequate for enabling effective responses to climate change and variability, most shared waters are not covered by agreements, and the majority of treaties lack vital mechanisms required for effective adaptation. The adequacy of

---

57 See Chapter 23, which puts forward the idea of a future Protocol to Govern Groundwater Resources of Relevance to International Law under the general framework of the UNWC.

58 With regard to the UNWC's future implementation, see Chapter 8, on factors that could limit the effectiveness of the Convention upon entry into force, and Chapter 22, on an institutional structure to support the implementation process.

legal and institutional arrangements for international river basins in supporting adaptation to climate change has been analyzed for more than 20 years.[59] The challenge now for the global community is to convert these assessments into adaptive institutions and legal instruments for practical action.

The UNWC is a key platform for such adaptive management. As discussed above, the Convention has workable provisions for equitable water- and benefit-sharing, harm prevention, information exchange and regular communication and dispute prevention and settlement. On the other hand, the Convention is largely silent on some key issues, notably funding, policy development and governance mechanisms to support its future implementation. Therefore, in preparation for entry into force, contracting states should start to consider the costs and benefits associated with, and options for, the creation of implementation bodies, procedures for the development of the Convention's text, institutional and funding mechanisms and so on.

Overall, the entry into force of the Convention would facilitate institutional attributes for adaptive management of international watercourses with climate change. The adoption of the appropriate governance mechanisms could further enhance the potential for the Convention to contribute to freshwater climate adaptation in a transboundary context, both on the ground and as a basis for the progressive development of international water law.[60]

59   Goldenman (note 39), at 741; Teclaff, L. A. 'The River Basin Concept and Global Climate Change', *Pace Environmental Law Review*, 1990; 8: 355–88.
60   See Chapters 8 and 14 for detailed discussions on the importance of governance mechanisms for the future implementation of the UNWC.

26 Benefit sharing in the UN Watercourses Convention and under international water law

Patricia Wouters and Ruby Moynihan

> Water knows no frontiers; as a common resource it demands international cooperation.[1]

More than half of the world's population depend upon the water resources of the 276 international watercourses shared by sovereign nation states around the world – connecting and unraveling societies in a myriad of ways. In less than 25 years, almost 70 percent of the global population will be living in water-stressed countries.[2] Africa, where an astounding 93 percent of the total available water is from shared river basins,[3] will be the worst affected. Some 75 million hectares of land currently suitable for rain-fed agriculture will be lost in sub-Saharan Africa by 2080, raising the threat of food insecurity across the continent. As nations seek to generate benefits from their increasingly scarce water resources in an attempt to achieve food, energy and economic security, the current world order grows vulnerable to the imminent adverse impacts from this brewing perfect storm of mounting insecurities.

Faced with this situation, 'it is essential for riparian countries to find ways of cooperating over the management of these transboundary water sources, if they are to maximize mutual benefits from the use of the resource'.[4] The recent examples of how national governments seek to address these challenges can be characterized (rather simplistically) into two categories: (i) unilateral response focused on the primacy of national sovereignty and self-help; and (ii) regional/global collective response, based upon the recognition of the interconnectedness of the problems and solutions required. This chapter examines how the rising discourse on 'benefit

---

1 European Water Charter (adopted May 26, 1967) Council of Europe, Committee of Ministers Resolution (67) 10 at Principle XII.

2 UN Educational, Scientific and Cultural Organization (UNESCO) International Hydrological Programme, *The Impact of Global Change on Water Resources: The Response of UNESCO's IHP* (UNESCO, 2011), at 1.

3 Phillips, D., Daoudy, M., MacCaffrey, S., Ojendal, J. and Turton, A. *Trans-boundary Water Cooperation as a Tool for Conflict Prevention and for Broader Benefit Sharing* (EGDI, Ministry for Foreign Affairs, 2006), at 8.

4 UN Commission on Sustainable Development (UNCSD), 'Report of the Secretary General on its 5th Session' (April 7–25, 1997) UN Doc E/CN.17/1997/9, at para. 17. Available online at www.un.org/esa/documents/ecosoc/cn17/1997/ecn171997-9.htm (accessed March 28, 2013).

sharing', as a collective response, is consistent with and supported by the rules of international water law, including those advanced under the UN Watercourses Convention (UNWC).

We begin by considering benefit sharing as it exists and is most firmly established within other areas of international law. This is followed by an examination of the concept's application in the specific area of international watercourses, including exploring how key principles of international water law provide the enabling environment for benefit sharing to succeed. Within this discussion, we examine how the concept of benefit-sharing relates to the governing principle of equitable and reasonable use, with a particular focus on this relationship under the UNWC, including attempting to dispel the misperception that these notions are adversaries. The chapter then examines how equitable utilization/participation and benefit sharing are supported by the cornerstone principle of international law – the duty to cooperate.[5] State practice on benefit sharing is considered, revealing some evidence of a similar understanding and application of these principles in treaty practice. The chapter concludes by summarizing some of the key remaining challenges regarding the collective and peaceful management of the world's international waters as they relate to benefit sharing.

Benefit sharing

Benefit sharing 'has been a recurrent theme in international debates, but the concept has never been satisfactorily defined'.[6] The concept at hand has its legal origins in a variety of sources – perhaps now most firmly established in international environmental law agreements. Instruments such as the UN Convention on Biological Diversity (CBD),[7] the Food and Agriculture Organization (FAO) International Treaty on Plant Genetic Resources for Food and Agriculture[8] and the UN Convention on the Law of the Sea (UNCLOS)[9] all refer to the notion.

---

5   The duty to cooperate urges all nation states to 'practice tolerance and live together in peace with one another as good neighbours...to maintain international peace and security and promote the fundamental freedoms of all'. Charter of the UN (adopted June 26, 1945, entered into force October 24, 1945), Preamble, Article 1(1) and (3) (UN Charter).

6   Schroeder, D. 'Benefit Sharing: It's Time for A Definition', *Journal of Medical Ethics*, 2007; 33: 205 at 209–9.

7   The CBD pursues the following objectives: the conservation of biological diversity, the sustainable use of its components and the fair and equitable sharing of the benefits arising out of the utilization of genetic resources, including by appropriate access to genetic resources and by appropriate transfer of relevant technologies, taking into account all rights over those resources and to technologies, and by appropriate funding. UN Convention on Biological Diversity (adopted June 5, 1992, entered into force December 29, 1993) 1760 UNTS 79, Article 1 (CBD). See also Morgera, E. and Tsioumani, E. 'The Evolution of Benefit Sharing: Linking Biodiversity and Community Livelihoods', *Review of European Community and International Environmental Law*, 2010; 19: 150–73.

8   International Treaty on Plant Genetic Resources for Food and Agriculture' (adopted November 3, 2001, entered into force June 29, 2004). Available online at ftp://ftp.fao.org/docrep/fao/011/i0510e/i0510e.pdf (accessed March 28, 2013).

9   UN Convention on the Law of the Sea (adopted December 10, 1982, entered into force November 16, 1994) 1833 UNTS 3, Articles 1(1) and 140 (UNCLOS).

The justification for benefit sharing, according to the CBD, relies on a mutually beneficial instrumental approach: '[i]n Aristotelian terms, we are dealing with "commutative justice", where each party gives one thing and receives another, with a focus on the equivalence of the exchange'.[10] In the case of biological diversity, the exchange is 'between the provision of access for bioprospecting and compensation, be it monetary or non-monetary'.[11] Benefit sharing in the context of international watercourses has developed in other ways, but the basic justifications for this approach align with the development of the concept under the CBD. A particularly important similarity is the role of equity in both cases to protect global resources – where both fields of international law may be seen as implementing the principle of intergenerational equity.[12]

The management and utilization of scarce international watercourses may stimulate cooperation or conflict. The benefit sharing discourse has recently emerged as one approach to overcoming this dichotomy, encouraging collective action through the sharing of benefits from the use of transboundary waters. Benefit sharing is sometimes framed as a preferable alternative to the purely volumetric allocation of water resources. The concept focuses on optimizing the values (economic, social, cultural, political and environmental) generated from water in its different uses and through the equitable distribution of the benefits arising.[13] However, from a broader disciplinary perspective, the benefit sharing discourse often omits a critical first step: effectively providing and sustaining the enabling environment that is required for its success. Robust legal frameworks are crucial for effective benefit sharing arrangements. This chapter thus explains how key principles of international water law, including those advanced under the UNWC, create the framework needed to promote cooperation and support the implementation of the benefit sharing approach.

Benefit sharing and equitable and reasonable use and participation

There is literature comparing and contrasting the principle of 'equitable and reasonable use' and the concept of benefit sharing, attempting to advance each as the preferred approach in the development and management of international watercourses.[14] This chapter disagrees with this juxtaposition, arguing that these two

---

10 Schroeder (note 6), at 207.

11 Ibid.

12 Essentially equity among those presently living on the planet. See McCaffrey, S. C. *The Law of International Watercourses* (2nd edn, Oxford University Press, 2007), at 404.

13 White, D., Wester, F., Huber-Lee, A., Hoanh, C. T. and Gichuki, F. *Water Benefits Sharing for Poverty Alleviation and Conflict Management, CPWF Topic 3 Synthesis Paper* (Consultative Group on International Agricultural Research Challenge Programme on Water and Food, 2008), at 4.

14 It is argued that one dilemma arising in the attempted harmonization of the concept of benefit sharing with the principle of equitable utilization is that 'optimal water-usage solutions may not always be congruent with the principle of equitable utilization'. Grey, D. 'Sharing Benefits of Transboundary Waters through Cooperation' (International Conference on Freshwater, Bonn, Germany, 2001).

notions are not alternatives – the first is a principle of international water law, and the second is an approach for managing water, which is supported by and contributes to the attainment of equitable and reasonable use in a transboundary river basin. It seems that the crux of the misunderstanding that benefit sharing is an adversary to equitable and reasonable utilization stems from a perception that the legal principle at hand is only about quantitative water allocation,[15] rather than its true meaning, which is much broader. According to the principle of equitable and reasonable utilization/participation, each state has a legally protected interest in an equitable share of the *uses* and *benefits* of an *international watercourse* and a correlated duty to participate equitably in the resource's development and protection.[16] This principle has been progressively developed in Article 5 of the UNWC, which provides:

1   Watercourse States shall in their respective territories utilize an international watercourse in an equitable and reasonable manner. In particular, an international watercourse shall be used and developed by watercourse States with a view to attaining optimal and sustainable utilization thereof and *benefits* therefrom, taking into account the interests of the watercourse States concerned, consistent with adequate protection of the watercourse.

2   Watercourse States shall participate in the use, development and protection of an international watercourse in an equitable and reasonable manner. Such participation includes both the right to utilize the watercourse and the duty to cooperate in the protection and development thereof.

The reference to 'utilization', 'participation' and 'benefits' here indicates that the UNWC anticipates the possibility that riparians may allocate rights to the water resource per se *and/or* agree to focus on the sharing of benefits arising from its use and protection.[17] What is to be considered a reasonable and equitable share of the uses/protection and benefits of an international watercourse in any given case is determined by taking into account a diverse range of relevant (non-exhaustive) factors, which focus on much more than just water allocation. These factors are listed in Article 6 of the UNWC and can be divided into two broad categories: factors of a natural character (hydrographic, hydrological, climatic, ecological, etc.) and economic, environmental and social factors (economic needs, population dependent on the watercourse, effects of use on co-riparians, existing and potential uses, conservation measures and availability of alternatives).[18]

15   'To negotiate the management and development of international shared rivers, riparians can focus their negotiations on the allocation of water rights or on the distribution of benefits derived from the use of water'. Sadoff, C. W. and Grey, D. 'Cooperation on International Rivers: A Continuum for Securing and Sharing Benefits', *Water International*, 2005; 30 (1): 420–8 at 422.

16   McCaffrey (note 12), at 388.

17   Phillips *et al.* (note 3), at 11.

18   Rieu-Clarke, A., Moynihan, R. and Magsig, B. O. *UN Watercourses Convention User's Guide* (IHP-HELP Centre for Water Law, Policy and Science, 2012), at 78. Available online at www.gwptoolbox.org/images/stories/Docs/unwaterconventionuseguide2012.pdf (accessed March 28, 2013).

Additionally, Article 3 of the UNWC preserves the contractual freedom of watercourse states. The Convention does not impose a duty on states to adopt future basin-specific treaties compatible with its provisions. Instead, the UNWC encourages states to consider harmonizing existing agreements with its basic provisions, as well as to adopt new agreements that apply and adjust the general principles of the Convention to the characteristics and uses of a particular watercourse. This means states can negotiate their own bilateral or multilateral watercourse agreements applying a basin-specific methodology for benefit sharing. Such a methodology can be compatible with and will ultimately be supported by the principle of equitable and reasonable use/participation, if the distribution of the benefits (and costs) is equitable. This is a key point – moving from a (misguided) focus on water allocation to a focus on sharing the *uses* and *benefits* of an *international watercourse* in a way that is truly fair is no easy task – the challenge to decide what exactly is equitable remains.

Evidence of broader consensus within international water law on the relationship between these notions is found in the International Law Association's (ILA) 1966 Helsinki Rules, which states that: '[e]ach basin State is entitled, within its territory, to a reasonable and equitable share in the beneficial uses of the waters of an international drainage basin'.[19] The Commentary continues:

> [a]ny use of water by a basin State, whether upper or lower, that denies an equitable sharing of uses by a co-basin State, conflicts with the community of interests of all basin States in obtaining maximum benefit from the common resource.[20]

The concept of benefit sharing, within the framework of equitable and reasonable use/ participation, was further supported in the *Gabčíkovo-Nagymaros* case. In its decision, the International Court of Justice (ICJ) found that the 1977 Treaty, signed between Hungary and Czechoslovakia, not only contains a joint investment programme, but also establishes a regime for the sharing of benefits. According to the Treaty, 'the main structures of the system of locks are the joint property of the Parties; their operation will take the form of a co-ordinated single unit; and the benefits of the projects shall be equally shared'.[21] Essentially, the ICJ confirmed a 'basic right to an equitable and reasonable sharing of the resources of an international watercourse', where these 'resources' include not only the water but also the capacity of the water to produce hydroelectric power, the ecological integrity of the water course system, and benefits from the stream.[22]

19　ILA, 'Helsinki Rules on the Uses of the Waters of International Rivers' in Report of the 52nd Conference (Helsinki 1966), Article IV (1966 ILA Report).
20　1966 ILA Report (note 19), at 114.
21　*Case Concerning the Gabčíkovo-Nagymaros Project (Hungary v Slovakia)* (Judgment) [1997] ICJ Rep 7 at 79 (*Gabčíkovo*).
22　McCaffrey (note 12), at 391 (quoting from *Gabčíkovo* [note 21], at 54).

Therefore, literature that focuses on debating the advantages of the concept of benefit sharing over the principle of equitable and reasonable use misunderstands the true meaning of the latter principle. The principle of equitable and reasonable utilization/participation is not a static rule; it is a dynamic process[23] that can support a benefit sharing approach to managing transboundary waters. This process depends on a strong underlying commitment from states to cooperate, which, as discussed in the next section, arguably stems from a general obligation to cooperate.

Incidentally, such literature also diverts from the core challenge facing transboundary water resources management in a world of sovereign nation states: how to move beyond the barriers of state sovereignty to a collective community of interests-based approach; i.e. one that facilitates and enables the sustained peaceful and orderly management of the world's internationally shared waters? And taking this further – how to define 'community of interests' within a cooperative framework? Considering the international watercourse from a regional and/or basin-wide perspective enables a more geographically functional designation of 'community' and often increases the diversity of potential basket of benefits. This broader perspective is also consistent with the principle of equitable utilization/participation, as codified in the UNWC.[24]

The duty to cooperate as the platform for equitable utilization and benefit sharing

Optimal and sustainable development of an international watercourse, including its protection and preservation, is dependent upon cooperation in good faith between the riparian states.[25] Taking a step out into general international law, the duty to cooperate lies at the heart of international relations and forms the foundation of the principles espoused under the UN Charter.[26] Concluded in 1945, the Charter begins, '[w]e the peoples of the UN', and lists its fundamental purposes:

1   To maintain international peace and security, to take effective collective measures for the prevention and removal of threats to the peace, and for the suppression of acts of aggression or other breaches of the peace, and to bring about by peaceful means, and in conformity with the principles of justice and international law, adjustment or settlement of international disputes or situations which might lead to a breach of the peace;
2   To develop friendly relations among nations based on respect for the principle of equal rights and self-determination of peoples, and to take other appropriate measures to strengthen universal peace;

23   Ibid., at 405.
24   Magsig, B. O. 'Overcoming State-Centrism in International Water Law: "Regional Common Concern" as the Normative Foundation of Water Security', *Goettingen Journal of International Law*, 2011; 1: 317–44, at 343.
25   McCaffrey (note 12), at 465.
26   Charter of the UN (adopted June 26, 1945, entered into force October 24, 1945) 1 UNTS XVI, Article 2(1).

3   To achieve international co-operation in solving international problems of an economic, social, cultural, or humanitarian character, and in promoting and encouraging respect for human rights and for fundamental freedoms for all without distinction as to race, sex, language, or religion; and

4   To be a centre for harmonizing the actions of nations in the attainment of these common ends.

The UN is 'based on the principle of the sovereign equality of all its Members',[27] who agree to 'settle their international disputes by peaceful means in such a manner that international peace and security, and justice, are not endangered'.[28] Reading these points together, it is clear that the law of nations, espoused by the UN, revolves around the fundamental principle of collective peace and security – in short, the duty to cooperate in this mission.

Translating this into the domain of international watercourses, the UNWC codifies and progressively develops the fundamental principles of international law, generally, and the obligation to cooperate, in particular. Consistent with, and as an operational bridge for the principle of equitable and reasonable use in Article 5, Article 8(1) of the UNWC sets forth the general obligation to cooperate: '[w]atercourse States shall cooperate on the basis of sovereign equality, territorial integrity and *mutual benefit* in order to attain optimal utilization and adequate protection of an international watercourse'.

According to the ILC, 'cooperation between watercourse States with regard to their utilization of an international watercourse is an important basis for the attainment and maintenance of an equitable allocation of the uses and *benefits* of the watercourse and for the smooth functioning of the procedural rules'.[29] Article 8(1) is the foundation for many of the UNWC's procedural rights and obligations contained primarily in Parts III–VI, including prior notification and consultation on works that may affect co-riparians in international watercourses. In this sense, the obligation to cooperate contains procedural duties which could be said to provide a platform for the operationalization of equitable and reasonable utilization. Moreover, as mentioned above, Article 5(2) of the UNWC introduces the obligation to 'participate in the use, development and protection of an international watercourse in an equitable and reasonable manner' – an obligation that 'includes both the right to utilize the watercourse *and the duty to cooperate* in the protection and development thereof'. This procedural component supports in a significant way the substantive rule of equitable and reasonable use, contained in Article 5(1), and must be read in that context.

---

27   Ibid.
28   Ibid., at Article 2(3).
29   International Law Commission, 'Draft Articles on the Law of the Non-Navigational Uses of International Watercourses and Commentaries thereto and Resolution on Transboundary Confined Groundwater' in ILC, 'Report on the Work of its Forty-sixth Session' (May 2–July 22, 1994) UN Doc A/49/10(Supp) (1994), at 105. Available online at http://untreaty.un.org/ilc/documentation/english/A_49_10.pdf (accessed March 28, 2013) (1994 ILC Report).

This articulation throughout the UNWC of the substantive duty, and the corresponding procedural duty of cooperation, recognizes that cooperative action by watercourse states is necessary to produce *maximum benefits for each of them*, while helping to maintain an equitable allocation of uses and affording adequate protection to the watercourse states and the international watercourse itself. Thus, watercourse states *have a right to the cooperation* of other watercourse states with regard to such matters as, inter alia, flood control, pollution abatement, drought mitigation, erosion control, disease vector control, river regulation, the safeguarding of hydraulic works and environmental protection, as appropriate under the circumstances. Of course, for maximum effectiveness, the details of such cooperative efforts should be provided for in watercourse agreements – an approach permitted and indeed encouraged under Article 3 of the UNWC. Yet, the obligation and correlative right provided for in Article 5(2) are not dependent upon a specific agreement for their implementation.[30] Thus, the cornerstone operational principle at the heart of the UNWC is arguably the *duty to cooperate*, in substance and in procedure.

The next section examines state practice on benefit sharing as it relates to the principles of equitable and reasonable use/participation, and cooperation. Our analysis reveals evidence of interpretation and application of these principles consistent with their meanings as articulated under the UNWC.

Selected state practice on benefit sharing

A growing number of international agreements provide for the sharing of the benefits from the use of transboundary waters. In these examples, benefit sharing is often viewed as a corollary to equitable utilization; i.e. it is often articulated as an approach to managing transboundary waters, supported by and stemming from the principle of equitable utilization/participation; in some instances, there is a looser connection between those two notions. In most cases, the obligation to cooperate is articulated as the fundamental basis for the attainment and maintenance of an equitable allocation of the uses and benefits of the watercourse. The reverse articulation is also sometimes made, in the sense that equitable and reasonable utilization/participation and benefit sharing are necessary building blocks to enable cooperation.

Indeed, all these elements integrate the broader process of transboundary water cooperation, which finds its foundation on the principle of equitable and reasonable use/participation, and can be implemented through benefit-sharing initiatives and on the basis of accepted and well-defined procedures of information exchange, notification, consultation and negotiations.

The concept of benefit sharing was first used in state practice in the 1961 Canada–United States Columbia River Treaty.[31] The notion was later cited with

---

30   Ibid., at 97.
31   Treaty Between Canada and the United States of America Relating to Cooperative Development of the Water Resources of the Columbia River Basin (adopted January 17, 1961, entered into force September 16, 1964) 542 UNTS 244 (Columbia River Treaty).

favor in Special Rapporteur Kearney's first Report, as part of the 30-year ILC study on the law non-navigational uses of international watercourses, which preceded negotiations on and the eventual adoption of the UNWC:

> in a two-state situation... whenever issues engendered by modern technology are involved, such as benefit-sharing from co-ordinated river regulation for hydroelectric production, the river has to be dealt with as a whole. The Canadian-United States Columbia River Treaty illustrates this requirement.[32]

The final text of the UNWC favors such an approach. Under Article 2, the Convention applies to the entire watercourse system, including the main-stem and any connected tributaries, lakes and aquifers. Furthermore, even though the UNWC foresees the possibility of agreements not applicable to an entire basin, Article 3 creates a solid framework by which the rights and interests of all states within a basin (and thus the watershed as a whole) must be duly considered.

Returning to the Columbia River case, Canada and the USA each wanted to develop the resource for a range of uses, including hydropower generation and flood control. Locating the dams downstream in the USA was not optimal and would have deprived Canada of opportunities for power generation. Locating the dams in Canada required the creation of significant storage reservoirs and would affect local communities. On balance, however, the second choice appeared the preferred solution, providing hydropower generation opportunities and also substantial flood control benefits, to the advantage especially of downstream US communities. After considerable studies and international and subnational consultations and negotiations, Canada and the USA agreed to the Columbia River Treaty. The Treaty served as the basis for the generation and sale of (excess) hydropower by Canada to the USA, and also the provision and compensation for downstream flood control to Canada – in short, a downstream-benefits arrangement where the division of benefits was based on equity and an underlying obligation to cooperate.[33]

Since then, the benefit sharing concept has evolved and been applied in different river basins around the world. For example, recent literature looking at the Nile defines the concept as the 'process where riparians cooperate in optimizing and equitably dividing goods, products and services connected *directly* or *indirectly* to the watercourse, or arising from the use of its waters'.[34] As already established throughout this chapter, the basic idea is to allocate the benefits of the uses of the water and water resources, as opposed to distributing quantities of water. The method focuses on optimizing the values (economic, social, cultural, political and

---

32 Kearney, R. (Special Rapporteur), '1st Report on the Law of the Non-Navigational Uses of International Watercourses' (May 7, 1976), (1976) II(1) Yearbook of the ILC 184 at 188, UN Doc A/CN.4/295. Available online at http://untreaty.un.org/ilc/publications/yearbooks/Ybkvolumes(e)/ILC_1976_v2_p1_e.pdf (accessed March 28, 2013).

33 Tarlock, D. and Wouters, P. 'Are Shared Benefits of International Waters an Equitable Apportionment?', *Colorado Journal of International Environmental Law and Policy*, 2007; 18: 523–36, at 527.

34 Phillips, D. and Woodhouse, M. 'Benefit Sharing in the Nile River Basin', (2013, forthcoming).

environmental) generated from water in its different uses and on the equitable distribution of the benefits among water users and suppliers.[35] From here, complex basin-specific benefit sharing mechanisms can be developed, which might be 'monetary or non-monetary and can be classified as ways to: (a) compensate for lost assets or loss of access, (b) restore and enhance livelihoods, (c) develop communities, (d) develop basins, and (e) share benefits'.[36]

In the Southern African region, where most of the available freshwater is transboundary, the Revised Southern African Development Community (SADC) Protocol[37] provides a legal framework for cooperation on water and development that includes the advancement of equitable and reasonable utilization, and guidelines for benefit sharing. In its Article 3(7)(a), the Protocol provides:

> Watercourse States shall in their respective territories utilize a shared watercourse in an equitable and reasonable manner. In particular, a shared watercourse shall be used and developed by Watercourse States with a view to attain optimal and sustainable utilization thereof *and benefits* therefrom, taking into account the interests of the Watercourse States concerned, consistent with adequate protection of the watercourse for the benefit of current and future generations.

Following on from this, the Article 7(3)(b) of the Protocol makes explicit that equitable and reasonable participation 'includes both the right to utilize the watercourse and the duty to co-operate in the protection and development thereof'. Additionally, the central underlying objective for the Protocol is to 'foster closer cooperation', which is necessary as a basis for equitable and reasonable utilization; reciprocally, advancing equitable and reasonable utilization is viewed as supporting greater cooperation.[38]

Recent developments in this region go further: an SADC Guideline on benefit sharing, currently being developed, aims to increase supply through reducing losses, reusing wastewater, accepting inter-basin transfers, promoting virtual water and reusing transferred water supply.[39] The Guideline study produces elaborate 'Benefit Wheels', which generate patterns of potential benefit sharing between upper and lower basin countries aimed at broadening the basket of benefits. According to this work, downstream riparians would support dam construction and hydropower development by upstream neighbors and receive benefits of seasonal flows that protect and enhance agriculture and tourism. Lower riparians could then trade stable crops back to upstream parties in exchange for energy, etc.[40]

---

35   White *et al.* (note 13), at 4.

36   Ibid., at 6.

37   SADC Revised Protocol on Shared International Watercourses (adopted August 7, 2000, entered into force September 22, 2003) (2001) 40 ILM 321 (Revised SADC Protocol).

38   Ibid., at Article 2(b).

39   SADC, *SADC Concept Paper on Benefit Sharing and Transboundary Water Management and Development* (SADC, 2011). Available online at www.orangesenqurak.org/UserFiles/File/SADC/SADC %20concept%20paper_benefit%20sharing.pdf (accessed March 28, 2013).

40   Ibid., at 6.

This approach is claimed to move the watercourse states towards optimal and higher value use of shared water resources.

This rapidly advancing and well-researched concept of benefit sharing in the SADC region is a welcome development. At the same time, this poses a significant challenge for international legal scholars to continue to explain how international law relates to such evolving policy guidelines and the role that law can play in providing the enabling environment for such frameworks to be sustainable, transparent, effective and legitimate.

Elsewhere in Africa, Article 4 of the 2003 Lake Tanganyika Convention[41] provides that 'States *shall co-operate* in good faith in the management of the Lake and its Basin *in a manner that . . . gives effect* to the general principles set out in Article 5', which states: 'the natural resources of the Lake shall be protected, conserved, managed, and used for sustainable development to meet the needs of present and future generations in an *equitable* manner'. Benefit sharing is stated as a concept to be applied to achieve equitable use, i.e. according to 'fair and equitable benefit sharing[,] . . . local communities are entitled to share in the benefits derived from local natural resources'.[42] In addition, per Article 11(c), states are

> required to cooperate in order to share in a fair and equitable way the results of research and development and the benefits arising from the utilization of the genetic and biochemical resources of the Lake and its Basin in accordance with the [CBD].

This example is illustrative of an underlying duty to cooperate, which then provides the basis for achieving equitable use/participation, which, in turn, supports the development of benefit sharing arrangements. Finally, such arrangements can contribute to the promotion of equitable use/participation.

The Nile Basin Initiative also 'seeks to develop the [Nile] river in a cooperative manner, sharing substantial socioeconomic benefits, and promote regional peace and security and to provide an institutional mechanism, a shared vision, and a set of agreed policy guidelines to provide a basin wide framework for cooperative action'.[43] The Initiative has recently launched the Nile Basin Sustainability Framework,[44] endorsed by all Nile states.[45] The case of the Nile

41 Convention on the Sustainable Management of Lake Tanganyika (adopted June 12, 2003, entered into force September 2005). Available online at www.ltbp.org/FTP/LAKECONV.pdf (accessed March 28, 2013) (2003 Lake Tanganyika Convention).

42 Ibid., at Article 5(f).

43 'Nile Basin Initiative'. Available online at www.nilebasin.org/newsite/ (accessed March 28, 2013).

44 'Rwanda's President of the Senate Opens the 3rd Nile Basin Development Forum' (Nile Basin Initiative, 27 Oct 2011). Available online at www.nilebasin.org/newsite/index.php?option= com_content&view=article&id=114%3Anbdf-opened-in-kigali&catid=40%3Alatest-news&Itemid=84&lang=en (accessed March 28, 2013). See 'Kigali Declaration', 3rd Nile Basin Development Forum (Kigali, Rwanda, October 26–28, 2011), (Kigali Declaration). Available online at www.nilebasin.org/newsite/attachments/article/115/3rd%20NBDF%20Kigali%20Declaration_ FNL.pdf (accessed March 28, 2013).

45 See Kigali Declaration (note 44).

thus demonstrates how the obligation to cooperate provides a basis upon which to implement equitable and reasonable utilization in order to share equitably the benefits arising.[46]

In Central Asia, the agreement between Kazakhstan, the Kyrgyz Republic, Uzbekistan and Tajikistan, concerning the Syr Darya and Aral Sea Basins, involves an arrangement for bartering hydropower, gas, coal and oil.[47]

In the Middle East, the Treaty of Peace between Israel and Jordan includes the understanding that cooperation on water-related subjects is to the benefit of both parties, in that it can help to alleviate water shortages. The Treaty also recognizes that water issues along their entire boundary must be dealt with in their totality, including the possibility of transboundary water transfers.[48] Arguably, an improved benefit sharing arrangement in that basin could include desalination as a viable option for several co-riparians to improve water availability. A further potential positive-sum outcome exists between Israel and Palestine involving the instigation of a coherent trade regime on low-cost agricultural products traded from Palestine to Israel.[49] In this example, the obligation to cooperate provides the starting point and the enabling environment for benefit sharing to be achieved.

In Southern Asia, the Preamble to the 1995 Mekong Agreement reaffirms 'the determination to continue to cooperate and promote in a constructive and mutually beneficial manner in the sustainable development, utilization, conservation, and management of the Mekong River Basin water and related resources'. The Preamble also refers to

> the promotion of interdependent subregional growth and cooperation among the community of Mekong nations, taking into account the regional benefits that could be derived and/or detriments that could be avoided or mitigated from activities within the Mekong River Basin undertaken by this framework of cooperation.[50]

A more significant and legally binding demonstration of the links between the principles of cooperation, equitable and reasonable use/participation and benefit

---

46   Abseno, M. 'The Concepts of Equitable Utilization, No Significant Harm and Benefit Sharing under the Nile River Basin Cooperative Framework Agreement: Some Highlights on Theory and Practice', *Journal of Water Law*, 2009; 20: 86–95.

47   Agreement on Cooperation in the Field of Joint Management of the Use and Conservation of Water Resources of Interstate Sources (adopted February 18, 1992). An English translation can be found online at www.icwc-aral.uz/statute1.htm (accessed March 28, 2013).

48   Treaty of Peace between the State of Israel and the Hashemite Kingdom of Jordan (signed October 26, 1994). Available online at www.mfa.gov.il/MFA/Peace%20Process/Guide%20to%20 the%20Peace%20Process/Israel-Jordan%20Peace%20Treaty (accessed March 28, 2013) (Treaty of Peace).

49   Phillips *et al.* (note 3), at xiv.

50   Agreement on the Cooperation for the Sustainable Development of the Mekong River Basin (adopted April 5, 1995) (1995) 34 ILM 864 [emphasis added] (1995 Mekong Agreement).

sharing is found in Articles 1 and 5 of the Agreement. Those provisions establish, respectively, an obligation 'to cooperate in both the utilization of water and its related resources ... in a manner to optimize the multiple-use and mutual benefits of all riparians' (Article 1), and an obligation to 'utilize the waters of the Mekong River system in a reasonable and equitable manner' (Article 5). The Mekong River Commission (MRC) is also actively promoting sharing of benefits through its other activities, including policy documents, such as the MRC 2011–15 Strategic Plan.[51]

The brief overview of selected existing legal agreements and policy guidelines above demonstrates that the concept of benefit sharing is often supported by and exists as corollary to the principle of equitable and reasonable use/participation, as the substantive norm. In its turn, the obligation to cooperate provides a platform in support of ongoing interstate interactions that are vital for enabling equitable benefit sharing.[52] The precise details of the benefits to be shared and how this is accomplished is often left to the parties to determine, usually with the assistance of joint governance bodies and on a case-by-case basis.

Challenges moving forward

Today, mankind is teetering on the brink of several possible catastrophes of its own making ... This should give us pause for thought. Now, more than ever, we need to cooperate and on a global scale. Although we are teetering on the brink of disaster, we are also on the brink of advancing to the next level of cooperation.[53]

The concept of benefit sharing will continue to evolve into a more elaborate approach, with increased input from across a broad interdisciplinary spectrum. A particularly strong contribution comes from the economics perspective, focused on the extent to which issue linkages within the water sector can be considered as underpinning the concept of benefit sharing.[54] The challenge thus remains for scholars to examine and explain how international legal frameworks remain part of the solution to the success of such an innovative approach. What is welcomed is that, in state practice, the benefit sharing concept and the principle of equitable and reasonable use/participation have often been combined as a powerful tool to achieving sustainable cooperation.

Furthermore, according to the Nobel Laureate Elinor Ostrom, in her analysis of

---

51 Mekong River Commission, *2011–2015 Strategic Plan* (MRC, 2011), at 41. Available online at www.mrcmekong.org/assets/Publications/strategies-workprog/Stratigic-Plan-2011-2015-council-approved25012011-final-.pdf (accessed March 28, 2013).

52 Abseno (note 46).

53 Nowak, M. and Highfield, R. *Supercooperators: Evolution, Altruism and Human Behaviour or, Why We Need Each Other to Succeed* (Cannongate, 2011), at 276–7.

54 Dombrowsky, I. 'Revisiting the Potential for Benefit Sharing in the Management of Trans-Boundary Rivers', *Water Policy*, 2009; 11: 125–140, at 126.

common pool resources,[55] the move from unilateral self-interest based on national sovereignty to the collective community-based approach is beneficial, but only if certain conditions are met – inter alia, all stakeholders need to be involved, and national governments have an important role to play at the subnational and regional levels. She further states that:

> The key problems to be solved are how to ensure that those using a common-pool resource share a similar and relatively accurate view of the problems they need to solve, how to devise rules to which most can contingently agree and how to monitor activities sufficiently so that those who break agreements (through error or succumbing to the continued temptations that exist in all such situations) are sanctioned, ensuring that trust and reciprocity are supported rather than undermined.[56]

From an international law perspective, this calls for improved governance, including strengthening the substantive and procedural aspects of the duty to cooperate within the context of international watercourses development and management, with an in-built system of monitoring compliance. However, addressing common pool resources problems calls for more targeted efforts. These matters must find clear expression and concrete articulation in current global initiatives that convene large communities in endeavors to tackle the global water crisis, including in debates around the UNWC, as a *global* legal instrument.

Conclusion

> We live at a time of immense opportunities. But it is also a world of great dangers, where a threat to one is a threat to all. The scale and complexity of these shared challenges often seems to have outgrown the framework through which they must be tackled.[57]

55 Common pool resources include natural and human-constructed resources in which: (i) exclusion of beneficiaries through physical and institutional means is especially costly; and, (ii) exploitation by one user reduces resource availability for others. These two characteristics – difficulty of exclusion and subtractability – create potential dilemmas in which people following their own short-term interests produce outcomes that are not in anyone's long-term interest. When resource users interact without the benefit of effective rules limiting access and defining rights and duties, substantial free-riding is likely, either through overuse without concern for the negative effects on others or a lack of contributed resources for maintaining and improving the common pool resource itself. Ostrom, E., Burger, J., Field, C. B., Norgaard, R. B. and Policansky, D. 'Revisiting the Commons: Local Lessons, Global Challenges', *Science*, 1999; 284: 278–82.
56 Ostrom, E. 'Coping with Tragedies of the Commons', *Annual Review of Political Science*, 1999; 2: 493–535.
57 Annan, K. 'Restoring Global Trust and Confidence' (Global Horizons Oxford Analytica's Conference, Oxford, England, September 16, 2011). Kofi Annan Foundation. Available online at www.kofiannanfoundation.org/newsroom/speeches/2011/09/%E2%80%9Crestoring-global-trust-and-confidence%E2%80%9D (accessed March 28, 2013).

How we develop and manage our international watercourses will have a significant impact on the social, economic and environmental welfare of communities around the world at all levels. Global problems linked to water-food-energy securities grow more complex within the current financial turmoil that continues to spread around the world. Faced with this reality, we need added impetus to our collective 'community of interests' approach to the effective management of the world's shared freshwaters. There are foundations for building such an approach: equitable and reasonable use/participation as a substantive principle and dynamic process that, among other things, supports benefit sharing – the success of which depends on a strong underlying commitment from states to the duty to cooperate. As discussed above, the UNWC codifies and develops such foundations.

However, more attention must be accorded to the substantive and procedural components of the duty to cooperate – the hallmark principle of international law. Furthermore, agreeing that the concept of benefit sharing is a useful approach should not detract from the critical examination of this concept as it is enabled and implemented in practice – generally an international agreement is the foundation for such an endeavour. Additionally, agreeing on what constitutes an equitable share of such benefits remains a complex challenge in each case. And, in some cases, despite the existence of cooperative frameworks unilateral measures nonetheless occur. This would call for the application, at outset, of the principles of equitable and reasonable use/participation and cooperation as the basis for interstate discussions on potential benefit sharing; or, at a minimum, to maintain the fair and peaceful relations between them.

Lessons learned from state practice in managing common pool resources (such as international watercourses) – such as the need for aligned political will, national and regional leadership and continued targeted efforts at capacity enhancement – are integral elements of overall good governance and form the building blocks for the future. The eventual entry into force of the UNWC will benefit from this and provide a legal framework for benefit sharing.

27 Water security – legal frameworks and the UN Watercourses Convention

Patricia Wouters and Ruby Moynihan

Water security is integral to human development and the prospects for peace.[1]

The propensity for conflict over water disappears where institutional arrangements such as treaties or river commissions exist to mitigate those pressures.[2]

International water security is a recent concept, emerging as one of the key challenges facing the global community. Public[3] and private sector[4] reports have seized upon the topic in research aimed at elucidating the scope, scale, substantive content and pathways towards resolution of this pervasive challenge. The absence of law in much of the emerging dialogue highlights the necessity for further examination, especially from a legal perspective. This chapter considers 'water security' through the prism of international law, with a particular focus on the law that governs international (transboundary)[5] watercourses and the contribution of the 1997 UN

---

1 Zimmerman, R. (ed.), *Afghanistan Human Development Report 2011. The Forgotten Front: Water Security and the Crisis in the Sanitation* (Centre for Policy and Human Development, 2011), at 2.

2 US Government Committee on Foreign Relations, *Avoiding Water Wars: Water Scarcity and Central Asia's Growing Importance for Stability in Afghanistan and Pakistan* (Report prepared for the 112th Congress First Session, US Government Printing Office, 2011), at 20.

3 UNSGAB, 'Hashimoto Action Plan: Compendium of Actions' (UN DESA, March 2006), at 9. Available online at www.unsgab.org/content/documents/HAP_en.pdf (accessed March 29, 2013) (Hashimoto Action Plan); UNSGAB, 'Hashimoto Action Plan II: Strategy and Objectives Through 2012' (UN DESA, January 2010), at 5, 15. Available online at www.preventionweb.net/files/12657_HAPIIen.pdf (accessed March 29, 2013) (Hashimoto Action Plan II). See also Global Water Partnership, *Water Security for Development: Insights from African Partnerships in Action* (GWP, 2010); Dannenmaier, E. 'Water Security: Identifying Governance Issues and Engaging Stakeholders', in Scozzari, A. and El Mansouri, B. (eds), *Water Security in the Mediterranean Region: An International Evaluation of Management, Control, and Governance Approaches* (Springer, 2011), at 11–20.

4 UBS, *The Rush for Resources Challenges Emerging Markets* (UBS, 2010); Maplecroft, 'New Maplecroft Index Rates Pakistan and Egypt Among Nations Facing 'Extreme' Water Security Risks' (Maplecroft, June 24, 2010). Available online at www.maplecroft.com/about/news/water-security.html (accessed March 29, 2013).

5 This chapter uses the term 'international' as opposed to 'transboundary' to characterize the freshwater resources shared across national borders. Other disciplinary work on water security tends to refer to transboundary watercourses, which does not refer exclusively to international watercourses. In international law, there is an important distinction whereby the term 'international' focuses on relations between sovereign nation states.

Watercourses Convention (UNWC) to the achievement of, or progress towards, water security. The global mission of securing 'water for all', prompted in large part by the collective sign-up to the Millennium Development Goals, lies at the heart of water security. Progress towards meaningful solutions in this regard requires unprecedented comprehensive understanding of the interdependencies between water and almost everything else. Such a context demands innovation from all sides and requires an integrated approach, which must include law.

The first part of this chapter sets out the context for the study – a world with growing water insecurity, at the local, national, regional, international and global levels.[6] Special attention is given to the international and regional levels, with the Nile case study providing a current example of where the legal dimensions of water security might be clarified. The Water Security Analytical Framework (WSAF) is then introduced, which comprises three core elements – availability, access and addressing conflicts of use – as an operational tool for identifying the central issues in this area. Each of these key elements is reviewed in detail, primarily through a legal perspective, focusing on how the provisions of the UNWC support the concept of water security and the three elements of the WSAF, including examining some shortcomings of the UNWC in this regard. The chapter concludes with observations and recommendations related to the role of the UNWC in addressing water security issues, and how the current global momentum around the water security concept might support greater endorsement of the UNWC and simultaneously provide an opportunity for international water law to rise to the new challenges raised by the concept of water security.

Global water security – emerging measures

Water – which maintains an unparalleled connectivity to all aspects of life on earth – can become scarce with climate change, population growth and urbanization, often resulting in deforestation and exacerbating erosion and soil degradation, aridity, desertification and salinization, challenging the peaceful and sustainable development of shared freshwater resources. There is already evidence that developing and managing water resources has led to conflicts of use, constraining an entire myriad of securities – human, environmental, food, economic, energy – at various levels of society.[7] Since water lies at the heart of all of these challenges, tackling water insecurity yields benefits beyond the water sector; indeed, it provides the fundamental basis for sustainable security and development.

---

6   The distinction between 'international' and 'global' from an international legal perspective relates to the rules of state responsibility – international rules relate to inter-state relations (i.e. nation state vis-à-vis another nation state), whereas global rules concern the global community, sometimes introducing obligations *erga omnes* and norms of *jus cogens*.

7   UN-Water/Africa, *The Africa Water Vision for 2025: Equitable and Sustainable Use of Water for Socioeconomic Development* (UN Economic Commission for Africa, 2009). Available online at www.afdb.org/fileadmin/uploads/afdb/Documents/Generic-Documents/african%20 water%20vision%202025%20to%20be%20sent%20to%20wwf5.pdf (accessed March 29, 2013).

Traditional perspectives on security have been conceived of primarily in terms of neutralizing military threats to the territorial integrity and political independence of the state. Since the end of the 1990s and Cold War, the concept of security has transformed significantly, with a widening and deepening of the concept to include cross-sectoral threats, i.e. across economic, societal and environmental dimensions.[8] The notion has also expanded across scales to extend beyond the original state-centric approach, to include not only threats to the 'state level' but also to human, national, regional and international levels.[9] Recognizing that water is a key component of 'ultimate security', the securitization of freshwater resources not only seems justified, but also inevitable.[10]

The 2011 *World Bank Development Report* calls for a new approach to solving conflicts and providing security, supporting a system which more accurately reflects twenty-first-century risks, identifying water issues as central to future conflict and supporting efforts to foster cross-border or subregional water management arrangements to ease regional tensions.[11] From an international law perspective, the response to such conflict is founded on the fundamental tenets of the law of nations, including those principles espoused under the UN Charter – the promotion of regional peace and security and the fundamental freedoms of all. Legal responses to water security challenges are considered within this paradigm.

Around the world, threats of water insecurity compromise the development aspirations of all nations, but especially newly emerging and developing countries. Already 80 percent of the world's population is exposed to high levels of risk to their water security.[12] Public international organizations,[13] nation states,[14] and private institutions[15] integrate water security risk into their threat analyses, and control over water

8   Brauch, H. G. 'Introduction: Globalization and Environmental Challenges: Reconceptualizing Security in the 21st Century', in Brauch, H. G., Spring, U. O., Mesjasz, C., Grin, J., Dunay, P., Behera, N. C., Chourou, B., Kameri-Mbote, P. and Liotta, P. H. (eds), *Globalization and Environmental Challenges: Reconceptualizing Security in the 21st Century* (Springer, 2008), at 1–24.

9   Brauch, H. G. 'Introduction: Facing Global Environmental Change and Sectorialization of Security' in Brauch, H. G., Spring, U. O., Grin, J., Mesjasz, C., Kameri-Mbote, P., Behera, N. C., Chourou, B. and Krummenacher, H. (eds), *Facing Global Environmental Change: Environmental, Human, Energy, Food, Health and Water Security Concepts* (Springer, 2009), at 21–42.

10 Magsig, B. O. 'Overcoming State-Centrism in International Water Law: "Regional Common Concern" as the Normative Foundation of Water Security', *Göttingen Journal of International Law 3*, 2011; 1: 317–44, at 331.

11 World Bank, *World Development Report 2011: Conflict, Security, and Development* (World Bank, 2011), at 2, 8 and 35.

12 Vörösmarty, C. J. and others, 'Global Threats to Human Water Security and River Biodiversity', *Nature*, 2010; 467: 555.

13 UN Environment Programme, *Water Security and Ecosystem Services: The Critical Connection* (UNEP, 2009); Global Water Partnership, *Water Security for Development: Insights from African Partnerships in Action* (GWP, 2010). Available online at www.gwp.org/Global/About%20GWP/Publications/Water%20Security%20for%20Development_report__final_2010.pdf (accessed March 29, 2013).

14 US Government Committee on Foreign Relations (note 2), at 20; Institute for Defence Studies and Analyses (IDSA), *Water Security for India: The External Dynamics* (IDSA, 2010); Zimmerman (note 1), at 2.

15 Strategic Foresight Group, *The Himalayan Challenge: Water Security in Emerging Asia* (Strategic Foresight Group, 2010); Maplecroft (note 4).

resources has become a strategic goal of national security policies and a significant consideration for international investment risk analyses – but the role of law is largely absent from this analysis. The UN Economic and Social Commission for the Asia and Pacific (UNESCAP) developed a Water Insecurity Index measuring 'the region's capacity to deliver the expected outcomes from investments and management in water resources for socially inclusive, environmentally sustainable economic development'.[16] Effective legal frameworks are essential for the delivery of more reliable, accountable and sustainable management and investment frameworks, especially concerning transboundary water resources, yet the Water Insecurity Index does not examine the integral role of law in achieving water security. Another index which offers to explain this emerging concept was released in March 2010, by Global Risk Institute – Maplecroft, which created the Water Security Risk Index. The index evaluates water security risk in 165 countries, finding that there are currently ten countries facing 'extreme water risk', including the emerging economies of Pakistan, Egypt and Uzbekistan, which are already experiencing internal and transboundary tensions from pressured water resources. The report predicts that future impacts of climate change on water stress will cause tensions and increase the potential to threaten global and regional stability.[17] The factors used to assess this risk in both the Water Insecurity Index and the Water Security Risk Index are evidence of a broader public-private consensus as to the multi-sectoral, multi-tiered nature of water security, but still the legal dimensions of this problem need to be better represented.

Further consideration is warranted – of the specific contribution that international law can make to achieving water security, including examining how legal variables interact with other scientific, socio-political and economic variables commonly related to conflict over water resources. More must be done to critically evaluate the role that international law can and does play in shaping state behavior over transboundary water sharing,[18] including examining how law has influenced 'informal interaction' among the basin states, such as nurturing epistemic communities or building intergovernmental networks around water security.[19]

It is the challenge of the international legal community to develop and articulate the role that law can play in this discourse, and to develop and articulate in readily accessible ways normative principles and legal frameworks to enable the legal perspective to be better integrated into future solutions in this field. This chapter offers one of several possible frameworks for legal analysis and for the provision of the legal dimensions of water security, and is a result of continuing legal research in this field.[20]

---

16 UNEP, Asian Development Bank and UNESCAP, *Green Growth, Resources and Resilience: Environmental Sustainability in Asia and the Pacific* (UN and Asian Development Bank, 2012). Available online at www.unescap.org/esd/environment/flagpubs/ggrap/documents/G2R2-web-20121121.pdf (March 29, 2013).
17 Maplecroft (note 4).
18 Hathaway, O. A. 'Between Power and Principle: An Integrated Theory of International Law', *University of Chicago Law Review*, 2005; 72: 469–536.
19 Brunnée, J. and Toope, S. J. 'The Changing Nile Basin Regime: Does Law Matter?', *Harvard International Law Journal*, 2002; 43: 105–59.
20 Magsig (note 8).

Regional water security – legal developments

Management of international freshwater resources at the regional level allows for the river basin to be the focal management unit. This method finds support in legal[21] and scientific[22] approaches. Maplecroft identifies the Middle East and North Africa region as the most at risk of extreme water insecurity, where some 15 countries have been identified as facing an 'extreme risk'.[23] The facts of this region demonstrate the nexus between energy and water security, where future water insecurity may lead to further increases in global oil prices.[24] Two examples of historic and continuing transboundary water disputes in this region that exemplify how conflict over water resources constrains a myriad of securities – human, environmental, food, economic and energy – include the Helmand River Basin, shared between Afghanistan and Iran, and the River Jordan, shared by Israel and Jordan. Another high-risk region is Southern Asia,[25] with a particular focus on the Indus and Ganges River Basins, inhabited by more than one billion people.[26] India sits at the center of many of the transboundary water risks in this region, where the securitization of water is unraveling, with many of India's internal and external security concerns directly correlated to water resource issues. The major land-based flashpoints for India–China tensions, particularly in Arunachal Pradesh state, are in the headwaters of Tibetan Plateau. Tensions between India and Pakistan over the Indus are causing the Indian Government to accept that India's security future will likely be not just focused on counterinsurgency, arms procurements and naval modernization, but also resource equity and efficiency.[27]

In East Africa, the Nile River Basin provides extensive evidence of the devastating impacts of water insecurity, with the four most water insecure countries on this planet situated in Africa.[28] The experience of this region demonstrates progressive and thwarted attempts to understand the pathways to achieve water security

---

21 McCaffrey, S. C. *The Law of International Watercourses* (2nd edn, Oxford University Press, 2007), at 35–7.

22 Falkenmark, M. 'Freshwater as Shared between Society and Ecosystems: From Divided Approaches to Integrated Challenges', *Philosophical Transactions: Biological Sciences*, 2003; 358: 2037–49.

23 These include: Mauritania (1), Kuwait (2), Jordan (3), Egypt (4), Israel (5), Niger (6), Iraq (7), Oman (8), United Arab Emirates (9), Syria (10), Saudi Arabia (11), Libya (14), Djibouti (16), Tunisia (17) and Algeria (18). Of the 12 Organization of the Petroleum Exporting Countries members, six are in the extreme risk category, while a further two – including Iran and Qatar – are rated as 'high risk'. Collectively, these countries produced approximately 30 percent of global oil production in 2009. See Maplecroft (note 4).

24 The correlation between water and energy security in this instance is due to the common practice of using 'lift water' in oil production in this region.

25 Bangladesh, Bhutan, India, Nepal, Pakistan and Sri Lanka comprise this region.

26 Sharma, B., Amarasinghe, U., Xuellang, C., de Condappa, D., Shah, T., Mukherji, A., Bharati, L., Ambill, G., Qureshi, A., Pant, D., Xenarlos, S., Singh, R. and Smakhtin, V. 'The Indus and the Ganges: River Basins under Extreme Pressure', *Water International*, 2010; 35 (5): 493–521.

27 Kugelman, M. (ed.), *India's Contemporary Security Challenges* (Woodrow Wilson International Center for Scholars, 2011), at 30.

28 The top four countries with the least secure supplies of water are 1. Somalia, 2. Mauritania, 3. Sudan, 4. Niger.

and provides important lessons necessary for understanding this evolving concept, particularly from a legal perspective. The Nile Basin Cooperative Framework Agreement (CFA) negotiated between Burundi, the Democratic Republic of Congo, Egypt, Ethiopia, Eritrea, Kenya, Rwanda, Sudan, Tanzania and Uganda is the first treaty to explicitly provide for water security.[29] Central to this Agreement is the definition of water security in Article 2(f), which 'means the right of all Nile Basin States to reliable access to and use of the Nile River system for health, agriculture, livelihoods, production and environment'. Further, the concept of water security is a substantive principle of the Agreement, which provides, in Article 14:

> [h]aving due regard to the provisions of Articles 4 and 5, Nile Basin States recognize the vital importance of water security to each of them. The States also recognize that the cooperation, management and development of waters of the Nile River System will facilitate achievement of water security and other benefits. Nile Basin States therefore agree, in a spirit of cooperation: (a) to work together to ensure that all states achieve and sustain water security.

While all Nile Basin States agree on the inclusion of the principle of water security, there remains a lingering misunderstanding, misinterpretation and disagreement on what this concept truly means from a legal perspective and how this concept might be implemented in practice, especially in the case of conflicts of use.[30] Some opponents of the use of water security in the CFA argue that it was

---

29 The Nile (comprising the Blue Nile and the White Nile) is shared by ten riparians. Negotiations on the Nile Basin Cooperation Framework Agreement commenced in 1999, resulting in an instrument now ready for ratification across the basin. At April 2012, the following countries had signed the CFA: Burundi, Ethiopia, Kenya, Uganda, Tanzania and Rwanda. See Nile Basin Initiative, 'Burundi Signs the Nile Cooperative Framework Agreement' (NBI, 2010). Available online at www.nilebasin.org/newsite/index.php?option=com_content&view=article&id=70%3Aburundi-signs-the-nile-cooperative-framework-agreement-pdf&catid=40%3Alatest-news&Itemid=84&lang=en (accessed March 29, 2013); Agreement on the Nile River Basin Cooperative Framework (opened for signature May 14, 2010). Available online at www.internationalwaterlaw.org/documents/regionaldocs/Nile_River_Basin_Cooperative_Framework_2010.pdf (accessed March 29, 2013) (CFA).

30 During discussions at the 15th Nile Council of Ministers (Nile-COM) meeting in Uganda, June 2007, over the draft CFA, the issue of 'water security' created an impasse in negotiations and the issue was referred for resolution by the Heads of State and Governments of the riparian countries. NBI, 'Minutes of the 15th Nile-COM' (Entebbe, Uganda, June 24–25, 2007). During the 16th Nile-COM meeting held in July 2008, the issue of water security again was deferred to the Nile River Basin Commission; during the Nile-COM meeting of April 13, 2010, in Sharm El-Sheikh, agreement on this concept still could not be reached, but it was agreed to open the CFA for signature including the concept of water security as it was. See NBI, 'Ministers of Water Affairs End Extraordinary Meeting over the Cooperative Framework Agreement' (NBI, April 14, 2010). Available online at www.nilebasin.org/newsite/index.php?option=com_content&view=frontpage&Itemid=1&lang=en (accessed March 29, 2013). See also Mekonnen, D. Z. 'Between the Scylla of Water Security and Charybdis of Benefit Sharing: The Nile Basin Cooperative Framework Agreement – Failed or Just Teetering on the Brink?', *Göttingen Journal of International Law*, 2011; 3: 345

deployed by the hydro-hegemons (Egypt and Sudan) in order to justify a traditional security response based on neutralizing threats to integrity and upholding
the political independence of the state; it is argued that this has derailed progress
on the CFA, and led to claims that the concept of water security, and its legal definition, is flawed and problematic.[31] While there is certainly a need for a more robust
examination of this topic, we would disagree with the position that attempting to
define the legal dimensions of water security is flawed or unhelpful. We would
agree that first attempts to define this concept have not been without difficulty and
states have perhaps chosen to interpret this concept, and in particular their
responses to the concept, in a way that suits their interests.

The case of the Nile shows us that the legal dimensions, appropriate responses
and measures to achieve water security need to be further clarified. Encouraging
states to address water security through a range of legal, policy, economic, scientific
and social mechanisms (moving away from traditional security responses), including through the utilization of international water law, particularly the UNWC,
should form the basis of such a response. Addressing water security on the Nile
should also focus on re-orienting the discourse to recognize the considerable
history of cooperation on the Nile, which appears now to have real potential for
deepening further. The third Nile Basin Development Forum, convened in Kigali
in October 2011, brought together high-level representatives from across the Nile
and resulted in an agreed Declaration and Recommendations enhancing joint
action around issues of climate change and sustainability – steps that provide a solid
(and growing) platform for cooperation based upon agreed processes and the
involvement of institutional mechanisms.[32] The meeting included the launch of
The Nile Basin Sustainability Framework, which identifies the broad range of institutional mechanisms responsible for implementing the framework.[33] From a water
security perspective, this approach of fostering regional dialogue is most welcomed
– it brings clarity to the roles of institutions and processes in addressing core issues,
and builds regional goodwill, trust and confidence, which form the foundation for
regional peace and security. Thus, our view is that the notion of water security, as
expressed in the CFA, should be seen to be consistent with the rule of equitable
and reasonable use, with its operation in practice determined through rules of
procedure (i.e. for development and management of the shared resource) and
through institutional mechanisms – as has been shown to evolve in positive ways
on the Nile.

It is now acknowledged that water is a security issue, but of a very unique kind
– water insecurity constrains the entire spectrum of securities: human, environmental, food, economic and energy. Thus, addressing water security at the national

---

31 Mekonnen (note 30), at 345, 361–2.
32 NBI, 'The 3rd Nile Basin Development Forum Closes in Kigali'. Available online at
    www.nilebasin.org/newsite/index.php?option=com_content&view=article&id=115%3Athe-3rd-
    nile-basin-development-forum-closes-in-kigali&catid=40%3Alatest-news&Itemid=84&lang=en
    (accessed March 29, 2013).
33 Ibid.

level requires embracing a range of interconnected issues that focus on minimizing risk by developing cooperative institutional, legal and political mechanisms to promote water security within and beyond sovereign boundaries as opposed to eliminating obstacles to gain control of scarce resources through military might.[34] Cooperation, through regional institutions, processes and rules of procedure – such as seems to be the case in many international basins across the globe, including the Nile – is the lynchpin of international water security.

The Water Security Analytical Framework: How international water law and the UNWC promote water security

The complex concept of water security, by definition, cuts across borders – sectoral, disciplinary, national and political. The WSAF is devised and continues to be developed to understand the concept of water security from an integrated perspective, with law as the central pivot, allowing for the coordination and unification of the surrounding disciplines and competing sectors. The WSAF is essentially an operational methodology for identifying the core legal elements required to address water security challenges across international watercourses. The three constituent elements of the WSAF are: (i) availability: a controlled supply of quality and safe water; (ii) access: enforceable rights to water for a range of stakeholders with priority for vital human needs and ecosystems; and (iii) addressing conflicts of use: where competing uses occur, mechanisms to avoid and/or address disputes.[35]

International water law contributes to understanding and addressing the notion of water security by: (i) defining and identifying legal rights and obligations tied to water use and provides the prescriptive parameters for resource development and management; and (ii) providing a framework to agree on processes and mechanisms that assist with ensuring the continuous integrity of the regime – that is, through cooperative platforms that encourage consultations and joint management procedures; exchange of information; ongoing monitoring and assessment of compliance and implementation of the treaty regime; promote dispute prevention, and peaceful settlement of disputes that may arise; and facilitate modifications of the existing regime, to be able to adapt to changing needs and circumstances.[36] The UNWC – a comprehensive framework treaty – provides rules and guidelines to achieve each of these objectives and thus supports the ability of states to achieve (or progress towards) water security. The UNWC provides also for the specific elements of WSAF – availability, access and addressing conflicts of use – which are reviewed in the next section.

---

34 The recent US Government Committee on Foreign Relations Report shows this paradigm shift. For more information, see US Government Committee on Foreign Relations (note 2), at 16.

35 For more details on this analytical framework, see Wouters, P., Vinogradov, S. and Magsig, B. O. 'Water Security, Hydrosolidarity and International Law: A River Runs Through It ...', *Yearbook of International Environmental Law*, 2008; 19: 97–134; Wouters, P. 'The International Law of Watercourses: New Dimensions', *Collected Courses of the Xiamen Academy of International Law*, 2010; 3: 349–541.

36 Wouters *et al.* (note 35), at 107.

The WSAF: Availability

From a legal perspective, the first element of the WSAF, availability, refers primarily to the geophysical aspects of water – the quantity, quality and control of the resource. These aspects of water resources management are regulated in law through rules related to the availability (quantity and quality) of water and associated flow regimes, in order to prevent or mitigate physical scarcity resulting from climatic or geographical factors, or unsustainable consumption and maintain the natural integrity of the water body. This includes laws and regulations that provide not only for reactionary mitigating measures but also for progressive protection-orientated rules to preserve ecosystems and enhance the availability of quality water, utilizing scientific understanding of hydrology and the complex relationship between land use, surface water and groundwater recharge, in order to avoid pollution and cope with the known and potential effects of new hydro-development, flooding or droughts. Quality and quality determinates are basin specific and involve a range of disciplinary expertise; an example of leading regional state practice in this field continues to evolve under the 1992 UN Economic Commission for Europe Convention on the Protection and Use of Transboundary Watercourses and International Lakes.[37] The Intergovernmental Panel on Climate Change report referred to availability quotients used as a measure to determine water-stressed basins.[38]

State practice of implementing different rules to achieve 'availability' continues to evolve[39] and the UNWC contains several articles supporting this element. The UNWC adopts a basin-wide approach to managing transboundary waters by designating the 'international watercourse' as the type of waters to which the UNWC applies (Articles 2 (a)–(b)), and highlights the need for an integrated approach to *systems* of surface and underground waters; this territorial scope recognizes the hydrological realities linked to the issue of the *availability* of water in a watershed.[40]

---

37   UNECE Convention on the Protection and Use of Transboundary Watercourses and International Lakes (adopted March 17, 1992, entered into force October 6, 1996) 1936 UNTS 269.

38   The IPCC defines water-stressed basins as those having either a per capita water availability below 1,000 m³ per year (based on long-term average runoff) or a ratio of withdrawals to long-term average annual runoff above 0.4. A water volume of 1,000 m3 per capita per year is typically more than is required for domestic, industrial and agricultural water uses.

39   Some examples of state practice include Article 1 of the 1996 Mahakali Treaty between Nepal and India concerning the Mahakali River, which stipulates that a specific quantity of water must be made available to Nepal during particular seasons. Treaty between His Majesty's Government of Nepal and the Government of India Concerning the Integrated Development of the Mahakali River including Sarada Barrage, Tanakpur Barrage and Pancheshwar Project (signed February 12, 1996, entered into force June 5, 1997) (1997) 36 ILM 531.

40   Discussion on the problems with the UNWC's inclusion of groundwater is discussed further in a separate chapter. For commentary, see International Law Commission, 'Draft Articles on the Law of the Non-Navigational Uses of International Watercourses and Commentaries thereto and Resolution on Transboundary Confined Groundwater' in ILC, 'Report on the Work of its Forty-sixth Session' (May 2–July 22, 1994) UN Doc A/49/10(Supp) (1994). Available online at http://untreaty.un.org/ilc/documentation/english/A_49_10.pdf (accessed March 29, 2013) (1994 ILC Report).

The UNWC contains rules setting out requirements related to quantity and quality issues. Article 20 codifies and develops the general obligation to protect and preserve the ecosystems of international watercourses where

> adequate protection' covers 'not only measures such as those relating to conservation, security and water-related disease, but also measures of 'control'... such as those taken to regulate flow, to control floods, pollution and erosion, to mitigate drought and to control saline intrusion.[41]

The obligation to protect ecosystems 'requires that watercourse States shield the ecosystems... from harm or damage'.[42] Article 21(2) obliges watercourse States to 'prevent, reduce and control the pollution of an international watercourse that may cause significant harm to other watercourse States or their environment'.[43]

Mitigation of water-related natural disasters is another element of 'availability', which can be covered by legal rules of emergency preparedness and response. Articles 27 and 28 of the UNWC provide for harmful conditions and emergency response. In this situation, watercourse states shall, per Article 28(2)–(3), 'without delay... notify other potentially affected States... of any emergency originating within its territory... and take all practicable measures... to prevent, mitigate and eliminate harmful effects of the emergency'. The Convention, in Articles 27–28, also encourages states to develop contingency plans for responding to emergencies – such an important consideration given the growing incidence and serious adverse impacts of floods and droughts. These rules set forth in the UNWC, considered together, provide a framework for managing issues related to the 'availability' issues linked to water security.

The WSAF: Access

The issue of 'access' concerns the extent to which existing and potential users are able to use the water resource which is available. At the international level, where the key actors are states, implementing the rights and duties related to access to water remains the responsibility of national governments, both externally (vis-à-vis other states) and internally (at the domestic level).[44] In a transboundary context, the discourse revolves around 'equitable and reasonable utilization', where allocation is concerned, and around the human right to water, in the specific area of human rights.[45]

---

41 Ibid., at 97.
42 Ibid., at 119.
43 This provision is clearly linked to the obligation contained in Article 7 to not cause significant harm, and is conditional on potential significant harm being caused to other riparian states. In addition, Article 23 includes the obligation to protect and preserve the marine environment.
44 The water security framework continues to examine the interface between national and international law, because the integration is vital to achieving water security. For further discussion, see Abseno, M. *How Can the Work of the General Assembly and the ILC on the Law of the Non-Navigational Uses of International Watercourses Contribute towards Basin-Wide Legal Framework for the Nile Basin* (University of Dundee, 2009).
45 UN General Assembly (UNGA), Resolution 64/294 'Right to Water and Sanitation' (August 3, 2010) UN Doc A/RES/64/294.

How are these issues of 'access' addressed in international law? In essence, the matter of access is dealt with through the rules for water allocation, reallocation, use and transfer of the water resources of shared international fresh waters. In the first instance, basic rights ensuring access to water for human survival and for ecosystems are protected through the governing international water law rule of 'equitable and reasonable use'. This rule of law, confirmed in both treaty and customary law, is codified in Article 5 of the UNWC and determines the right of a state to use the transboundary water resources in two distinct ways.[46] First, it lays out the objective to be achieved (an equitable and reasonable use), which then determines the legitimacy of the new or increased utilization. Second, it entails an important operational function, since it requires that all relevant factors (including scientific, economic and social factors) and circumstances to be identified and duly considered when evaluating the legality of the proposed new or increased use (i.e. does it qualify as an equitable and reasonable use?).

What level of access is to be enjoyed? The rule of equitable and reasonable use does not require equal water allocation or equal benefit sharing among all watercourse states – such a strict prescription is not what is meant by the concept. Instead, as provided for under Article 6 of the UNWC, all relevant factors are to be considered together and a decision made on the basis of the whole. Article 6 provides a non-exhaustive list of factors to be considered in this regard; each factor is of equal weight, with a priority of use given to vital 'human' and 'environmental' needs, under Articles 10 and 23 of the UNWC.[47] Considering all relevant factors is central to the operation of the governing rule of equitable and reasonable use and provides inherent relevance of process. Article 6 makes it clear that the application of these principles is not an isolated step, but is a process to be carried out continuously in the context of the overall cooperative arrangements among watercourse states – an ongoing assessment of changing circumstances and potential revision of allocation rights to bring them back in conformity with the principle of equitable and reasonable use. While the UNWC has its critics and some call for increased clarity around the governing rule of equitable and reasonable use,[48] it is clear that it provides a flexible approach that recognizes and addresses the evolving nature of the development of the uses of international watercourses. Importantly, the rule is supported by a body of procedural rules that assist with the cooperative management of the basin.

Water security concerns relating to *access* arise also within the context of changing environmental conditions, including the needs of ecosystems and the

---

46   Article 5(1) of the UNWC determines that 'watercourse States shall in their respective territories utilize an international watercourse in an equitable and reasonable manner. In particular, an international watercourse shall be used and developed by watercourse States with a view to attaining optimal and sustainable utilization thereof and benefits there from, taking into account the interests of the watercourse States concerned, consistent with adequate protection of the watercourse'.

47   Article 10 establishes that, in the case of conflicting uses across international borders, watercourse states must give special regard to vital human needs in solving such a conflict.

48   McIntyre, O. *Environmental Protection of International Watercourses under International Law* (Ashgate, 2007), at 155–89.

(re)allocation requirements to be met in order to deal with fluctuations in precipitation, droughts and floods and emergency situations, under which changing circumstances must be identified and evaluated in order to determine the lawfulness of various uses. Existing international practice shows that states rely upon rules of procedure and the establishment of joint institutional mechanisms to provide for adaptability of allocation and access rights in these circumstances.

The UNWC provides these rules, including the duty to give prior notice of new or increased uses (Articles 12–16 and 18–19),[49] consult on changes to the regime (Article 17), and to exchange information on a regular basis (generally), and in the event of planned measures (more specifically) (Article 11). This is generally accomplished through the involvement of a joint institutional mechanism, such as a river basin organization or Meeting of the Parties.[50] One shortcoming of the UNWC on this point is that it does not require the mandatory establishment of such joint institutions which could be considered a shortcoming in its ability to provide for the 'access' element of water security.

Information sharing and regular and meaningful communication form the bedrock of cooperation – a global study of 287 transboundary water agreements found that close to 90 percent contain formal communication mechanisms, including procedures for prior notification; only 40 percent of the treaties surveyed included make direct reference to exchange of water resources data and information, leaving a large gap to be filled, but providing a solid example of consistent state practice in this regard.[51] The propensity to exchange data and information appears to find greater favor in multilateral agreements, with just over 45 percent having a data and information exchange provision (compared with only 35 percent of bilateral agreements).[52]

---

49 These provisions cover a range of issues – from timing of notification, response to notification and what should occur in the absence of notification, or where there is need for urgent implementation of planned measures.

50 Article 24 provides: 'Watercourse States shall, at the request of any of them, enter into consultations concerning the management of an international watercourse, which may include the establishment of a joint management mechanism. For the purposes of this article, "management" refers, in particular, to: (a) Planning the sustainable development of an international watercourse and providing for the implementation of any plans adopted; and (b) Otherwise promoting the rational and optimal utilization, protection and control of the watercourse'.

51 A study that examined all available transboundary water agreements signed between 1900 and 2007 to determine the degree to which water resources data and information is exchanged in the world's regions, how the level of exchange has developed over time and the different ways in which data and information sharing has been codified in practice concluded that data and information exchange is now more widespread than may commonly be recognized. The results revealed key linkages between democracy and data and information exchange. The study predicted 'exchange provisions to be present in more democratic basins, where common values are expected to reduce transaction costs in seeking solutions to collective action problems'. Gerlak, A., Lautze, J. and Giordano, M. 'Water Resources Data and Information Exchange in Transboundary Water Treaties', *International Environmental Agreements: Politics, Law and Economics*, 2011; 11: 179–99, at 185.

52 Ibid., at 192.

The WSAF: Addressing conflict of use

Where all of the demands for use of the fresh water resources of an international watercourse cannot be satisfied (for reasons of quantity and/or quality), then a conflict of use occurs. Under the WSAF, this scenario is dealt with through the suite of mechanisms that address disputes, from dispute prevention to dispute resolution. The UNWC responds to this element of water security in two ways: first, by including a number of provisions that support dispute prevention (generally); and second, through its inclusion of dispute resolution procedures, which include traditional methods and also includes the fact-finding approach. The UNWC embraces dispute prevention by promoting cooperation and thus avoiding conflict, throughout the text. Article 8 is the central legal provision on cooperation, but numerous other articles also call for cooperation, e.g. the obligation to regularly exchange data and information under Article 9 directly flows from the general obligation to cooperate in Article 8 of the Convention. Additionally, cooperation is included in the provisions encouraging the establishment of joint institutions (Articles 4 and 8(2)) and through the procedures for consultation and negotiation under Planned Measures (Articles 11–19). Cooperation is also essential to equitable participation under Article 5.

Article 33 of the Convention offers mechanisms for dispute resolution for watercourse states, including negotiation, good offices, mediation, conciliation, joint watercourse institutions or submission of the dispute to arbitration or to the International Court of Justice. These procedures require the consent of all parties concerned, consistent with the rules of international law, generally. Where these methods of dispute settlement are unsuccessful, the Convention provides for unilateral action – any watercourse state that is party to the dispute can unilaterally invoke the compulsory fact-finding procedure provided for under Article 33(3)–(9). The disputing states are required to consider the recommendations of the fact-finding commission in good faith, but these are not binding. Thus, through a considerable portfolio of provisions, the UNWC offers a range of opportunities for the peaceful management of international watercourses, including instances where disputes might arise. From a water security perspective, this provides a peaceful mechanism for addressing conflicts over water, offering a platform, instead, for cooperation.

Water security and the UNWC: Complementary or contradictory?

In considering the relative synergies (or lack thereof) between the emerging concept of water security and the UNWC, several questions can be posed:

*   How could the concept of water security affect the influence of the UNWC?
*   Is the concept of water security consistent with the rules espoused by the UNWC?
*   What are some of the challenges remaining in this field – given the emergence of the concept of water security and the current non-entry into force of the UNWC?

While the UNWC does not define or refer to the concept of water security, this should not be considered a shortcoming on either side of the equation. If we examine this fact further, we can see that the UNWC, as a framework instrument, leaves considerable latitude to watercourse states to draft their own agreements that 'apply and adjust' its provisions (Article 3). The UNWC contains numerous provisions that address the cornerstones of water security – availability, access and addressing conflicts of use – but challenges remain. The closed definition of 'international watercourse' adopted under the UNWC (Article 2) limits its reach in terms of 'availability' issues; thus, issues related to confined aquifers (which are important sources of fresh water) remain unregulated under this instrument. The ILC has sought to respond to this outstanding issue of confined transboundary groundwater in the 2008 ILC Draft Articles on the Law of Transboundary Aquifers,[53] and although these Draft Articles do provide coverage of confined transboundary aquifers there are many criticisms of the scope and substantive principles of this work (e.g. it does not cover domestically located confined aquifers even if they are remotely connected to a transboundary surface water resource).[54] A further concern relates to the state-centric approach adopted by the UNWC, compared with the more progressive approach adopted under the UN Framework Convention on Climate Change (UNFCCC). This state-centric focus raises the hackles of state sovereignty.[55] Ensuring 'water for all', whereby access to available water is agreed on a basin-wide basis (i.e. a water-secure approach), must transcend national borders and could be based upon 'regional common concern'.[56] It remains to be seen whether the UNWC goes far enough in supporting such an orientation – which seems watered down in its provision on the duty to cooperate as expressed in Article 8. Nonetheless, we do welcome this formal recognition of the duty to cooperate, which is now supported by the UN Resolution declaring 2013 the International Year of Water Cooperation.[57]

The global common concern approach fostered in the UNFCCC might be an alternative approach – but one must question the transaction costs and actual benefits from a fresh start in negotiating a new universal instrument in the transboundary watercourse arena. Given the vast scope of the global water challenge and the numerous inter-linkages with other crises (e.g. energy, food and health), it is possible to develop an analogous understanding for transboundary

53　ILC, 'Draft Articles on the Law of Transboundary Aquifers, with Commentaries' in ILC, 'Report of the Work of its 63rd Session' (May 5–June 6 and July 7–August 8, 2008) UN Doc A/63/10 (Supp) (2008) (2008 ILC Report).

54　McIntyre, O. 'International Water Resources Law and the International Law Commission Draft Articles on Transboundary Aquifers: A Missed Opportunity for Cross-Fertilization?', *International Community Law Review*, 2011; 13: at 237–54.

55　Falk, R. 'The Coming Global Civilization: Neo-Liberal or Humanist?', in Anghie, A. and Sturgess, G. (eds), *Legal Visions of the 21st Century: Essays in Honour of Judge Christopher Weeramantry* (Kluwer Law International, 1998), at 15.

56　Magsig (note 10), at 341.

57　UNGA, Resolution 65/154 'International Year of Water Cooperation, 2013' (February 11, 2011) UN Doc A/RES/65/154.

freshwater management. However, there is no evidence of an international consensus on whether water security is of 'global common concern'. It would be better to scale down and focus on implementation of the current body of rules and procedures at the regional (basin-wide) level, which is an avenue promoted intrinsically by the UNWC. Considering water security as a 'regional common concern' could strengthen the emergence of choate rules of international law in this field and thus counterbalance state-centered behavior focused on sovereign self-interest.[58] This new perspective requires further examination and should include deeper enquiry into the role of law in this field, including the contribution that the UNWC might play in the evolution of the concept of collective security. Such an examination may also expose some of the shortcomings of international water law and, at the same time, provide the catalyst for international water law to adapt and better address the challenges posed by water security and global change.

Addressing water security at the regional level is being suggested to Central Asia by external countries, although this has not yet translated into acceptance by the states themselves. A US Foreign Policy Report on water scarcity and water management in Central and South Asia highlighted the importance of water security across this region,[59] and a related report confirms that water security in this region and other regions has been explicitly identified as a security interest to the USA.[60]

> The national security implications of this looming water shortage – exacerbated and directly caused by agriculture demands, hydroelectric power generation, and climate instability – will be felt all over the world.... [We] have recognized the threat of conflict stemming from ineffective water management.[61]

The report called for a quickening pace to focus efforts on the provision of water security to achieve USA foreign policy objectives and directly acknowledged the vital role of international law in this task, stating that

> the USA government calls for an increased focus on supporting programmes that build the institutional capacity of government agencies and universities in areas such as international water law, dispute resolution, mediation, and arbitration and greater investment in institutions that support developing transboundary water sharing agreements.[62]

---

58   Magsig, B. O. 'Rising to the Challenge of Water Security: International (Water) Law in Need of Refinement', *International Journal of Sustainable Society*, 2012; 4: 28–44, at 38.
59   See US Government Committee on Foreign Relations (note 2).
60   National Intelligence Council, *Global Water Security: Intelligence Community Assessment* (Office of the Director of National Intelligence, 2012).
61   Ibid., at 10.
62   Ibid., at 3.
63   Resolution 65/154 (note 57).

Water security is a global issue, connecting regions and an interdependent global community.

Conclusion

We live in uncertain times. The critical importance of water to every facet of our lives, including the sustenance of ecosystems, cannot be overstated. International watercourses crisscross sovereign nations around the globe and offer opportunities for regional cooperation, peace and security, as recognized by the UN, which has declared 2013 the International Year for Water Cooperation.[63] While the particular contours of water security have yet to be fully identified and examined, the concept has emerged as an approach which attempts to comprehend the complexity and interconnected nature of the global water challenge. This chapter has identified three constituent elements of water security – availability, access and addressing conflicts of use – as the hallmark parameters of this new approach. The UNWC provides a framework instrument that is consistent with the cornerstone elements of water security, as presented in this chapter. However, the UNWC is a framework instrument – it remains for nation states to agree to basin-wide agreements that meet regional water security imperatives. Such an approach, based on the duty to cooperate, which is at the heart of international relations, would articulate more clearly the rights and obligations of watercourse states as they work together to meet needs related to availability, access and addressing conflicts of use on a case-by-case basis.

28 Transboundary water interactions and the UN Watercourses Convention

Allocating waters and implementing principles[1]

Naho Mirumachi, Mark Zeitoun and Jeroen Warner

In the 1990s and 2000s, institution building for the management and governance of international transboundary waters has become a prominent agenda item both at the basin and global level. There has been an increase in the number of international agreements over shared waters.[2] Global water policy initiatives, including international organizations such as the Global Water Partnership and forums such as the World Water Week, have also increased in number,[3] with many focusing on transboundary issues. In the process of institution building, many discussions and deliberations between basin states and international organizations refer to legal principles. The purpose of this chapter is to explore and discuss the political context in which legal principles – such as those enshrined in the UN Watercourses Convention (UNWC) – are acknowledged and implemented in the allocation and management of shared waters. The chapter argues that an assessment of the way in which such principles may be used for improved water resources management is possible only through an understanding of the context and nature of transboundary interactions between sovereign basin states.

The chapter first clarifies base assumptions, taking the perspective that legal principles inform the institutional structure of transboundary water management and governance through political processes. These processes help to explain why legal principles are effective or ineffective. In this regard, the concept of transboundary water interaction as one of coexisting conflict and cooperation is

---

1 This chapter draws upon previous works by the authors, notably Zeitoun, M. and Mirumachi, N. 'Transboundary Water Interaction I: Reconsidering Conflict and Cooperation', *International Environmental Agreements: Politics, Law and Economics*, 2008; 8: 297–316; and Zeitoun, M., Mirumachi, N. and Warner, J. 'Transboundary Water Interaction II: Soft Power Underlying Conflict and Cooperation', *International Environmental Agreements: Politics, Law and Economics*, 2011; 11: 159–78.

2 Gerlak, A. K. and Grant, K. A. 'The Correlates of Cooperative Institutions for International Rivers', in Volgy, T. J. Šabič, Z., Roter, P. and Gerlak, A. K. (eds), *Mapping the New World Order* (Wiley-Blackwell, 2009), at 114–47.

3 Varady, R. G., Meehan, K., Rodda, J., McGovern, E. and Iles-Shih, M. 'Strengthening Global Water Initiatives', *Environment: Science and Policy for Sustainable Development*, 2008; 50 (2): 18.

introduced. Second, power asymmetry and 'hydro-hegemony' are introduced as a key lens to help interpret how legal frameworks, such as that of the UNWC, are subject to deliberations in which 'soft' power is utilized sometimes alongside 'harder', coercive power. Third, in such asymmetric political conditions, the role of UNWC principles is examined using examples from the Ganges-Brahmaputra-Meghna river system, Nile and Euphrates-Tigris River Basins. The chapter concludes with a summary argument and policy implications, notably the challenge of applying UNWC principles at the basin level.

Transboundary water interaction at the interface of international politics and law

Legal frameworks, such as the UNWC, are part of the institutional structure of transboundary water resources management and governance between basin states. Basin-wide agreements for river basin management, bilateral agreements on specific water resources development projects and river basin organizations also make up the institutional structure. A constructivist perspective of international relations sees this institutional structure as constructed by actors such as basin states and international organizations focusing on water resources governance, as well as donor agencies funding river basin management or development projects. This institutional structure can in turn influence the kinds of actions taken by the states.[4] Reus-Smit has argued that examining the interface of international politics and international law is important to understand how legal principles form institutions.[5] With this argument underpinning this chapter's analysis, the main focus is to examine, not the legal framework or the individual agreements between basin states, but the transboundary water interactions that establish and shape the institutions.

Thanks to the groundbreaking development of the Transboundary Freshwater Dispute Database,[6] studies have identified and analyzed specific events relating to water allocation, development and management in international river basins around the world. However, if we are to understand the interface of politics and legal institutions, focus on the political process of deliberation and negotiation, or on transboundary water interaction, is the crucial next step.

As Zeitoun and Mirumachi pointed out, interaction between states over international transboundary river basins has typically been described as one of either conflict *or* cooperation.[7] Factors that induce acute conflict or cooperation have been examined. For example, quantitative water scarcity has often been associated

---

4 Finnemore, M. J. and Toope, S. 'Alternatives to "Legalization": Richer Views of Law and Politics', *International Organization*, 2001; 55: 743–58.

5 Reus-Smit, C. 'The Politics of International Law', in Reus-Smit, C. (ed.), *The Politics of International Law*, Cambridge Studies in International Relations, No. 96 (Cambridge University Press, 2004), at 14–44.

6 'Transboundary Freshwater Dispute Database' (Oregon State University). Available online at www.transboundarywaters.orst.edu/database (accessed March 29, 2013).

7 Zeitoun and Mirumachi (note 1), at 298.

with conflict, as seen in the media hype of 'water wars'.[8] However, a broader perspective on the factors of conflict or cooperation has been discussed, including geographical conditions, economic interdependency and the type of political regime.[9] Moreover, policy-oriented studies have indicated that different types of benefits and disbenefits from the use and management of shared waters can lead to conflict and cooperation.[10] However, the conceptual foundation of these studies relies on a linear juxtaposition of conflict and cooperation as two main states of interaction over international transboundary waters. Put differently, conflict and cooperation are seen as opposite ends of a linear scale, which leads to the status of interaction over international transboundary waters being described as either conflictive or cooperative, or more conflictive than cooperative (and vice versa). For example, Sadoff and Grey proposed a linear progression of the state of basins from unilateral action to coordination, to collaboration and ultimately to joint action.[11] The Transboundary Freshwater Dispute Database also uses a linear 15 point scale to describe events relating to water in shared basins from 'formal declaration of war' to 'voluntary unification into one nation'.[12]

There are also studies which attempt to provide more nuance to this conceptualization. For example, Zawahri added a third category – 'unstable cooperation' – as one between conflict and cooperation to exemplify how negotiations over water resources may lead to adverse political situations.[13] To illustrate that interstate cooperation is a means to ends of issues beyond environmental ones (not to mention water allocation), basin states may engage in 'coerced cooperation'[14] or 'tactical functional cooperation'.[15]

While these analyses begin to unpack the politics of water resources management at the international level, the linear conceptualization of conflict and cooperation poses two main analytical pitfalls, with subsequent implications for

8 Vidal, J. 'Water Wars Loom as Demand Grows' (*The Guardian*, June 26, 2010).

9 Song, J. and Whittington, D. 'Why Have Some Countries on International Rivers Been Successful Negotiating Treaties? A Global Perspective', *Water Resources Research*, 2004; 40: W05S06; Gerlak and Grant (note 2); Tir, J. and Ackerman, J. T. 'Politics of Formalized River Cooperation', *Journal of Peace Research*, 2009; 46: 623–40.

10 Sadoff, C. W. and Grey, D. 'Beyond the River: The Benefits of Cooperation on International Rivers', *Water Policy*, 2002; 4: 389–403; Phillips, D., Daoudy, M., McCaffrey, S., Ojendal, J. and Turton, A. *Trans-boundary Water Cooperation as a Tool for Conflict Prevention and for Broader Benefit-Sharing: Prepared for the Ministry for Foreign Affairs, Sweden* (EGDI Secretariat, Ministry for Foreign Affairs, 2006).

11 Sadoff, C. W. and Grey, D. 'Cooperation on International Rivers: A Continuum for Securing and Sharing Benefits', *Water International*, 2005; 30: 420–4.

12 Yoffe, S. B., Wolf, A. T. and Giordano, M. 'Conflict and Cooperation over International Freshwater Resources: Indicators of Basins at Risk', *Journal of the American Water Resources Association*, 2003; 39: 1109–26, at 1112.

13 Zawahri, N. A. 'Capturing the Nature of Cooperation, Unstable Cooperation, and Conflict over International Rivers: The Story of the Indus, Yarmouk, Euphrates, and Tigris Rivers', *International Journal of Global Environmental Issues*, 2008; 8: 286–310.

14 Weinthal, E. *State Making and Environmental Cooperation: Linking Domestic and International Politics in Central Asia* (MIT Press, 2002), at 35.

15 Sosland, J. K. *Cooperating Rivals: The Riparian Politics of the Jordan River Basin* (State University of New York Press, 2007), at 9.

policy particularly when concerned about the interface between politics and international law. The first analytical pitfall of this conceptualization is that the scale aggregates environmental conditions, political relationships and histories, economic capacities, geographical and hydrological features and institutions into incremental measurements. Such aggregation may oversimplify the more complex reality, and importantly, fall short of reflecting how negotiations and deliberations over transboundary water resources do not progress in a linear fashion, and certainly not equally on all issues.

The second pitfall of the 'either/or' approach is the inherent normative assumption about conflict and cooperation. The conflict–cooperation scale associates the state of conflict as undesirable and the state of cooperation as desirable. By implicitly assuming that cooperation is the sole target, or in some way 'better' than non-cooperation, analysis obfuscates whether the cooperative arrangement is actually improving water governance[16] and/or the stated goal of the arrangement. The 'either/or' conceptualization may evaluate the legal framework as a successful output, but not scrutinize whether it is effectively contributing to the sustainable management of shared waters, obscuring the understanding of the interface between politics and institutional structures.

The policy implication resulting from use of this conflict-cooperation scale should be considered. Policy derived from the 'either/or' analysis may only focus on: (a) addressing conflict without specifying what cooperation would entail for the specific basin; or (b) supporting cooperative initiatives without addressing the issues that may be the source of conflict and tensions. For example, the UN Development Programme (UNDP) published a report that asserted the importance of 'promot[ing] and support[ing] cooperation of any sort, no matter how slight'.[17] This approach risks overlooking the fundamental reason water resources have come on to the international agenda between the basin states. Thus, even if policy promotes the principles of the UNWC, such as equitable and reasonable utilization, no significant harm and prior notification, as a guideline for cooperation, discussion on why and how these principles will address the fundamental issues of water allocation and sustainable water resources use in a specific project or specific case may be sidestepped. As such, a policy initiative that undermines the political context in which cooperation or conflict resolution is promoted does not guarantee optimal, improved water use and allocation. It may, in fact, simply serve to delay or perpetuate the conflict – as in the Nile River Basin, as we shall see.

The interface of international politics and international law over shared waters may be more usefully examined through the understanding of *coexisting conflict and cooperation*. Within transboundary water interaction, conflict and cooperation coexist in differing intensities. For example, there can be interstate interaction where

16 Kistin, E. J. and Phillips, D. 'A Critique of Existing Agreements on Transboundary Waters, and Proposals for Creating Effective Cooperation between Co-riparians' (Third International Workshop on Hydro-Hegemony, London, May 12–13, 2007).
17 UNDP, *Human Development Report 2006: Beyond Scarcity: Power, Poverty and the Global Water Crisis* (Palgrave Macmillan, 2006), at 228.

low intensity conflict coexists alongside high intensity conflict, as in a case where there is an international river basin agreement in place but dispute over parameters of a particular river development project. Conflict does not exclude cooperation and vice versa. While it is not within the scope of this chapter to describe them in detail, new approaches have been proposed to examine coexisting conflict and cooperation. The Transboundary Waters Interaction Nexus provides a historical analysis of the process in which basin states have engaged in both conflict and cooperation over time, thus painting a more nuanced and broader picture of trans-boundary water interactions within a basin.[18] The perspective of coexisting conflict and cooperation helps to analyze situations where institution-building efforts are made but without much change in the way water is allocated.

Hydro-hegemony in transboundary water interactions

The focus on transboundary water interaction between states reveals that the polit-ical process of developing an institutional structure for water resources management and governance is influenced by actor-specific characteristics. While material conditions such as physical scarcity of water indeed matter, the types of rules and norms (that is, the institutions) that are deliberated, negotiated and accepted are shaped by the basin states themselves. Zeitoun and Warner have argued that power plays an important role in this political process.[19] The authors used the term 'hydro-hegemony' to show that different types of power can be used to determine the process and outcomes of negotiations regarding water resources management and governance. Within basins where there are clear hegemonic rela-tionships,[20] a state that has relative power over others, or 'the hydro-hegemon', can enable preferential water allocation through three main strategies: resource capture, integration and containment. The ways in which these strategies are achieved is through a combination of riparian position, capacity for hydraulic development and, importantly, different types of power: coercive (or 'hard') power, bargaining (i.e. agenda-setting power) and ideational (i.e. power over ideas). It is this last aspect of power in particular that concerns this chapter.

18 See Sojamo, S. 'Illustrating Co-Existing Conflict and Cooperation in the Aral Sea Basin with TWINS Approach', in Rahaman, M. M. and Varis, O. (eds), *Central Asian Waters: Social, Economic, Environmental and Governance Puzzle* (Water and Development Publications, Helsinki University of Technology, 2008), at 75–88; Zeitoun and Mirumachi (note 1); Warner, J. and van Buuren, A. 'Multi-Stakeholder Learning and Fighting on the River Scheldt', *International Negotiation*, 2009; 14: 419–40; Allan, J. A. and Mirumachi, N. 'Why Negotiate? Asymmetric Endowments, Asymmetric Power and the Invisible Nexus of Water, Trade and Power that Brings Apparent Water Security', in Earle. A,, Jägerskog, A. and Öjendal, J. (eds), *Transboundary Water Management: Principles and Practice* (Earthscan, 2010), at 13–26; Mirumachi, N. 'Study of Conflict and Cooperation in International Transboundary River Basins: The TWINS Framework' (PhD thesis, King's College, London, 2010).
19 Zeitoun, M. and Warner, J. 'Hydro-hegemony: A Framework for Analysis of Trans-boundary Water Conflicts', *Water Policy*, 2006; 8: 435–60.
20 As opposed to contexts where there is considerably more parity on paper and in practice, such as the EU, or contexts where formal parity is not even pretended, such as those established by impe-rial countries upon their empire members – see Zeitoun *et al.* (note 1).

An example of coercive power influencing transboundary interstate interaction is the military force to secure dams and forcibly allocate water resources. Bargaining and ideational power, which can be loosely described as 'soft' power, become useful in deliberations and negotiations of water resources allocation and management. Coercive power such as economic sanctions are used, but 'soft' power is an equally, if not more, effective way of ensuring compliance through the use of discourse and development of ideas. 'Soft' power is employed during formal and informal negotiations, declarations, media reports and rumors, and through the development of persuasive 'storylines'[21] that justify the maintenance of the status quo in the way water resources are allocated or managed.

More specifically, 'soft' power can be used for distributive or integrative ends.[22] Both types of arrangements entail compliance from the other states to the hydro-hegemon's pre-eminence, rule or wishes. However, 'soft' power exercised for distributive ends is based upon resistance of or resignation after resistance of the non-hydro-hegemon to the constructed understanding of the status quo of water allocation and management. On the other hand, 'soft' power exercised for integrative ends is based upon the non-hydro-hegemon's explicit or implicit *consent* to the proposed water allocation and management. Consent is achieved in this case because the non-hydro-hegemon seemingly benefits, or at least is not worse off compared to challenging the hydro-hegemon.[23] Whether distributive or integrative, 'soft' power can change intensities of coexisting conflict and cooperation. Moreover, it can shape the institutional structure through the adoption of certain norms and principles and the rejection of certain rules to suit the interests of the hydro-hegemon.

The Ganges-Brahmaputra-Meghna river system in South Asia is useful to illustrate the use of 'soft' power in hegemonic transboundary interactions. There have been landmark agreements on river basin development between the basin's main states: Nepal, India and Bangladesh. Project specific bilateral agreements between India and Nepal have existed since the 1950s. The year 1996 marked a significant milestone in the institutionalization of water resources management in this basin. The Mahakali Treaty between India and Nepal formally acknowledged the 'equal partnership' of the two basin states to develop tributary waters for mutual economic benefit.[24] In the same year, the Ganges Treaty formally established water allocation between upstream India and downstream Bangladesh. Here, it is clear that India has been the central player in the institutionalization of the shared waters in the South Asian region. For India, coercive power was not the main means to

---

21 Hajer, M. A. *The Politics of Environmental Discourse: Ecological Modernization and the Policy Process* (Oxford University Press, 1995).
22 Zeitoun *et al.* (note 1).
23 Haugaard, M. and Lentner, H. H. *Hegemony and Power. Consensus and Coercion in Contemporary Politics* (Lexington Books, 2006).
24 Treaty between His Majesty's Government of Nepal and the Government of India Concerning the Integrated Development of the Mahakali River including Sarada Barrage, Tanakpur Barrage and Pancheshwar Project (signed February 12, 1996, entered into force June 5, 1997) (1997) 36 ILM 531, at Preamble.

achieve development on the rivers; the consensual means of establishing bilateral agreements has proved useful.

While the integrative use of 'soft' power has established cooperative forms of interaction between the basin states, there are, of course, elements of coexisting conflict. A feature of the institution-building in the Ganges-Brahmaputra-Meghna river system is the development of bilateral treaties as opposed to basin-wide treaties. India has selectively signed bilateral agreements with its upstream and downstream neighbors, and avoided the establishment of a formal basin-wide institutional structure.[25] These bilateral agreements have tended to benefit India, to the extent that in the case of Indo-Nepal relations, agreements have been renegotiated. Specifically, the Kosi and Gandak Agreements have been so controversial that clauses were amended to ensure a more explicit acknowledgement of Nepal's share of economic benefit from hydraulic infrastructure development. Moreover, these bilateral agreements have different approaches to water resources allocation and management. For example, the Mahakali Treaty between India and Nepal specifies an integrated approach to water resources management, while the Ganges Treaty between India and Bangladesh is narrower in scope, focusing on water allocation.[26]

The South Asian example illustrates that the basin state with the relative power advantage has the choice in determining when to sign up to cooperative initiatives, what kind of institutional structure to promote and who to include and exclude in such institutional structure. This is not to say that the non-hydro-hegemonic states are without any means to challenge the actions of the hydro-hegemon. For example, in an attempt to open up more opportunities for basin-wide dialogue, representatives from Pakistan, Nepal and Bangladesh and India have engaged in track two diplomacy or unofficial deliberative platforms between non-governmental organizations.[27] What is important to bear in mind, however, is that the power asymmetry between basin states enables the hydro-hegemon to open up and close down discursive spaces for achieving more sustainable water use or challenging the status quo of water allocation.[28] Sovereign states of an international river basin are recognized as being equal under international law, but some of their actions may occur as if unseen by it. Within hegemonic political conditions, certain basin states may thus act as 'first among equals' in pursuing their interests in water resources management and governance, selectively utilizing or ignoring the implementation of international law principles such as those of the UNWC, as we shall

25 Crow, B. and Singh, N. 'Impediments and Innovation in International Rivers: The Waters of South Asia', *World Development*, 2000; 28: 1907–25; Brichieri-Colombi, S. and Bradnock, R. W. 'Geopolitics, Water and Development in South Asia: Cooperative Development in the Ganges-Brahmaputra Delta', *Geographical Journal*, 2003; 169: 43–64.

26 Salman, S. M. A. and Uprety, K. 'Hydro-Politics in South Asia: A Comparative Analysis of the Mahakali and the Ganges Treaties', *Natural Resources Journal*, 1999; 39: 295–9.

27 Nishat, A. and Faisal, I. M. 'An Assessment of the Institutional Mechanisms for Water Negotiations in the Ganges-Brahmaputra-Meghna System', *International Negotiation*, 2000; 5: 289–310; Swain, A. 'Environmental Cooperation in South Asia', in Conca, K. and Dabelko, G. D. (eds), *Environmental Peacemaking* (Woodrow Wilson Center Press and Johns Hopkins University Press, 2002), at 61.

28 Zeitoun *et al.* (note 1).

see. This understanding of the 'hegemon's prerogative' helps with the interpretation of the effectiveness of institutional structures for water resources management and governance and the role of international water law principles, as is shown in the next section.

The role of legal principles in asymmetric transboundary water interactions

Zeitoun and Jägerskog have suggested that in hegemonic political contexts, there are two ways of achieving sustainable and equitable water resources use and management: leveling the grounds for deliberation and negotiation, and leveling the capacity of states to influence water resources management and governance.[29] The former specifically points to ensuring discursive space that allows interests of both the hydro-hegemons and non-hydro-hegemons to be expressed and deliberated. International law and legal frameworks, such as the UNWC, have been suggested as an effective way of 'levelling the playing field'.[30] This is because the UNWC highlights key principles that inform the foundations of an institutional structure for water resources management and governance: equitable and reasonable utilization, no significant harm and prior notification (and the concept of 'shared' or 'limited' sovereignty that accompanies them).[31] While the ratification of the UNWC by member states has been slow and it has yet to come into force, the UNWC has in some cases had the effect of underlining customary international law in the context of shared natural resources (see Chapters 2 and 5 of this book).

Three countries voted against the UNWC at the UN General Assembly in 1997: China, Turkey and Burundi. While geography is not the sole determinant of water allocation,[32] the interests expressed through votes of each of these upstream states (in the Mekong River Basin, Euphrates-Tigris River Basin and the Nile River Basin, respectively) reflects their concerns of being disadvantaged through changed water-use allocation based on the UNWC's principles. Put differently, existing water allocation may be challenged and upstream plans for water resources development may be criticized as potentially harmful downstream.[33]

For Turkey, water resources have been perceived as a way of becoming more self-sufficient by developing the hydropower capacity of the Euphrates-Tigris River

29 Zeitoun, M. and Jägerskog, A. 'Confronting Power: Strategies to Support Less Powerful States', in Jägerskog, A. and Zeitoun, M. (eds), *Getting Transboundary Water Right: Theory and Practice for Effective Cooperation* (Stockholm International Water Institute, 2009), at 9.

30 Ibid.

31 McCaffrey, S. C. *The Law of International Watercourses* (2nd edn, Oxford University Press, 2007); McIntyre, O. 'International Water Law: Concepts, Evolution and Development' in Earle *et al.* (note 18).

32 Zeitoun, M. and Warner, J. 'Hydro-hegemony: A Framework for Analysis of Trans-boundary Water Conflicts', *Water Policy*, 2006; 8 (5): 435–60.

33 Salman, S. M. A. 'The UN Watercourses Convention Ten Years Later: Why Has its Entry into Force Proven Difficult?', *Water International*, 2007; 32: 1–15.

Basin. The Southeast Anatolia Project (*Güneydoğu Anadolu Projesi* – GAP), consisting of 22 dam projects in the southeast of Turkey, is a prime example of
infrastructure development for securing water and hydropower. Successive Turkish
governments deny their interventions are doing harm downstream, maintaining
that Syria has plenty of water resources but which are not managed effectively,[34]
and that Turkish projects in fact regulate flood risk. The government of Turkey cited
Article 7 of the UNWC regarding 'no significant harm' as the reason for its objection to the Convention.[35] Like Egypt, Turkey claimed it would not accept the
formation of international custom on the basis of the UNWC.[36] Successive Turkish
administrations have maintained that the Euphrates and Tigris are Turkish rivers, as
they originate in Turkey, and therefore argued that Turkey has absolute sovereignty
over them. From these Turkish administrations' perspective, the implicit endorsement of 'limited sovereignty' on transboundary water resources utilization by the
UNWC is at odds with its perception of the Euphrates-Tigris waters.

Even if states have voted in favor of the UNWC and international agreements
incorporate those principles, power asymmetry acting upon coexisting conflict and
cooperation has ensured that their implementation has been a major challenge. For
example, drawing on the aforementioned example of Ganges-Brahmaputra-
Meghna river system, it has been argued that the UNWC has not been applied by
the basin states to provide a coherent understanding of equitable and reasonable
utilization.[37] The Mahakali Treaty between Nepal and India does incorporate the
notion of equitable and reasonable utilization through the operation of the
Mahakali River Commission, an organization responsible for the implementation
of the treaty.[38] In addition, Article 7 of the treaty specifies the maintenance of the
natural flow and level of the river and the need for prior agreement to any such
changes. This article facilitates the obligation to cause no significant harm,[39] and
thus in accordance with the spirit, if not letter, of the UNWC and its principles.

However, the Mahakali Treaty has not facilitated a swift conclusion to the
prolonged decision making on the Pancheshwar Multipurpose Dam project. This
project was planned as a multi-purposed dam on the Mahakali River, a border river
between Nepal and India. This project had been at the center of debate since the
1970s, when both states were developing large-scale infrastructure projects as part
of their hydraulic mission.[40] Water allocation was a contentious issue during the
phase of institution building, with the Nepali government focusing on quantitative

34   Yesilkaya, T. 'Hydropolitics: Searching for a Solution for the Water Dispute in the Euphrates-Tigris
      River Basin' (International Studies Association South Conference, Miami, November 2005).
35   Ibid.
36   Tanzi, A. 'The Completion of the Preparatory Work for the UN Convention on the Law of
      International Watercourses', *Natural Resources Forum*, 1997; 21: 239–45.
37   Salman, S. M. A. and Uprety, K. *Conflict and Cooperation on South Asia's International Rivers: A Legal
      Perspective* (World Bank, 2002).
38   Rahaman, M. M. 'Principles of Transboundary Water Resources Management and Ganges Treaties:
      An Analysis', *International Journal of Water Resources Development* , 2009; 25: 159–73.
39   Ibid.
40   Mirumachi (note 18).

allocation while the Indian government argued for sharing benefits, rather than the actual water. While the treaty established an institution to govern the shared waters, project-specific negotiations between the Nepali and Indian governments have been at a deadlock over the determination of 'mutual benefits'.

Why are global norms such as the UNWC difficult to implement at the basin level? Giordano and Wolf pointed out that local-specific arrangements, rather than global principles, are the preferred path of establishing agreement in international transboundary river basins.[41] Woodhouse and Zeitoun call our attention to further issues with interstate treaties trumping international law. While treaty law explicitly addresses the use of coercion between states in concluding a treaty – Articles 51 and 52 of the Vienna Convention on the Law of Treaties cover 'Coercion of a representative of a State' and 'Coercion of a State by the threat or use of force' – it does not recognize any expressions of 'soft' power.[42] For this chapter's exploration, this is a fundamental flaw with international law more generally – it inadequately takes power asymmetry between basin states into account. In hegemonic political conditions where the hydro-hegemon uses 'soft' power to determine and establish water use and management practices, international water law is put to the test. As previously discussed, non-hydro-hegemonic states may attempt to use their ability to level the playing field, or to develop their own strategies to counter the hegemony outside of law's reach.[43]

Thus, the UNWC principles may or may not be the tools with which basin states establish institutions for water resources management and allocation. Put differently, upstream hegemonic states can use their available water resources to assert power, thereby relying less on legal principles to establish order within the basin. In such cases, soft power is 'invisible' to international law (international water law as well as international treaty law), which is equipped to deal only with covert expressions of power. For instance, while the GAP is a Turkish project, its development is bound in wider geopolitical considerations. The project site is in the Kurdish-dominated territory of Turkey and the dams became the target of Kurdish independence struggles. The Syrian government supported the independence struggles of Kurdish insurgents PKK to force Turkish concessions on water allocation. To counter such strategy, Turkey created alliances with Israel to strengthen military capacity.[44] The resulting effect is a political climate where water and security issues are linked, with both countries leveraging power of obstruction in one domain to secure concessions from the other. This situation can be seen as an example where the upstream hegemonic state has used its available water resources to gain compliance.

---

41   Giordano, M. A. and Wolf, A. T. 'Incorporating Equity into International Water Agreements', *Social Justice Research*, 2001; 14: 349–66.

42   Woodhouse, M. and Zeitoun, M. 'Hydro-Hegemony and International Water Law: Grappling with the Gaps of Power and Law', *Water Policy*, 2008; 10 (Supplement 2): 103–19.

43   See, e.g. Cascão, A. E. 'Political Economy of Water Resources Management and Allocation in the Eastern Nile River Basin' (PhD thesis, King's College, London, 2009).

44   Daoudy, M. 'Asymmetric Power: Negotiating Water in the Euphrates and Tigris', *International Negotiation*, 2009; 14: 361–9.

Compliance is important for Turkey because ever since the dissolution of the Ottoman Empire and the foundation of a truncated Republic in 1922, Turkey finds itself in a 'rough neighbourhood'.[45] From the Turkish perspective, while maintaining interests when confronting neighbors, it is worth trying to be or be perceived to be a peacemaker of the region rather than exerting power through coercive means. While mooted, Turkey had proposed the 'Peace Pipeline' in the late 1980s to transfer water beyond the Euphrates-Tigris River Basin to water scarce countries in the Gulf and Eastern Mediterranean region. This international water grid would not only exploit Turkey's hydro-strategic position in the basin, but also serve to contain water-related claims by neighbors. Water resources are presented as a technical issue, and 'technical cooperation' with Syria and Iraq has been stepped up since 1998. Turkey and Syria have increased technical exchange with their respective GAP and GOLD (General Organization for Land Development) projects, both of which involve utilizing water and land resources of the Euphrates River Basin. In recent years, Turkey has successfully used its 'soft' power in an attempt to gain more influence over water resources governance, organizing the World Water Forum and encouraging exchange and cooperation within the basin.

While Turkey voted against the UNWC, it has proposed water resources management principles at the basin level that are not incompatible with the principles of the UNWC. In the 1980s, the Turkish government actively proposed a so-called three-staged plan to assess and evaluate land and water resources so that 'optimum, equitable and reasonable use' of the Euphrates-Tigris waters could be achieved. While not explicitly adopting the UNWC principle of 'equitable and reasonable use', the Turkish government referred to international law conventions to seek consent from Syria and Iraq.[46] Although Syria and Turkey have disputed over the Kurdish independence movements, there has been a security and economic cooperation treaty in 1987, with a note for Turkey to guarantee an average water flow of 500 cubic meters per second to Syria. Turkey has made considerable effort to ensure promised amounts crossed the Syrian border, making up for shortages in later periods. In 1990, when Turkey started to fill its enormous Atatürk Dam as part of the GAP, the country did give downstream Syria and Iraq prior notice. Crucially, however, the Turkish government did not confer with them about how to reduce the impact of interrupting the flow of the Euphrates River for a month. While the GAP is now an integrated dam project incorporating environmental and participatory issues, downstream demands are carefully managed on Turkey's terms. These upstream actions reflect international law principles used in a discursive manner to control and manage water resources in an institutional structure that is similar to but different from one established squarely on UNWC principles.

---

45 Aydin, M. 'Securitization of History and Geography: Understanding of Security in Turkey', *Southeast European and Black Sea Studies*, 2003; 3: 163.

46 Tomanbay, M. 'Turkey's Approach to Utilization of the Euphrates and Tigris Rivers', *Arab Studies Quarterly*, 2000; 22 (2): 79–101.

In the case where the hydro-hegemon is downstream, it can use power to secure water resources. Here the UNWC can be used in the process of discursive institution building; downstream hegemonic states can use 'soft' power legitimized through the legal framework to influence water resources use and management. In the case of the Nile River Basin, for over a decade downstream Egypt has been able to maintain its position on the established water allocation throughout related deliberations with eight of the nine upstream states through the Nile Basin Initiative (NBI). In this case, the legal framework has served to maintain the status quo of water allocations that favors Egypt. The 1959 Nile Waters Treaty signed by the Egyptian and Sudanese government allocates flow volumes roughly equivalent to 75 percent of the Nile flow vis-à-vis Sudan (and consequently excluding an allocation to the rest of the eight states).[47] Egypt participated in NBI from its establishment in 1999, and was part of the deliberations that were directly guided by the UNWC.[48] However, the Egyptian government has repeatedly stalled on developing a basin-specific legal framework that would guide water resources utilization.[49] Indeed, the final stages of negotiations over the text of the Cooperative Framework Agreement between the Nile Basin states came to a halt when Egypt raised its concerns about reallocation of waters (Article 14b).[50] This example serves to show that legal principles can be one part of the institutional structure, which may justify the hydraulic control of the hydro-hegemon: law 'unlevelling the playing field', in other words.

In situations where water is used to secure power and vice versa, 'soft' power can shape prioritization or de-prioritization of legal principles in allocating and managing water resources management. While non-hydro-hegemons can also promote UNWC principles for institution building, the examples from the Ganges-Brahmaputra-Meghna river system, Euphrates-Tigris and Nile River Basins illustrated the asymmetric transboundary water interactions that determine institution-building. These empirical observations relate to the claims made by Giordano and Wolf about the geographical scale at which water resources are managed: at the basin scale, and not at the global scale.[51] While the UNWC has advanced the global normative understanding of shared waters, the challenge of the UNWC principles guiding equitable and sustainable water resources use for both the hydro-hegemon and non-hydro-hegemon states remains.

Conclusion

This chapter has focused on the interface between politics and the principles of international water law, to gain a better understanding on the development of

47 Tvedt, T. *The River Nile in the Age of the British: Political Ecology and the Quest for Economic Power* (I. B. Tauris, 2004).
48 Cascão, A. E. 'Changing Power Relations in the Nile River Basin: Unilateralism vs. Cooperation?', *Water Alternatives*, 2009; 2: 245–68.
49 Ibid.
50 Cascão, A. E. and Zeitoun, M. 'Power, Hegemony and Critical Hydropolitics', in Earle *et al.* (note 18).
51 Giordano and Wolf (note 41).

institutional structures to manage and govern transboundary waters. It argued that without the understanding of the political context, which gives rise to changing transboundary water interactions between basin states, it is not possible to even begin discussing the utility and effectiveness of the UNWC. Especially as the UNWC is designed as a global, rather than basin-specific, legal framework there are challenges in reconciling the global expectations and local concerns of water resources allocation and management. The chapter has highlighted the importance of considering the asymmetric transboundary water interactions in which hydro-hegemons have various strategies in place to influence the decision making of water resources management and governance. Specifically, this influence is supported by the use of 'soft' power. This discursive way of gaining compliance impacts both the process and outcome of water negotiations. The principles of international water law are seen as one of several tools by which institutions are built, but, as with other such tools, subject to the asymmetric power relations of basin states. It was shown that in asymmetric transboundary interactions even where the principles of the UNWC are acknowledged in cooperative initiatives and basin agreements, the *practice* of these principles remains a very elusive goal. Empirical examples showed that hydro-hegemons have the option to exercise or ignore these principles in practice. Rather than blindly promoting the UNWC as a template for cooperation, policy would benefit from first assessing the political context of transboundary water interactions and then evaluating the ways in which institutions can be, or have been, established. In this context, policy would be providing a range of tools, including the UNWC's principles, upon which negotiations and deliberations can be made between basin states and within river basin organizations and other actors involved in institution-building.

Index

Page numbers in bold refer to figures and tables